Pandas and People

Pandas and People

Coupling Human and Natural Systems for Sustainability

EDITED BY

Jianguo Liu, Vanessa Hull, Wu Yang, Andrés Viña, Xiaodong Chen, Zhiyun Ouyang, and Hemin Zhang

OXFORD
UNIVERSITY PRESS

OXFORD
UNIVERSITY PRESS

Great Clarendon Street, Oxford, OX2 6DP,
United Kingdom

Oxford University Press is a department of the University of Oxford.
It furthers the University's objective of excellence in research, scholarship,
and education by publishing worldwide. Oxford is a registered trade mark of
Oxford University Press in the UK and in certain other countries

First Edition published in 2016

Impression: 1

Published in the United States of America by Oxford University Press
198 Madison Avenue, New York, NY 10016, United States of America

British Library Cataloguing in Publication Data
Data available

Library of Congress Control Number: 2015955005

ISBN 978–0–19–870354–9 (hbk.)
ISBN 978–0–19–870355–6 (pbk.)

Printed and bound by
CPI Group (UK) Ltd, Croydon, CR0 4YY

Preface

One of the greatest global challenges is to understand and manage the complex human–nature interactions that affect sustainability. Humans and nature have interacted inextricably with each other since the dawn of humanity. On one hand, humans have directly or indirectly altered every inch of the Earth's surface. On the other hand, changes in natural systems have profoundly shaped human behavior and well-being. Although researchers have made much progress in uncovering these endless interactions, many mysteries remain due to a lack of effective integrated frameworks and dearth of long-term systematic and in-depth inquiries.

The emerging coupled human and natural systems framework provides an integrated approach to explicate intricate human–nature interactions. It explicitly recognizes human–nature interactions and feedbacks among organizational levels, across space, and over time. As many feedbacks take a long time to emerge and sustainability is inherently a long-term issue, long-term research is warranted.

Since 1996, our interdisciplinary and international team has been studying the complexity of a globally important coupled human and natural system—Wolong Nature Reserve in southwestern China. As a UNESCO biosphere reserve, Wolong is 2000 km^2 in size and lies within a global biodiversity hotspot and a World Heritage site. It contains one of the largest populations of the world-famous and endangered giant panda, together with several thousand other animal and plant species. Like most nature reserves in China and many other developing countries, Wolong is also home to several thousand local residents who undertake a variety of activities that involve interactions with the natural system.

Over the past two decades we have learned a great deal about Wolong. We have come to know Wolong as a vibrant and dynamic place. We also have appreciated Wolong as a challenging object of study with seemingly countless connections between humans and nature. Our long-term interdisciplinary research has generated a series of results. Many of the findings and methods have been applied to studies on other coupled systems at local, regional, national, and global levels. These findings have not only advanced the science of coupled human and natural systems, but also helped improve the development and implementation of policy and management programs. Additionally, they have been featured in the global news media and have been widely used to educate students and the general public about conservation, sustainability, and human–nature interactions.

The purpose of this book is to synthesize some of our team's work in the past and to provide research directions for the future. The synthesis is further enhanced by analyzing various types of data from a long-term lens to glean new information about coupled systems. While publishing our results in separate papers in various outlets has generated a variety of impacts, this book offers a more comprehensive and overall picture by connecting different human and natural components and integrating some major results on theory, methodology, and applications of coupled systems.

This book contains 19 interrelated chapters in four major parts. Part I provides empirical and theoretical bases for the rest of the book. It offers a multifaceted peek into the dynamic world of people and the iconic pandas (Chapter 1), a theme that is subsequently interwoven throughout the book. The backdrop of pandas–people interactions and feedbacks sets the stage for framing sustainability of coupled human and natural systems (Chapter 2).

As intensive long-term research on coupled systems requires much manpower and is very time-consuming, it is impossible to study every coupled system for a long time and in a detailed manner. Thus, we present a novel approach of model coupled systems, analogous to widely used model organisms and model ecosystems that involve in-depth and long-term studies. Since many findings from Wolong have proven applicable to other coupled systems, we view Wolong as such a model coupled system. Part II addresses various aspects of the model coupled system. It begins with an overview (Chapter 3), followed by the central challenges and mechanisms behind coexistence and competition between people and the endangered pandas (Chapter 4), and quantification of human dependence on ecosystem services (Chapter 5). This part also documents both gradual and drastic dynamics of various system components and processes, including landscape change in space and time (Chapter 6), panda habitat transition from decades of degradation to recovery (Chapter 7), demographic decisions and cascading consequences (Chapter 8), economic transformation from a subsistence-based community to a society with diverse livelihood strategies (Chapter 9), and energy transition from fuelwood to electricity (Chapter 10). These dynamics are driven by a range of factors such as social norms and social capital (Chapter 11), natural disasters (Chapter 12), and policy interventions (Chapter 13). Building on what happened in the past, the final chapter (Chapter 14) of Part II lays out a number of scenarios for a sustainable future in Wolong in an uncertain world.

Going beyond Wolong, Part III highlights a few applications of findings from Wolong to other coupled systems across the globe. It scales up methods developed in Wolong to the entire geographic range of pandas (Chapter 15), reveals similar complexities to those in another high-profile coupled system—Chitwan National Park for tigers in Nepal (Chapter 16), illustrates socioeconomic and environmental interactions over distances (Chapter 17), and presents the top ten lessons we learned from our previous work (Chapter 18).

Although substantial progress has been made, many challenges and opportunities are in front of us. In the final part (Part IV), we outline several key topics for future research that would help elevate the understanding and management of coupled systems to new heights (Chapter 19).

By looking both back and forward, we hope this book illustrates the value of the coupled human and natural systems framework in exploring mechanisms for improving global sustainability and of long-term interdisciplinary approaches in understanding human–nature interactions. While we are pleased that Wolong is an effective model coupled system for discerning human–nature interactions and for addressing sustainability challenges, much more needs to be done. Hopefully this book can serve as a stepping stone for future endeavors on research and management of coupled systems. We invite fellow researchers, students, and stakeholders to join us in further extending and expanding coupled systems research and management in Wolong and elsewhere for sustainability across local to global scales.

Acknowledgments

This book is the result of generous support and help from numerous marvelous people and organizations over the past two decades.

We are most grateful to Jared Diamond, Paul Ehrlich, Carl Folke, Simon Levin, Emilio Moran, and Stuart Pimm, who read earlier versions of the book manuscript and provided constructive comments and suggestions. We extend our deep appreciation to all contributors of the book, and to Joanna Broderick and Sue Nichols whose excellent editing skills substantially improved the writing and clarity of the book manuscript. We also appreciate the helpful comments on the initial book proposal from anonymous reviewers.

This book synthesizes some of the work that we conducted with former students, postdoctoral associates, and collaborators. Besides those who are contributors to this book, other collaborators are Sandra Batie, Rique Campa III, Betty Cheng, Angela Mertig, Kelly Millenbah, Jiaguo Qi, Shawn Riley, Gary Roloff, Mike Walters, Julie Winkler, and Runsheng Yin who served on the guidance committees of graduate students on various relevant projects. There are many other collaborators who applied insights from Wolong Nature Reserve to other coupled systems or compared results from Wolong with other coupled systems: Marina Alberti, William Axinn, Mateus Batistella, Stephen Carpenter, Manuel Colunga, Gretchen Daily, Stephen Davis, Jared Diamond, Gilberto de Miranda Rocha, Peter Deadman, Ruth DeFries, Paul Ehrlich, Carl Folke, Feng Fu, Stuart Gage, Joanne Gaskell, Dirgha Ghimire, Peter Gleick, Richard Groop, Nianyong Han, Thomas Hertel, R. Cesar Izaurralde, Timothy Kratz, Claire Kremen, Eric Lambin, Jane Lubchenco, Gary Luck, Luiz Martinelli, Hong Miao, Harold Mooney, Paul Moorcroft, Emilio Moran, Harini Nagendra, Rosamond Naylor, Elinor Ostrom, Alice Pell, Stuart Pimm, Karen Polenske, William Provencher, Peter Raven, Charles Redman, Anette Reenberg, Stephen Schneider, Karen Seto, Ryan Sheely, Cynthia Simmons, Simon Swaffield, Christine Tam, William Taylor, Billie L.Turner II, Peter Verburg, Peter Vitousek, Xiaoke Wang, Yi Xiao, Fuzuo Zhang, and Chunquan Zhu. Former students and postdoctoral associates include Weilin Chen, Yi Chen, Lily Cheng, Cathy Chung, Jayson Eageler, Guangming He, Clinton Jenkins, Aaron Kortenhoven, Carrie Lentz, Yu Li, Hongxia Lu, Yi Mou, Olympia Moy, Nathan Pfost, Brent Wheat, Eunice Yu, Yan Zhang, and Yi Zhang.

We are deeply indebted to many administrators, faculty members, and staff in our institutions for their strong support and generous help. Particularly, Doug Buhler, Ian Gray, Steven Hsu, Karen Klomparens, Fred Poston, Lou Anna K. Simon, William W. Taylor, Scott Winterstein, and June Youatt at Michigan State University have been wonderful cheerleaders and exceptional supporters throughout the past two decades. We also acknowledge the great support of the Chinese Academy of Sciences for our work, and the Chinese government agencies (State Forestry Administration, Forestry Department of Sichuan Province, and Wolong Administration Bureau) for permissions to conduct research in China. Also gratefully acknowledged are the timely support from Wolong staff such as Yan Huang, Desheng Li, Mingchong Liu, Xiaogang Shi, Chunxiang Tang, Hao Tang, Pengyan Wang, Rongping Wei, Yuanliang Xie, Guiquan Zhang, and Xiaoping Zhou.

We were fortunate to have so many dedicated field assistants and knowledgeable field guides who were essential for us to collect ecological and socioeconomic data through hiking in the rugged mountains and interviewing local households in

Wolong. They include: Dong Chen, Yanxi Cheng, Linhua Deng, Ke Dong, Shuming Fan, Datian He, Senlong Jin, Caiwu Li, Mao Li, Rengui Li, Bin Liu, Dian Liu, Jianping Liu, Jiang Ming, Meng Ming, Jie Shen, Shixian Song, Weihong Tan, Yang Xu, Fugui Wan, Bo Wang, Hongdan Wang, Jifu Wang, Lun Wang, Maolin Wang, Qunyin Wang, Xiangwen Wang, Youfu Wang, Han Xiao, Haibin Xu, Bihong Yang, Bijun Yang, Bing Yang, Biyou Yang, Changyou Yang, Fan Yang, Kai Yang, Sen Yang, Wenbing Yang, Liang Zhang, Kaiqiang Zhang, Yahui Zhang, Zhijun Zhang, Sha Zhou, Quan Zhou, and Yongguang Zhou. We are especially grateful to the numerous interviewees in Wolong and the adjacent Sanjiang Township for their cooperation over the years. Likewise, we thank the field assistants and collaborators for the work in Chitwan National Park (e.g., Bhim Gurung, Jhamak B. Karki, Narendra Man Babu Pradhan, and Binoj K. Shrestha) and the interviewees in Chitwan, Nepal.

We also thank the following organizations and individuals for their help with the logistics of field work conducted across other areas of the panda's range: the Forestry Departments of Gansu, Shaanxi, and Sichuan Provinces, and administrations of many nature reserves beyond Wolong; and Gaodi Dang, Xichun Du, Junzheng Meng, and Yong Wang for their expertise with plant species identification. Similarly, we are grateful to the staff at the Institute for Social and Environmental Research—Nepal for the logistic support during field work in Chitwan National Park.

This book would not have been possible without the financial support from funding agencies for collecting, analyzing, and integrating various sources of data. The major funding agencies were the US National Science Foundation, National Institutes of Health, National Aeronautics and Space Administration, the John Simon Guggenheim Foundation, the American Association for the Advancement of Sciences, The John D. and Catherine T. MacArthur Foundation, Michigan State University, Michigan AgBioResearch, the William W. and Evelyn M. Taylor Endowed Fellowship for International Engagement in Coupled Human and Natural Systems, United States Fish and Wildlife Service, the St. Louis Zoo, the International Association for Bear Research and Management Research and Conservation Grant, the Giant Panda International Collaboration Fund, China Bridges International, State Key Laboratory of Urban and Regional Ecology, Research Center for Eco-Environmental Sciences, the Chinese Academy of Sciences, the National Natural Science Foundation of China, and the Ministry of Science and Technology of China. The program managers at these agencies such as Thomas Baerwald, Elizabeth Blood, William Chang, Rebecca Clark, Henry Gholz, Garik Gutman, Sarah Ruth, Alan Tessier, Woody Turner, Saran Twombly, and Min-Ying Wei, were excellent in managing the projects and providing useful feedback.

We thank Shuxin Li who drew several figures, compiled some data, and found literature for the book, and Junyan Luo who organized the Wolong household survey data. The input from Li An and Mao-Ning Tuanmu to Chapter 14, comments from Thomas Connor and William McConnell on Chapter 17, and Thomas Dietz's comments on the introduction of Part II are greatly acknowledged. We are very grateful to those publishers who granted us permissions to reproduce some figures (see the specific sources at the end of relevant figure legends). Also, Chapter 5 is largely based on Yang et al. 2013 (*PLoS ONE*), and Chapter 17 is largely based on Liu et al. (2015).

We would like to express our sincere appreciation to Lucy Nash and Ian Sherman at Oxford University Press for their guidance during the book writing process and their patience in waiting for the completion of the book manuscript, to Julian Thomas for skillful copyediting, to Kaarkuzhali Gunasekaran for overseeing the production, and to others who are responsible for various stages of the book production and distribution.

Finally, we cannot thank our families enough for their unconditional support and sacrifice when we were in the field collecting data, and spending endless hours during evenings and weekends for data analysis and writing.

Contents

Acronyms, Abbreviations, and Terms

ABM	agent-based model(ing)
AUC	area under the receiver operating characteristic curve
CBA	cost–benefit analyses
CCRCGP	China Conservation and Research Center for the Giant Panda
CERN	European Organization for Nuclear Research
CHANS	coupled human and natural systems
CITES	Convention on International Trade in Endangered Species
ESA	ecosystem service assessments
ESP	Electricity Subsidy Program
ES	ecosystem services
ETM +	Enhanced Thematic Mapper Plus, for Landsat
FAO	United Nations Food and Agriculture Organization
GHG	greenhouse gas emissions
GIS	geographic information system(s)
GPS	global positioning system
GTBP	Grain to Bamboo Program
GTGB	Grain to Green/Bamboo Program
GTGP	Grain to Green Program
HSI	habitat suitability index
HWB	human well-being
HWBI	human well-being index
IDES	index of dependence on ecosystem services
ILTER	International Long Term Ecological Research Network
IMSHED	Integrative Model for Simulating Household and Ecosystem Dynamics
InVEST	Integrated Valuation for Ecosystem Services and Tradeoffs
LTER	Long Term Ecological Research Network
LTSER	Long-Term Socio-Ecological Research
MA	Millennium Ecosystem Assessment
MAB	Man and Biosphere Programme
MODIS	Moderate Resolution Imaging Spectroradiometer
NFCP	Natural Forest Conservation Program
NSF	National Science Foundation
PES	payments for ecosystem services
SRTM	Shuttle Radar Topography Mission
TDP	Tourism Development Program
TM	Thematic Mapper (Landsat)
UNESCO	United Nations Educational, Scientific, and Cultural Organization
VCM	vicious circle model
VIF	variance inflation factor
WWF	World Wildlife Fund
yuan	Chinese currency, equal to 0.16 USD in May 2015

List of Contributors

Li An Center for Systems Integration and Sustainability, Department of Fisheries and Wildlife, Michigan State University, *present address:* Department of Geography, San Diego State University, 5500 Campanile Drive, San Diego, CA 92182-4493, USA

Scott Bearer Center for Systems Integration and Sustainability, Department of Fisheries and Wildlife, Michigan State University, *present address:* The Nature Conservancy, Community Arts Center, 220 West Fourth Street, Williamsport, PA 17701, USA

Neil Carter Center for Systems Integration and Sustainability, Department of Fisheries and Wildlife, Michigan State University *present address:* Human-Environment Systems Center, Boise State University, Environmental Research Building, 1910 University Dr., Boise, ID 83725, USA

Xiaodong Chen Center for Systems Integration and Sustainability, Department of Fisheries and Wildlife, Michigan State University, *present address:* Department of Geography, The University of North Carolina at Chapel Hill, Carolina Hall 319-C, Chapel Hill, NC 27599, USA

Thomas Dietz Department of Sociology, Environmental Science and Policy Program, Center for Systems Integration and Sustainability, Michigan State University, 6H Berkey Hall, East Lansing, Michigan 48824, USA

Ken Frank Department of Counseling, Educational Psychology and Special Education, Department of Fisheries and Wildlife, Center for Systems Integration and Sustainability, Michigan State University, Room 462 Erickson Hall, East Lansing, MI 48824-1034, USA

Jinyan Huang China Conservation and Research Center for the Giant Panda, Wolong Nature Reserve, Wenchuan County, Sichuan, 623006, China

Vanessa Hull Center for Systems Integration and Sustainability, Department of Fisheries and Wildlife, Michigan State University, 115 Manly Miles Building, 1405 S. Harrison Rd., East Lansing, MI 48823, USA

Daniel Kramer James Madison College and the Department of Fisheries and Wildlife, Michigan State University, Case Hall, 842 Chestnut Rd Room N370, East Lansing, MI 48825, USA

Zai Liang Department of Sociology, State University of New York at Albany, 1400 Washington Ave, Albany, NY 12222, USA

Shuxin Li Center for Systems Integration and Sustainability, Department of Fisheries and Wildlife, Michigan State University, 115 Manly Miles Building, 1405 S. Harrison Rd., East Lansing, MI 48823, USA

Marc Linderman Center for Systems Integration and Sustainability, Department of Fisheries and Wildlife, Michigan State University, *present address:* Department of Geographical and Sustainability Sciences, The University of Iowa, 316 Jessup Hall, Iowa City, Iowa 52242, USA

Jianguo Liu Center for Systems Integration and Sustainability, Department of Fisheries and Wildlife, Michigan State University, 115 Manly Miles Building, 1405 S. Harrison Rd., East Lansing, MI 48823, USA

Wei Liu Center for Systems Integration and Sustainability, Department of Fisheries and Wildlife, Michigan State University, *present address:* International Institute for Applied Systems Analysis, Schlossplatz 1, A-2361 Laxenburg, Austria

Frank Lupi Department of Agricultural, Food, and Resource Economics, Department of Fisheries and Wildlife, Center for Systems Integration and Sustainability, Michigan State University, 446 W. Circle, Dr., Rm 301B, Morrill Hall of Agriculture, East Lansing, Michigan 48824-1039, USA

Junyan Luo Center for Systems Integration and Sustainability, Department of Fisheries and Wildlife, Michigan State University, 115 Manly Miles Building, 1405 S. Harrison Rd., East Lansing, MI 48823, USA

William McConnell Center for Systems Integration and Sustainability, Department of Fisheries and Wildlife, Michigan State University, *present address:* Center for Global Change and Earth Observations, Michigan State University, 218 Manly Miles Building, 1405 S. Harrison Road, E, East Lansing, MI 48823-5423, USA

Zhiyun Ouyang State Key Laboratory of Regional and Urban Ecology, Research Center for Eco-Environmental Sciences, Chinese Academy of Sciences, Beijing, 100085, China

Ashton Shortridge Department of Geography, Michigan State University, 235 Geography Building, 673 Auditorium Road, East Lansing, MI, 48824 USA

Yingchun Tan Wolong Nature Reserve, Wenchuan County, Sichuan, 623006, China

Mao-Ning Tuanmu Center for Systems Integration and Sustainability, Department of Fisheries and Wildlife, Michigan State University, *present address:* Department of Ecology and Evolutionary Biology, Yale University, 165 Prospect St., Room 405, New Haven, CT 06520, USA

Andrés Viña Center for Systems Integration and Sustainability, Department of Fisheries and Wildlife, Michigan State University, 115 Manly Miles Building, 1405 S. Harrison Rd., East Lansing, MI 48823, USA

Christine Vogt Department of Community Sustainability, Michigan State University, *present address:* Center for Sustainable Tourism, Arizona State University, Mail Code 4020, 411 North Central Avenue, Suite 550, Phoenix, Arizona 85004-0690, USA

Weihua Xu State Key Laboratory of Regional and Urban Ecology, Research Center for Eco-Environmental Sciences, Chinese Academy of Sciences, Beijing, 100085, China

Zhenci Xu Center for Systems Integration and Sustainability, Department of Fisheries and Wildlife, Michigan State University, 115 Manly Miles Building, 1405 S. Harrison Rd., East Lansing, MI 48823, USA

Hongbo Yang Center for Systems Integration and Sustainability, Department of Fisheries and Wildlife, Michigan State University, 115 Manly Miles Building, 1405 S. Harrison Rd., East Lansing, MI 48823, USA

Jian Yang Wolong Nature Reserve, Wenchuan County, Sichuan, 623006, China

Wu Yang Center for Systems Integration and Sustainability, Department of Fisheries and Wildlife, Michigan State University, *present address:* Conservation International, 2011 Crystal Dr Suite #500, Arlington, VA 22202, USA

Hemin Zhang China Conservation and Research Center for the Giant Panda, Wolong Nature Reserve, Wenchuan County, Sichuan, 623006, China

Jindong Zhang Center for Systems Integration and Sustainability, Department of Fisheries and Wildlife, Michigan State University; Institute of Rare Animal and Plants, China West Normal University, Nanchong, 637002, China

Shiqiang Zhou China Conservation and Research Center for the Giant Panda, Wolong Nature Reserve, Wenchuan County, Sichuan, 623006, China

Currency exchange rates from US dollars to Chinese Yuan between 1995 and 2014

Year	Exchange rate (US$ to Yuan)*
1995	8.351
1996	8.314
1997	8.290
1998	8.279
1999	8.278
2000	8.279
2001	8.277
2002	8.277
2003	8.277
2004	8.277
2005	8.194
2006	7.973
2007	7.608
2008	6.949
2009	6.831
2010	6.770
2011	6.461
2012	6.312
2013	6.196
2014	6.143

*value given is yuan equivalent of US$1, calculated as an annual average by the World Bank (data.worldbank.org)

PART I

Empirical and Theoretical Foundations

Since the dawn of the human race, humans have interacted with nature. Humans rely on nature for survival, and nature in turn has responded to human presence in diverse and sometimes unpredictable ways. With human impacts on nearly every corner of the world, it is more important than ever before to understand how humans and nature interact with each other and how such interactions can be better managed to ensure that biodiversity and ecosystem services on the planet Earth can be sustained for the future while meeting today's human needs. Addressing such questions is the central theme of this book.

Part I lays empirical and theoretical foundations for the book. Chapter 1 provides a multifaceted glimpse into the dynamic world of people and the iconic giant pandas, a theme that is subsequently interwoven throughout the entire book. In this chapter, Hull and Liu give an overview of the charismatic and emblematic giant panda in the past and at present and outline the panda's geographic range and biological requirements. They highlight the influences of pandas on humans, impacts of myriad human activities on pandas, and pandas–people feedbacks through conservation policies.

The backdrop of pandas–people interactions and feedbacks sets the stage for framing sustainability of coupled human and natural systems, in which humans and natural components interact. Chapter 2 presents an overall framework that explicitly analyzes complex interactions and feedbacks between humans and nature in a holistic manner. Liu et al. outline key concepts and typology of coupled systems. They also provide an example application of the overall framework to coupled human and wildlife systems in which wildlife are a key component of nature, and give an overview of major relevant theories and methods for studying coupled systems. Liu et al. conclude by suggesting a novel model coupled system approach, analogous to widely used model organisms and model ecosystems that warrant in-depth and long-term studies. Such an approach may yield valuable insights that can be useful for studying complexities of human–nature relationships in other coupled systems across local to global scales.

CHAPTER 1

A Global Icon for Nature in the Human-Dominated World

Vanessa Hull and Jianguo Liu

1.1 Introduction

Humans depend on nature for ecosystem services, such as clean air, energy, food, spiritual inspiration, and clean water (Daily, 1997, Millennium Ecosystem Assessment, 2005, Yang et al., 2013). Biodiversity, the various forms of life and habitat that support life, is essential to provide ecosystem services for humans (Intergovernmental Platform on Biodiversity and Ecosystem Services, 2014). However, humans have created an unprecedented biodiversity crisis (Pimm et al., 2014). For example, many plant and wildlife species have become extinct or endangered due to human impacts. For planetary sustainability and long-term human well-being, it is crucial to transform human–nature conflicts into harmonious human–nature coexistence (Carter et al., 2012, Kareiva and Marvier, 2015). A global environmental icon—the giant panda (*Ailuropoda melanoleuca*)—holds valuable lessons that can help advance such transformation. We illustrate the complex interactions between humans and this global icon whose challenges and hopes are similar to those of numerous other wildlife worldwide.

The giant panda is a large mammal, with unmistakable black-and-white coloration, a unique pattern that has attracted curiosity and fascination from humans for centuries (Morris and Morris, 1966). A species currently found in bamboo forests in southwestern China (Figure 1.1), the giant panda's main food source is bamboo. Bamboo makes up over 99% of its diet (Schaller et al., 1985). Pandas are largely solitary, living in distinct but overlapping home ranges roughly 3–10 km^2 in area (Hull et al., 2015b, Pan et al., 2001, Schaller et al., 1985). Individuals convene in groups during the mating season in the spring of each year, when multiple males aggressively compete for access to estrous females (Pan et al., 2001). Females typically raise one young at a time and invest 1.5–2 years in raising their offspring (Schaller et al., 1985). Pandas spend much of the time physically apart from one another but monitor the whereabouts of their neighbors through an advanced scent-marking system (Schaller et al., 1985).

The *Ailuropoda* genus first appeared in the fossil record in the late Pliocene era (~2.4 million years ago) in caves of southern China (Jin et al., 2007). The genus evolved from the bear family, specifically the ursid *Ailurarctos lufengensis*. Fossils of this genus dates back to the late Miocene age (~7–8 million years ago) in the Yunnan province of southwestern China (Hunt, 2004). Profound climate instabilities during the early Pleistocene era caused a large-scale die-off of mammals, including the *Ailuropoda*. The genus later rebounded and records spread across China, Myanmar, Laos, Vietnam, and Thailand in the late Pleistocene era (15 000 years ago; Hunt, 2004, Jin et al., 2007). The extant species of giant panda (*Ailuropoda melanoleuca*) also appeared in the fossil record around the mid to late Pleistocene (Wang et al., 2007) throughout all of the aforementioned areas (Tougard, 2001).

By 1800, the distribution of giant pandas had shrunk to roughly 262 000 km^2 and was limited to five provinces in southwestern China (Reid

Pandas and People. Edited by Jianguo Liu, Vanessa Hull, Wu Yang, Andrés Viña, Xiaodong Chen, Zhiyun Ouyang, and Hemin Zhang. Published 2016 by Oxford University Press.

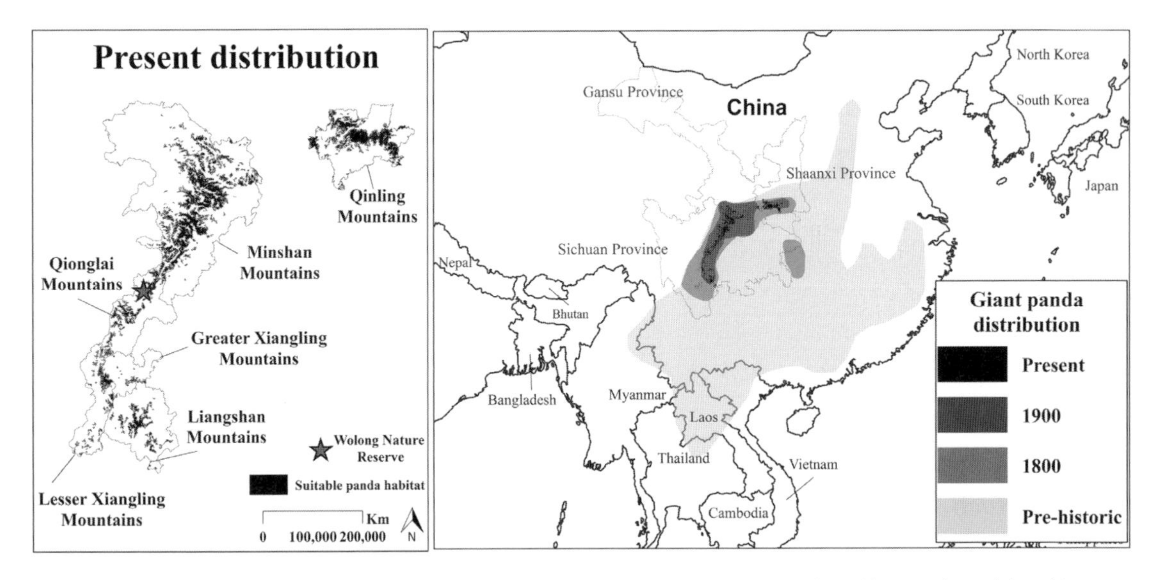

Figure 1.1 Distribution of giant pandas (*Ailuropoda melanoleuca*) over time. Prehistoric estimates are derived from Loucks et al. (2001) but adjusted for additional fossil evidence detailed in Jin et al. (2007). Estimates for 1980 and 1990 are derived from Reid and Gong (1999) that summarizes data in Zhu and Long (1983). Present estimates are derived from a supervised habitat classification by Viña et al. (2010). Map on the left depicts the six mountain ranges where giant pandas are currently found and is reprinted with permission from the Missouri Botanical Garden Press (adapted from Liu and Viña, 2014).

and Gong, 1999, Zhu and Long, 1983). The range then dropped to 124 000 km^2 in the year 1900, and to 21 300 km^2 today (Reid and Gong, 1999, Viña et al., 2010, Zhu and Long, 1983). The current giant panda range is fragmented into over 30 isolated populations in three provinces—Gansu, Sichuan, and Shaanxi—in southwestern China (Hu, 2001; Figure 1.1). These patterns of decline also reflect population trends for the species. There are no estimates of the total number of giant pandas prior to the twentieth century, but expert opinion suggests that the population numbered around 3000 individuals in 1950 (Hu, 2001). Rough estimates for the wild panda population obtained from panda surveys were subsequently changed: 2459 in 1974–1977, 1114 in 1985–1988, 1596 in 1999–2003 (Hu, 2001, Liu, 2015) and 1864 in 2011–2013 (State Council Information Office of China, 2015). These estimates may not be accurate because the methods used in the surveys were not fully consistent. Nonetheless, the panda habitat in many places has been recovering (Liu and Viña, 2014, Viña et al., 2007, 2011) and captive breeding programs have been remarkably successful (State Council Information Office of China, 2015). The growth of the human population and its expansion into the giant panda range and associated human impacts caused the earlier decline (Liu et al., 2001, Reid and Gong, 1999). Conservation efforts are believed to be responsible for improvements since the early 2000s (State Council Information Office of China, 2015).

Despite encouraging signs, pandas still face many threats such as climate change and habitat fragmentation (Tuanmu et al., 2013, Viña et al., 2010). Analyzing panda–people interactions in the past and at present can help better address these threats in the future. The experiences and lessons from panda research and conservation can also offer insights for understanding and managing interactions between people and other species worldwide. In this chapter, we highlight the influences of pandas on people and outline human impacts on pandas. We then illustrate the feedbacks between the two through policies. We also discuss how panda conservation can help other species and how research on pandas–people interactions can help understand and manage human–nature interactions on the planet through developing and quantifying an integrated framework (Chapter 2).

1.2 Influences of pandas on humans

The giant panda has appeared throughout Chinese historical texts under various names, including *pixiu*, *mo*, *zouyu*, white bear, flowery-bear, bamboo-bear, and iron-eating animal (Hu, 2001). The first purported appearance of the animal is in a text from over 4000 years ago documenting the activities of two warring tribes battling in the Henan province of China. One tribe was led by a warrior named Huangdi who strategically used trained animals in combat, including the *pixiu* (Hu, 2001). Around the same time, another well-known tribe leader named Dayu purportedly received a panda skin for his efforts in flood management (Hu, 2001).

The panda pelt became a symbol of status and achievement as it was later associated with ruling emperors. The Empress Dowager Bo (170 BC) chose to be buried along with a panda skull (Hu, 2001). The Emperor Hanwu (~100 BC) held the panda as one of the esteemed animals in his garden for hunting (Hu, 2001). The Emperor Tang Taizong (~600 AD) purportedly held a banquet and honored his 14 guests by giving them panda skins (Hu, 2001). The panda also developed a reputation as a symbol of peace and friendship. It became revered in historical texts as far back as 290 AD as a creature that kills no other animals or people and coexists peacefully with its neighbors. For example, in a historic gesture, an emperor of the Tang dynasty (618–907 AD) sent two live pandas and a collection of panda skins to Japan as a sign of friendship (Schaller, 1993).

Despite the long history of folklore, the panda was not known to the Western world until 1869, when French missionary and naturalist Père Armand David received a panda pelt from a hunter while stationed in China (Hu, 2001). A series of panda hunting expeditions by curious Westerners soon followed, beginning with the Roosevelt brothers Theodore Jr. and Kermit in 1929 (Schaller, 1993). The first live panda to reach the West was a cub named Su-Lin, who was captured in 1936 by Ruth Harkness, an American socialite (Schaller, 1993). The cub was brought back to Chicago's Brookfield Zoo and died less than two years later (Schaller, 1993). The arrival ushered in a new area of fascination with these rare animals in the West and led to a clamoring of attempts by zoos to obtain live pandas. Between 1936 and 1946 alone, 14 live pandas were taken from China by foreigners (Lü and Kemf, 2001). By 1949, at least 73 pandas (dead or alive) had left China (Schaller et al., 1985).

China also began to embrace the worldwide interest in giant pandas and revived the tradition of presenting pandas to other countries as signs of friendship and diplomacy. One of the earliest efforts was a diplomatic gift of two pandas to the Bronx Zoo in 1941. The Chinese government gave the pandas as an appreciation to the USA for providing aid during World War II and as a token of international friendship (Morris and Morris, 1966). China gifted a total of 24 pandas to a number of countries, including Mexico, Germany, France, and the United Kingdom between 1957 and 1983 (Lü and Kemf, 2001). One well-known example during this period involved two pandas named Ling Ling and Hsing Hsing received by President Richard Nixon and First Lady Pat Nixon during their historic trip to China in 1972. This trip paved the way for a new era of improved ties between the two countries.

As the giant panda's global presence grew, its symbolic presence in its home country also flourished as the Chinese embraced the now world-famous creature found only within their borders. Interestingly, despite the previously mentioned periodic appearances of pandas in ancient texts, pandas were largely absent from cultural relics such as artworks in ancient China and pre-modern books (Songster, 2004). Unlike the dragon or the tiger, which appear prominently throughout ancient Chinese literature and art, the panda did not become a prominent cultural symbol until modern times (Songster, 2004). Beginning in the 1950s, pandas appeared in artwork and on manufactured goods and advertising campaigns (Songster, 2004). The panda was a logo on a variety of mass-produced products including electronics, dairy products, plastics, and stamps (Songster, 2004). The species very quickly ascended to iconic status, now widely and affectionately known by the Chinese people as a *guobao*, or national treasure of China. By the 1960s, pandas came to symbolize China itself, whereby the rendering of the panda's image was seen as glorifying China (Songster, 2004). Pandas also gained popularity due to their appearances in zoos, most notably the Beijing Zoo beginning in 1955 (Songster, 2004).

The panda later became one of the official mascots of the 1990 Asian Games and the 2008 Beijing Olympics, events crucial in presenting China to the international community in the twenty-first century.

The relatively small number of foreign "ambassador" pandas quickly rose to the status of icons in each country they visited. The National Zoo in Washington, DC, attracted over one million visitors during the first four weeks of Ling Ling and Hsing Hsing's residence in 1972 (up 48% from the previous year; Wright, 2012). As of November 2011, 154 giant pandas had either traveled to or were born outside of China in 20 different countries worldwide (Xie and Gipps, 2011). At the same time, media attention grew markedly. Global news articles published on giant pandas grew from 20 in 1972 to over 2700 in 2011 (according to a search in LexisNexis® of all news media, Figure 1.2). This amounts to more articles than about any other species of conservation concern from China, including the snub-nosed monkey (62), blue sheep (106), clouded leopard (258), and Asian elephant (1224). Perhaps the most popular international media outlet for pandas has been the online panda cameras associated with zoos, which receive millions of visitors per year. The panda cam website of the National Zoo in Washington, DC, received more than 237 000 hits within one day after the birth of a panda (Associated Press, 2012).

Through its popularity, the giant panda has also become an icon for conservation as a whole, extending its influence far beyond its habitat. The giant panda was a key face of early efforts to set up nature reserves across China and a target species that helped keep wildlife a priority of emerging government policies on nature conservation (Songster, 2004). Internationally, the public appeal of the panda Chi-Chi gifted to the London Zoo in 1958 helped inspire the founders of the World Wildlife Fund (WWF) to select the giant panda as its logo in 1961 (Lü and Kemf, 2001). The panda has remained the trademark of this famous conservation organization, which has since grown to involve over 1300 conservation projects around the world (World Wildlife Fund, 2012). As a result, the giant panda has come to be known by the world as a universal symbol for the global conservation movement (Lü and Kemf, 2001).

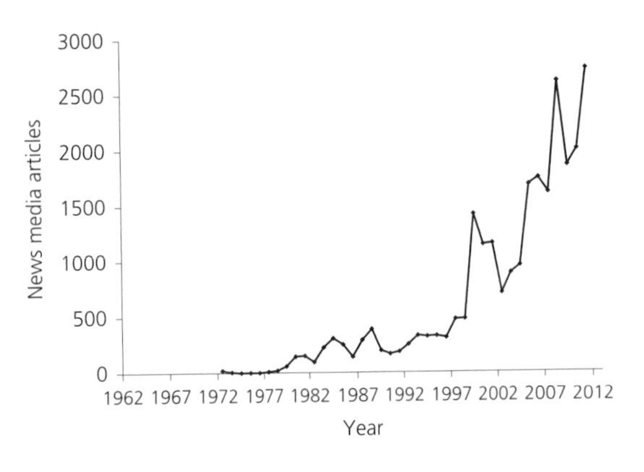

Figure 1.2 Number of world media news articles published in the English language on the giant panda by year (since 1972 and recorded in the LexusNexis® database).

At the same time, it is also important to appreciate the role of the giant panda as a flagship species, or one that has improved conservation efforts for countless other species that share the same habitat. Some areas of giant panda habitat rival the highest biodiversity of any ecosystem in the world. This is in part owing to the wide range in elevation, humid conditions, physical complexity, and diverse geological processes over time (Mackinnon, 2008). For example, the 9000 km^2 Qionglai Mountains support over 100 species of mammals (double that of the United Kingdom), 350 breeding bird species (more than Canada), and over 4000 plant species (more than both aforementioned countries; Mackinnon, 2008). Other threatened and endangered animal species whose ranges overlap with giant panda habitat include the golden snub-nosed monkey (*Rhinopithecus roxellana*), takin (*Budorcas taxicolor*), and red panda (*Ailurus fulgens*). Rare plants found throughout include the Chinese larch (*Larix chinensis*), Chinese yew (*Taxus chinensis*), and dove tree (*Davidia involucrata*).

The giant panda range also provides humans with many essential ecosystem services, such as fresh air, clean water, carbon sequestration, fuelwood, timber and non-timber forest products, and soil erosion prevention. Hundreds of thousands of people benefit from recreational tourism in panda habitat. Panda habitat also has cultural value for ethnic minority groups. For example, the estimated 10 000 remaining Baima people who occupy northern Sichuan and southern Gansu believe that mountains, trees, and animals are sacred (Luo et al., 2009).

Besides the benefits that pandas confer on people, those closest to panda habitat have incurred high costs from conservation efforts (Carter et al., 2014, He et al., 2008). These costs include restrictions on some human activities, such as farming, timber harvesting, fuelwood collection, and hunting (Section 1.4) as these activities negatively affect the panda and its habitat (Section 1.3). We expand upon the challenge of balancing panda and human needs in the sections that follow.

1.3 Human impacts on the giant panda and its habitat

The irony of the rising popularity of the giant panda on the international stage is that the wild giant panda population was dwindling and facing ever-increasing human threats during the same period. One of the driving forces behind such increases is a marked increase in the human population in China from roughly 400 million in 1900 to almost 1.4 billion today. Such a trend accompanies changes in human activities in the industrialized era. Examples include an exponential increase in gross domestic product (World Bank, 2007), and a rapid decline in household size and corresponding increase in per-individual resource use (Liu et al., 2003a).

The majority of disturbances in panda habitat in recent centuries have been anthropogenic. Nonetheless, natural disturbance events such as bamboo die-offs and earthquakes also have had impacts over short-term periods in some areas (State Forestry Administration, 2006). The main human threats include poaching, land cultivation, fuelwood collection, timber harvesting, road construction, livestock grazing, tourism, medicinal herb collection, bamboo harvesting, and mining (State Forestry Administration, 2006). Some of these threats relate to the conflict between panda conservation and meeting the livelihood needs of millions of low-income rural people living throughout panda habitat. Outside industrial and economic forces have caused other threats. The severity of each of these threats varies across space and has changed over time.

Historically, poaching of pandas for sport and fur was not an activity traditionally undertaken by villagers living in panda habitat. This is because of the large cost of acquiring a panda in the rough terrain relative to the minimal benefit of obtaining an animal with little medicinal, nutritional, or utilitarian use (Songster, 2004). Culturally, local hunters also view pandas as a taboo. Encountering a panda in the field during a hunting trip is said to bring bad luck, possibly because in Chinese culture people wear black and white during funerals. However, beginning in the mid-twentieth century international interest in obtaining pandas or their skins created a lucrative black market that increased the incidence of poaching substantially (Reid and Gong, 1999, Schaller, 1993). Poaching killed as many as 400 pandas in the late 1970s and 1980s (Hu, 1989), which was approximately 17% of the total population at that time (Hu, 2001). Poaching is believed to have decreased significantly since then due to strict punishments and increased awareness (State Forestry Administration, 2006). However, poaching still occurs, in part due to unintended harm to pandas while hunting for other animals such as musk deer and takin (Lü and Kemf, 2001).

Panda habitat has been disturbed by cultivation for several hundred years, with documented evidence dating back to at least the 1600s (Pan et al., 2001). This human activity is a threat to panda habitat because forests are cleared to cultivate crops such as rice, vegetables, and medicinal plants (State Forestry Administration, 2006). Land cultivation most severely affects low-elevation regions of panda habitat that are more accessible to humans. The elevational gradient has historically allowed people and pandas to coexist in their respective elevational ranges (Pan et al., 2001). But in the last several decades cultivation has extended up to prime panda habitat in the higher elevations (Schaller et al., 1985). For instance, in a roughly 1300 km^2 region in the mid-Minshan Mountains supporting approximately 115 pandas, the amount of cultivated land increased from 63 km^2 to 77 km^2 over the period 1994–2008. More importantly, new cultivation expanded to higher elevation areas, with the amount of cultivated land in the 1700–2400 m elevation range increasing by 65% (Wang, 2008). Cultivation contributed to the loss of 6% of panda habitat (~5.3 km^2) in this area during that period (Wang, 2008).

Timber harvesting was the most significant threat to the giant panda at one time, primarily from the 1960s to the early 1990s. Although a 1998 nationwide logging ban in natural forests in China has curbed this impact considerably, the legacy effect of past timber harvesting is still apparent in panda habitat (State Forestry Administration, 2006). Timber harvesting was the most frequently encountered disturbance in the third panda survey (2000–2004). It occurred in 28% of 34 187 plots across the panda range (more than double that of any other disturbance type and accounting for 41% of all disturbances; State Forestry Administration, 2006). In contrast to land cultivation driven by local communities living in panda habitat, timber harvesting was driven by outside forces, namely state- and privately owned logging corporations. Increases in timber harvesting in panda habitat reflect nationwide trends. Across China, harvest increased from 20 million m^3/year in the 1950s to 63 million m^3/year in the 1990s (Zhang et al., 2000). Pandas are affected by this activity because they cannot make good use of clear-cut stands. Bamboo will not grow, or grows in overly dense and non-nutritious patches, in clear-cuts (Bearer et al., 2008, Schaller et al., 1985). After the timber harvesting ban covering much of panda habitat in 1998, fuelwood collection in forests by locals to fulfill everyday cooking and heating needs continued (Bearer, 2005, Liu et al., 1999). Although less intensive, fuelwood collection can also disrupt forest structure and preclude panda inhabitance (Bearer et al., 2008). Despite these threats, pandas can use stands that are in the recovery stage after not having been cut for at least 37 years (Bearer et al., 2008).

Road construction is a major source of fragmentation of giant panda habitat. Within the giant panda range, there are over 1300 km of major roads (State Council Information Office of China, 2015). For example, the panda habitat in the Minshan Mountains, home to roughly 44% of the panda population, was once connected but is now divided into three distinct panda populations. This fragmentation is due to the creation of major highways and associated human activities surrounding them (State Forestry Administration, 2006). A similar pattern occurred in the Qionglai Mountains, home to 27% of the panda population. Here, the habitat has been fragmented into four main blocks, with further fragmentation occurring within each block (Xu et al., 2006). While there is some anecdotal documentation of pandas crossing roads (Zhu et al., 2011), there is also potential for roads to isolate subpopulations of pandas located on either side. Such isolation was shown in the significant genetic differentiation created by a main road in the Daxiangling and Xiaoxiangling mountains (Zhu et al., 2011).

Livestock grazing is another threat to the giant panda and its habitat. This threat was the most significant threat to giant pandas in the most recent fourth national giant panda survey (State Forestry Administration of China, 2015). Considering that most rural residents in giant panda habitat are farmers, livestock–panda interactions are expected. Livestock grazing was the most commonly observed disturbance to giant panda habitat across their entire range (34% of plots, 14% of all disturbances; State Forestry Administration of China, 2015). Across western China, livestock grazing has increased in recent years and threatens forest conservation goals because locals see animal husbandry as an alternative means to meet livelihood needs (Melick et al., 2007). Research in the Xiaoxiangling Mountains showed that giant pandas and livestock select similar habitats (Ran, 2003). Livestock grazing occurred at the same elevation as giant panda habitat nearly 50% of the time, and pandas appeared to avoid areas where livestock grazing occurred (Ran, 2004).

Tourism is another human activity that threatens giant pandas, a threat that has increased in recent years. Tourists across the nation of China reached over 3.09 billion in 2012 (including domestic and international visitors; CNTA, 2013). As of 2000, around 80% of China's nature reserves had developed tourism programs and nearly 16% attracted more than 100 000 tourists each year (Li and Han, 2001). For example, Wolong Nature Reserve (the major focus of this book) attracted over 200 000 tourists with more than US$6 million in revenue in 2006 (Liu et al., 2012). Wanglang Nature Reserve is another popular tourist destination and brings in over 20 000 visitors per year (Li et al., 2009). Tourism can impact giant pandas and their habitat through construction of tourism facilities and amenities, in addition to noise produced and garbage waste left behind by tourists (Liu, 2012, Liu et al., 2012).

Three other threats from medicinal herb collection, bamboo harvesting, and mining are less well studied but are nonetheless frequently observed across giant panda habitat (State Forestry Administration, 2006). New human impacts are also likely to emerge in the future as China and the rest of the world continue to undergo rapid economic development.

1.4 Policies as feedback to panda–people interactions

In the face of the dynamic and diverse human impacts occurring across panda habitat, the Chinese government launched an aggressive campaign to protect the species starting around the mid-twentieth century. This campaign mirrored global conservation efforts for other species over the last several decades (Ehrlich and Ehrlich, 1991). Across the world, diverse policies were put forth to reverse earlier destruction of nature and reduce threats to sustainability (Hull et al., 2015a, Liu et al., 2007a).

Perhaps the first documentation indicating a realization among the Chinese public of the decline in the panda population can be traced to 1946. At that time, the Chinese newspaper *Da Gong Bao* published an article entitled "Panda on the Brink of Extinction" (Schaller et al., 1985). The main concern at the time was excessive capturing of pandas by hunters. Government action came in 1959 when the Chinese government created its first wild animal protection policy. The government also listed the giant panda as one of nine "precious and rare" species, which prohibited the export of pandas (Songster, 2004). However, poaching of pandas was not officially outlawed until 1962 when the State Council of China called for new measures ensuring the protection of all wildlife (Schaller et al., 1985). The aforementioned frequency of poaching in the 1970s and 1980s indicate that the law was ineffective. However, gains were made when the government stepped up enforcement in the late 1980s with strict punishments for offenders, including capital punishment (Reid and Gong, 1999).

The Chinese government also took broader measures to curb wildlife trade. China joined the Convention on International Trade in Endangered Species (CITES) in 1981, which made international trade of pandas and their skins illegal (Lü and Kemf, 2001). The giant panda was then recognized by the International Union for Conservation of Nature as being rare (1986). The government later revised the status to endangered in 1990, as it was deemed to face "a very high risk of extinction in the wild" (IUCN, 2012). China also took steps to create broad legal protection for wildlife within the country by establishing the Chinese Wildlife Conservation Law in 1988. This law listed the giant panda as a category I species receiving the highest level of protection (Reid and Gong, 1999). The Chinese Ministry of Forestry and WWF (MacKinnon et al., 1989) later drafted the National Conservation Plan for the Giant Panda and its Habitat. This plan was adopted in 1992. It provided a set of guidelines for improving protection of this species. These included measures for reduction of human activities in panda habitat, removal of human settlements, modification of forestry operations, control of poaching, restoration of habitat, establishment of nature reserves, and captive breeding and reintroduction of pandas to the wild (summarized in Schaller, 1993).

Nature reserves emerged early on as one of the central strategies employed by the Chinese government to conserve giant pandas. This strategy was first proposed by a government order to establish forest reserves throughout China at the Third National People's Congress in 1957 (Schaller et al., 1985). The first nature reserves specifically for giant panda protection were created in 1963, a landmark year in which five nature reserves were established totaling 3550 km^2 (Ghimire, 1997). By 1983, there were 12 reserves covering nearly double that area (Schaller et al., 1985). By 1990, reserves encompassed around 35% of the wild giant panda population (Reid and Gong, 1999). As of the most recent giant panda survey conducted in 2011–2013, there were 67 reserves covering 54% of the predicted giant panda habitat (State Council Information Office of China, 2015; Figure 1.3). This trend reflects increases in nature reserve establishment throughout China as a whole. Nature reserves increased from 34 in 1978 to 2669 by the end of 2012, accounting for approximately 15% of the nation's total land area (SEPA, 2007, Xu et al., 2014).

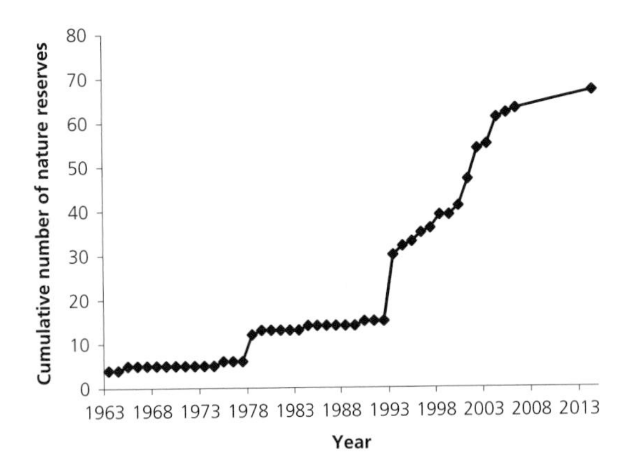

Figure 1.3 Cumulative number of nature reserves for giant pandas in China. Data are from the State Forestry Administration (2006) and the State Council Information Office of China (2015).

In an effort to improve effectiveness, all reserves are required to have a zoning scheme to contain human impacts. Reserves were required to delineate "experimental zones" for human development inside the protected area, with other areas set aside as "core zones" for biodiversity protection (The State Council, 1994). "Buffer zones" were required to soften neighboring human impacts (The State Council, 1994). However, zoning schemes have had their own limitations due to unclear requirements for each zone, poor design, and impracticality of enforcement on the ground (Hull et al., 2011a, Jim and Xu, 2004, Liu and Li, 2008; see also Chapters 4 and 13).

Much like many other protected areas (Hayes, 2006), panda reserves have struggled with lack of funding (Liu et al., 2003b), lack of enforcement capability, and lack of adequate monitoring (Lü and Kemf, 2001). As a result, the establishment of panda reserves has not always achieved successful conservation goals (Ghimire, 1997). For example, in Wolong Nature Reserve, the rate of habitat decline was more severe after the reserve was established than in the years before (Liu et al., 2001).

Perhaps the underlying challenge in panda reserve management comes from the fact that human communities reside within most of the reserves and have ancestors who lived in those communities for hundreds of years prior to reserve establishment. Historically, efforts to relocate people outside of reserves have not been successful (Schaller, 1993). Such efforts have also created ethical dilemmas since people living in reserves may already be disenfranchised due to their poor economic status and position as ethnic minorities (Johnson et al., 1996). In fact, as of 1999, residents in all of the counties that encompassed panda habitat had average per capita annual incomes that fell below the international poverty line (State Forestry Administration, 2006). Therefore, it is important to find creative ways to work together with local people to meet conservation goals without threatening human livelihoods.

Two of the best examples of conservation policies that accomplish this task are the Grain to Green Program (GTGP) and Natural Forest Conservation Program (NFCP). These two payments for ecosystem services (PES) programs were launched in the late 1990s at the national level in China to curb deforestation and the associated soil erosion (Liu et al., 2008). NFCP includes strict regulations on timber harvesting (including a harvesting ban in nature reserves) and payments for forest protection (Liu et al., 2008). GTGP provides farmers with subsidies for converting farmland to planted forest (Liu et al., 2008). The programs have helped protect panda habitat by discouraging illegal logging, improving local livelihoods, and prompting residents to harvest less fuelwood from forests (Chapters 6, 10, and 13; see also Chen et al., 2010, He et al., 2009, Viña et al., 2011).

Ecotourism appears on the surface to be another potential mechanism for improving both panda habitat and local livelihoods. Like many other countries, the Chinese government has actively promoted ecotourism (Liu et al., 2012). However, ecotourism has not functioned effectively in panda habitat due to a tendency for mass tourism development. There has so far been too little concern for environmental impacts and insufficient economic benefits have transferred to local communities (He et al., 2008, Liu et al., 2012).

Captive breeding has historically been a strong focus of interest in panda conservation efforts. Since the first successful breeding of a giant panda in the Beijing Zoo in 1963, the road to reaching this aim has been long and winding. In the first 26 years of breeding attempts, only 30% of captive pandas successfully reproduced and cub mortality was more than 60% (Zhang et al., 2006). Since then, improvements in maternal and infant care, captive breeding, and artificial insemination more than tripled the captive population between 1970 and 2000 (Zhang et al., 2006). In fact, the captive population reached almost 400 at the end of 2014 (State Council Information Office of China, 2015). The National Conservation Plan stated that the aim of captive breeding should be to bolster the wild giant panda population (Schaller, 1993). However, reintroduction is a complex process with a low overall success rate (~11%) across various animal species (Beck et al., 1994). The reintroduction of pandas has recently been attempted (Hull et al., 2011b, Zhou et al., 2005) and ultimately will only be successful if ongoing threats to the habitat are eliminated (Reading et al., 1999).

1.5 Concluding remarks

As a flagship for conservation, the fate of giant pandas may echo the success or failure of thousands of other endangered and threatened species worldwide (Barnosky et al., 2011). Timber harvesting, roads, livestock, and tourism also threaten numerous other mammals, birds, reptiles, and amphibians in the panda's range and in other ecosystems (Bowles et al., 1998, Gossling, 2002, Steinfeld et al., 2006, Trombulak and Frissell, 2000). Therefore, efforts to understand and prevent threats to the panda can help guide conservation of other less studied and less recognized species. For instance, studies demonstrating the limitations of nature reserves for the panda (Liu et al., 2001) have contributed to global dialogues on shortcomings of protected areas from marine environments (Jameson et al., 2002) to rainforests (Laurance et al., 2012). On the other hand, positive outcomes of PES programs on the recovery of panda habitat (Liu et al., 2008) have contributed to the worldwide appreciation of the utility of this method to conserve biodiversity while meeting basic livelihood needs of local people (Bulte et al., 2008, Farley and Costanza, 2010). The recent emerging recovery of panda population and habitat also provides hope for nature conservation in the face of mounting human impacts in the Anthropocene.

Research and management efforts on pandas–people interactions are also poised to help develop new approaches and generate new insights to achieve environmental sustainability and improve human well-being globally. They have also helped develop and quantify the framework of coupled human and natural systems, in which humans and natural components (e.g., pandas) interact (Chapters 2–17, Liu et al., 1999, 2007a, b, 2015). This chapter highlighted three major processes in coupled systems—influences of natural systems on humans, human impacts on natural systems, and human–nature feedbacks (see also Chapter 2). Further development and quantification of the coupled systems framework can help better understand and manage human–nature interactions around the planet. This can be accomplished by applying and scaling up methods and lessons from one coupled pandas–people system (Chapters 2–14,17, and 18) to other coupled human and natural systems across local to global scales (Chapters 15–19).

1.6 Summary

The giant panda is one of the world's most well-known endangered species. It has been our focal study species in the past two decades. In this chapter, we have explored impacts of pandas on people, impacts of people on pandas, and pandas–people feedbacks. These topics represent major processes in coupled human and natural systems. We began by highlighting references to pandas in Chinese historical texts as a symbol for peace and status and more recently as a national icon for China and an international icon for nature. The growth in popularity of the charismatic panda mirrors its decline to around 1864 remaining individuals in more than 30 isolated patches. Humans have threatened pandas and their habitat through diverse activities such as timber harvesting, livestock grazing, poaching, road construction, and tourism, all of which reflect the challenge of managing conflicting panda and human needs. The Chinese government has launched aggressive conservation policies to combat human impacts. These measures act as feedbacks geared toward reversing patterns of decline. Examples include strict punishments for poaching, the establishment of 67 nature reserves and implementation of payments for ecosystem services programs. We also illustrated how panda conservation has broad significance because panda habitat supports high biodiversity and supplies ecosystem services for human well-being. The pandas–people interactions and feedbacks represent many complex features of coupled systems and thus inform theoretical and applied research in this emerging field. Ultimately, the panda has provided, and will continue to provide, invaluable lessons about how to achieve sustainability in a human-dominated world.

References

Associated Press (2012) *After a Surprise Panda Birth in DC, Anxiety Awaits*. http://www.wivb.com/dpp/news/nation/After-a-surprise-panda-birth-in-DC-anxiety-awaits_79060825.

Barnosky, A.D., Matzke, N., Tomiya, S., et al. (2011) Has the Earth's sixth mass extinction already arrived? *Nature*, **471**, 51–57.

Bearer, S.L. (2005) *The Effects of Forest Harvesting on Giant Panda Habitat Use in Wolong Nature Reserve, China*. Doctoral Dissertation, Michigan State University, East Lansing, MI.

Bearer, S., Linderman, M., Huang, J., et al. (2008) Effects of fuelwood collection and timber harvesting on giant panda habitat use. *Biological Conservation*, **141**, 385–93.

Beck, B.B., Rapaport, L.G., Stanley-Price, M.R., and Wilson, A.C. (1994) Reintroduction in captive born animals. In P.J. Onley, G.M. Mace, and A.T.C. Feistner, eds, *Creative Conservation: Interactive Management of Wild and Captive Animals*, pp. 265–86. Chapman & Hall, London, UK.

Bowles, I.A., Rice, R.E., Mittermeier, R.A., and da Fonseca, G.A.B. (1998) Logging and tropical forest conservation. *Science*, **280**, 1899–900.

Bulte, E.H., Lipper, L., Stringer, R., and Zilberman, D. (2008) Payments for ecosystem services and poverty reduction: concepts, issues, and empirical perspectives. *Environment and Development Economics*, **13**, 245–54.

CNTA (2013) CNTA. *Statistical Bulletin of Tourism Industry in China on 2012*. Available online: http://www.cnta.gov.cn/html/2013–19/2013–19–12-%7B@hur%7D-39–08306.html (in Chinese).

Carter, N.H., Shrestha, B.K., Karki, J.B., et al. (2012) Coexistence between wildlife and humans at fine spatial scales. *Proceedings of the National Academy of Sciences of the United States of America*, **109**, 15360–65.

Carter, N.H., Viña, A., Hull, V., et al. (2014) Coupled human and natural systems approach to wildlife research and conservation. *Ecology and Society*, **19**, 43.

Chen, X., Lupi, F., Viña, A., et al. (2010) Using cost-effective targeting to enhance the efficiency of conservation investments in payments for ecosystem services. *Conservation Biology*, **24**, 1469–78.

Daily, G. (1997) *Nature's Services: Societal Dependence on Natural Ecosystems*. Island Press, Washington, DC.

Ehrlich, P.R. and Ehrlich, A.H. (1991) *Healing the Planet*. Addison-Wesley, New York, NY.

Farley, J. and Costanza, R. (2010) Payments for ecosystem services: from local to global. *Ecological Economics*, **69**, 2060–68.

Ghimire, K.B. (1997) Conservation and social development: an assessment of Wolong and other panda reserves in China. In K.B. Ghimire and M.P. Pimbert, eds, *Environmental Politics and Impacts of National Parks and Protected Areas*, pp. 187–213. Earthscan Publications, London, UK.

Gossling, S. (2002) Global environmental consequences of tourism. *Global Environmental Change-Human and Policy Dimensions*, **12**, 283–302.

Hayes, T.M. (2006) Parks, people, and forest protection: an institutional assessment of the effectiveness of protected areas. *World Development*, **34**, 2064–75.

He, G., Chen, X., Bearer, S., et al. (2009) Spatial and temporal patterns of fuelwood collection in Wolong Nature Reserve: implications for panda conservation. *Landscape and Urban Planning*, **92**, 1–9.

He, G., Chen, X., Liu, W., et al. (2008) Distribution of economic benefits from ecotourism: a case study of Wolong Nature Reserve for giant pandas in China. *Environmental Management*, **42**, 1017–25.

Hu, J. (1989) *Life of the Giant Panda*. Chongqing University Press, Chongqing, Sichuan (in Chinese).

Hu, J. (2001) *Research on the Giant Panda*. Shanghai Science and Education Publishing House, Shanghai, China (in Chinese).

Hull, V., Shortridge, A., Liu, B., et al. (2011b) The impact of giant panda foraging on bamboo dynamics in an isolated environment. *Plant Ecology*, **212**, 43–54.

Hull, V., Tuanmu, M.-N., and Liu, J. (2015a) Synthesis of human-nature feedbacks. *Ecology and Society* **20**(3), 17.

Hull, V., Xu, W.H., Liu, W., et al. (2011a) Evaluating the efficacy of zoning designations for protected area management. *Biological Conservation*, **144**, 3028–37.

Hull, V., Zhang, J., Zhou, S., et al. (2015b) Space use by endangered giant pandas. *Journal of Mammalogy*, **96**, 230–36.

Hunt, R.M. (2004) A paleontologist's perspective on the origin and relationships of the giant panda. In D. Lindburg and K. Baragona, eds, *Giant Pandas: Biology and Conservation*, pp. 45–52. University of California Press, Berkeley, CA.

IUCN (2012) *The IUCN Red List of Threatened Species*. Version 2012.2. http://www.iucnredlist.org. Downloaded on 17 October 2012.

Intergovernmental Platform on Biodiversity and Ecosystem Services (2014) http://www.ipbes.net/.

Jameson, S.C., Tupper, M.H., and Ridley, J.M. (2002) The three screen doors: can marine protected areas be effective? *Marine Pollution Bulletin*, **44**, 1177–83.

Jim, C.Y. and Xu, S.S.W. (2004) Recent protected-area designation in China: an evaluation of administrative and statutory procedures. *The Geographical Journal*, **170**, 39–50.

Jin, C., Ciochon, R.L., Dong, W., et al. (2007) The first skull of the earliest giant panda. *Proceedings of the National Academy of Sciences of the United States of America*, **104**, 10932–37.

Johnson, K.G., Yao, Y., You, C., et al. (1996) Human/carnivore interactions: conservation and management implications from China. In J.L. Gittleman, ed., *Carnivore Behavior, Ecology, and Evolution* (vol. **2**), pp. 337–70. Cornell University Press, Ithaca, NY.

Kareiva, P. and Marvier, M. (2015) *Conservation Science: Balancing the Needs of People and Nature, second edition*. Roberts & Company Publishers, Greenwood Village, CO.

Laurance, W.F., Useche, D.C., Rendeiro, J., et al. (2012) Averting biodiversity collapse in tropical forest protected areas. *Nature*, **489**, 290–94.

Li, Q., Feng, M., and Yan, Y. (2009) Analysis and forecast of the dynamic number of tourists about an ecotourism area on the basis of a logistic growth model: Wanglang Nature Reserve in Mianyang city. *Ecological Economy*, **1**, 234–38 (in Chinese).

Li, W. and Han, N. (2001) Ecotourism management in China's nature reserves. *Ambio*, **30**, 62–63.

Liu, W. (2012) *Patterns and Impacts of Tourism Development in a Coupled Human and Natural System*. Doctoral Dissertation, Michigan State University, East Lansing, MI.

Liu, J. (2015) Promises and perils for the panda. *Science* **348**, 642.

Liu, J., Daily, G.C., Ehrlich, P.R., and Luck G.W. (2003a) Effects of household dynamics on resource consumption and biodiversity. *Nature*, **421**, 530–33.

Liu, J., Dietz, T., Carpenter, S.R., et al. (2007a) Complexity of coupled human and natural systems. *Science*, **317**, 1513–16.

Liu, J., Dietz, T., Carpenter, S.R., et al. (2007b) Coupled human and natural systems. *Ambio*, **36**, 639–49.

Liu, J., Li, S., Ouyang, Z., et al. (2008) Ecological and socioeconomic effects of China's policies for ecosystem services. *Proceedings of the National Academy of Sciences of the United States of America*, **105**, 9477–82.

Liu, J., Linderman, M., Ouyang, Z., et al. (2001) Ecological degradation in protected areas: the case of Wolong Nature Reserve for giant pandas. *Science*, **292**, 98–101.

Liu, J., Mooney, H., Hull, V., et al. (2015) Systems integration for global sustainability. *Science*, **347**, 1258832.

Liu, J., Ouyang, Z., Pimm, S.L., et al. (2003b) Protecting China's biodiversity. *Science*, **300**, 1240–41.

Liu, J., Ouyang, Z., Taylor, W.W., et al. (1999) A framework for evaluating the effects of human factors on wildlife habitat: the case of giant pandas. *Conservation Biology*, **13**, 1360–70.

Liu, J. and Viña, A. (2014) Pandas, plants, and people. *Annals of the Missouri Botanical Garden*, **100**, 108–25.

Liu, W., Vogt, C.A., Luo, J., et al. (2012) Drivers and socioeconomic impacts of tourism participation in protected areas. *PLoS ONE*, **7**, e35420.

Liu, X. and Li, J. (2008) Scientific solutions for the functional zoning of nature reserves in China. *Ecological Modelling*, **215**, 237–46.

Loucks, C.J., Lü, Z., Dinerstein, E., et al. (2001) Giant pandas in a changing landscape. *Science*, **294**, 1465.

Lü, Z. and Kemf, E. (2001) *Giant pandas in the Wild: A WWF Species Status Report*. WWF International, Gland, Switzerland.

Luo, Y., Liu, J., and Zhang, D. (2009) Role of traditional beliefs of Baima Tibetans in biodiversity conservation in China. *Forest Ecology and Management*, **257**, 1995–2001.

MacKinnon, J. (2008) Species richness and adaptive capacity in animal communities: lessons from China. *Integrative Zoology*, **3**, 95–100.

MacKinnon, J.F., Bi, F., Qiu, M., et al. (1989) *National Conservation Plan for the Giant Panda and its Habitat*. Ministry of Forestry and WWF, Peoples Republic of China.

Melick, D., Yang, X., and Xu, J. (2007) Seeing the wood for the trees: how conservation policies can place greater pressure on village forests in southwest China. *Biodiversity and Conservation*, **16**, 1959–71.

Millennium Ecosystem Assessment (2005) *Ecosystems and Human Well-being*. Island Press, Washington, DC.

Morris, R. and Morris, D. (1966) *Men and Pandas*. McGraw-Hill Book Company, New York, NY.

Pan, W., Lü, Z., Zhu, X.J., et al. (2001) *A Chance for Lasting Survival*. Beijing University Press, Beijing, China (in Chinese).

Pimm, S., Jenkins, C., Abell, R., et al. (2014) The biodiversity of species and their rates of extinction, distribution, and protection. *Science*, **344**, 1246752.

Ran, J. (2003) Habitat selection by giant pandas and grazing livestock in the Xiaoxiangling Mountains of Sichuan Province. *Journal of Sun Yatsen University Social Science Edition*, **23**, 2253–59 (in Chinese).

Ran, J. (2004) A survey of disturbance of giant panda habitat in the Xiaoxiangling Mountains of Sichuan Province. *Acta Theriologica Sinica*, **24**, 277–81 (in Chinese).

Reading, R.P., Miller, B.J., and Price, M.R.S. (1999) Reintroducing animals into the wild: lessons for giant pandas. In S. Mainka and Z. Lü, eds, *International Workshop on the Feasibility of Giant Panda Re-introduction*, pp. 146–57.

China Forestry Publishing House, Wolong Nature Reserve, Sichuan, China.

Reid, D.G. and Gong, J. (1999) Giant panda conservation action plan. In C. Servheen, S. Herrero, and B. Peyton, eds, *Bears Status Survey and Conservation Action Plan*, pp. 241–45. IUCN/SSC Bear and Polar Bear Specialist Groups, Gland, Switzerland.

SEPA (2007) *China Environmental Statistics Bulletin*. State Environmental Protection Administration, Beijing,China (in Chinese).

Schaller, G.B. (1993) *The Last Panda*. University of Chicago Press, Chicago, IL.

Schaller, G.B., Hu, J., Pan, W., and Zhu, J. (1985) *The Giant Pandas of Wolong*. University of Chicago Press, Chicago, IL.

Songster, E. (2004) *A Natural Place for Nationalism: The Wanglang Nature Reserve and the Emergence of the Giant Panda as a National Icon*. Doctoral Dissertation, University of California, San Diego, San Diego, CA.

State Council Information Office of China (2015) *Press Conference on the Fourth National Panda Survey Results* http://www.scio.gov.cn/xwfbh/gbwxwfbh/fbh/Document/1395514/1395514.htm (in Chinese).

State Forestry Administration of China (2015) *The Fourth National Giant Panda Survey*. http://www.forestry.gov.cn/main/4462/content-743596.html (in Chinese).

State Forestry Administration (2006) *The Third National Survey Report on the Giant Panda in China*. Science Publishing House, Beijing,China (in Chinese).

Steinfeld, H., Gerber, P., Wassenaar, T., et al. (2006) *Livestock's Long Shadow: Environmental Issues and Options*. FAO, Rome.

The State Council (1994) *Bylaws of Nature Reserves in the People's Republic of China* (in Chinese).

Tougard, C. (2001) Biogeography and migration routes of large mammal faunas in Southeast Asia during the Late Middle Pleistocene: focus on the fossil and extant faunas from Thailand. *Palaeogeography, Palaeoclimatology, Palaeoecology*, **168**, 337–58.

Trombulak, S.C. and Frissell, C.A. (2000) Review of ecological effects of roads on terrestrial and aquatic communities. *Conservation Biology*, **14**, 18–30.

Tuanmu, M.-N., Viña, A., Winkler, J.A., et al. (2013) Climate-change impacts on understorey bamboo species and giant pandas in China's Qinling Mountains. *Nature Climate Change*, **3**, 249–53.

Viña, A., Bearer, S., Chen, X., et al. (2007) Temporal changes in giant panda habitat connectivity across boundaries of Wolong Nature Reserve, China. *Ecological Applications*, **17**, 1019–30.

Viña, A., Chen, X.D., McConnell, W.J., et al. (2011) Effects of natural disasters on conservation policies: the case of the 2008 Wenchuan Earthquake, China. *Ambio*, **40**, 274–84.

Viña, A., Tuanmu, M.-N., Xu, W., et al. (2010) Range-wide analysis of wildlife habitat: implications for conservation. *Biological Conservation*, **143**, 1960–69.

Wang, W., Potts, R., Baoyin, Y., et al. (2007) Sequence of mammalian fossils, including hominoid teeth, from the Bubing Basin caves, South China. *Journal of Human Evolution*, **52**, 370–79.

Wang, X. (2008) *The Impact of Human Disturbance on Giant Panda Habitat in the Minshan Mountains*. Doctoral Dissertation, Chinese Academy of Sciences, Beijing, China (in Chinese).

World Bank (2007) *World Development Indicators*. World Bank, Washington, DC, USA.

World Wildlife Fund (2012) *WWF: history, people, operations*. http://wwf.panda.org/who_we_are/.

Wright, J. (2012) *Panda-monium!* http://siarchives.si.edu/blog/panda-monium.

Xie, Z. and Gipps, J.(2011) *The International Studbook for Giant Panda (Ailuropoda melanoleuca)*. Updated November 2011. Beijing, China.

Xu, J., Sun, G., and Liu, Y. (2014) Diversity and complexity in the forms and functions of protected areas in China. *Journal of International Wildlife Law & Policy*, **17**, 102–14.

Xu, W., Ouyang, Z., Vina, A., et al. (2006) Designing a conservation plan for protecting the habitat for giant pandas in the Qionglai mountain range, China. *Diversity and Distributions*, **12**, 610–19.

Yang, W., Dietz, T., Liu, W., et al. (2013) Going beyond the Millennium Ecosystem Assessment: an index system of human dependence on ecosystem services. *PLoS ONE*, **8**, e64581.

Zhang, P., Shao, G., Zhao, G., et al. (2000) China's forest policy for the 21st century. *Science*, **288**, 2135.

Zhang, Z., Zhang, A., Hou, R., et al. (2006) Historical perspective of breeding giant pandas ex situ in China and high priorities for the future. In D.E. Wildt, A. Zhang, H. Zhang, D.L. Janssen, and S. Ellis, eds, *Giant Pandas: Biology, Veterinary Medicine and Management*, pp. 455–68. Cambridge University Press, Cambridge, UK.

Zhou, X., Tan, Y., Song, S., et al. (2005) Comparative study on behavior and ecology between captivity and seminature enclosure of giant panda. *Sichuan Journal of Zoology*, **24**, 143–46 (in Chinese).

Zhu, J. and Long, Z. (1983) The vicissitudes of the giant panda. *Acta Zoologica Sinica*, **29**, 93–104 (in Chinese).

Zhu, L., Zhang, S., Gu, X., and Wei, F. (2011) Significant genetic boundaries and spatial dynamics of giant pandas occupying fragmented habitat across southwest China. *Molecular Ecology*, **20**, 1122–32.

CHAPTER 2

Framing Sustainability of Coupled Human and Natural Systems

Jianguo Liu, Vanessa Hull, Neil Carter, Andrés Viña, and Wu Yang

2.1 Introduction

Human–nature interactions have existed since the beginning of human history, but their scope and intensity have increased drastically since the Industrial Revolution, especially in the past several decades. Historically, most human–nature interactions were at the local scale, although some large-scale activities such as migrations and trade did occur. However, human–nature interactions at large scales have become much more frequent and faster than before (Liu et al., 2007a). With more people and fewer resources on the planet, the world has entered the epoch of the Anthropocene (Steffen et al., 2007).

As the human population continues to increase, human impacts have spread to every corner of the earth (e.g., IPCC, 2008, Marsh, 1864, Millennium Ecosystem Assessment, 2005, Thomas Jr, 1956, Turner et al., 1990, Vitousek et al., 1997). Both human and natural systems have become more vulnerable, and the earth has crossed several planetary boundaries (Rockstrom et al., 2009, Steffen et al., 2015). Such consequences have largely resulted from the historical views of human–nature relationships—humans should and can conquer nature (Liu, 2010), or humans can utilize nature without limits (Simon, 1996). Some scholars have recognized that human development is constrained by natural resources (e.g., Meadows et al., 2004) and that humans should maintain harmonious relationships with nature (e.g., in ancient Chinese and Native American cultures; Diamond, 2005, Shapiro, 2001). However, these views were largely downplayed or ignored in practice (Ehrlich and Pringle, 2008). For example, the major development model around the world has been "pollute first and clean up later." Even though some areas were cleaned up later, improvements were often made at the cost of damaging other areas (Hird, 1993).

How to enable humans to prosper while sustaining natural systems is among the most challenging questions in the world today (Clark and Dickson, 2003, Kates et al., 2001, Matson, 2012). As human impacts continue to rise, more people have realized that the traditional human-dominant views cannot sustain life-supporting systems and new development paradigms are needed. A watershed event took place when the World Commission on Environment and Development published the historic document *Our Common Future* in which sustainable development was proposed (World Commission on Environment and Development, 1987). The research and management communities have widely accepted the concept of sustainable development (meeting the needs of the present society without compromising future generations), but the practice of sustainable development has lagged behind.

A major barrier to effective implementation of sustainable development is the lack of sufficient knowledge on the complex relationships between humans and nature. To help generate such knowledge, an integrated concept—Coupled Human and Natural Systems (CHANS)—has emerged. Coupled systems are those in which human and natural components interact (Liu et al., 2007b), emphasizing reciprocal interactions and feedbacks. These not

Pandas and People. Edited by Jianguo Liu, Vanessa Hull, Wu Yang, Andrés Viña, Xiaodong Chen, Zhiyun Ouyang, and Hemin Zhang. Published 2016 by Oxford University Press.

only include human impacts on nature, but also nature's influences on humans. Furthermore, substantial progress has been made in the development and quantification of integrated conceptual frameworks (Liu et al., 2015). In this chapter, we set the stage for the remainder of the book by presenting an overview of a coupled human and natural system framework. We also demonstrate typology of coupled systems and apply the overall integrated framework to coupled human and wildlife systems. In addition, we highlight theoretical foundations and relevant methods for studying coupled systems and propose a model coupled system approach.

2.2 Overall conceptual framework

Like all other studies, conceptual frameworks are essential for effective research on coupled human and natural systems. Such frameworks provide a shared set of variables for the design of data collection instruments, the conduct of fieldwork or experiments, and the analysis and integration of results (Ostrom, 2007). They help identify variables and relationships that may influence how policies enhance the sustainability of some components of a coupled system while potentially compromising the sustainability of other components. Frameworks connecting relevant variables can help integrate isolated knowledge obtained from separate studies conducted in different places at different times. When presented in the form of diagrams (e.g., Figure 2.1), they also present a visual representation of various components and their relationships.

The core of the conceptual framework of coupled human and natural systems lies in the interactions and feedbacks between the human subsystem and

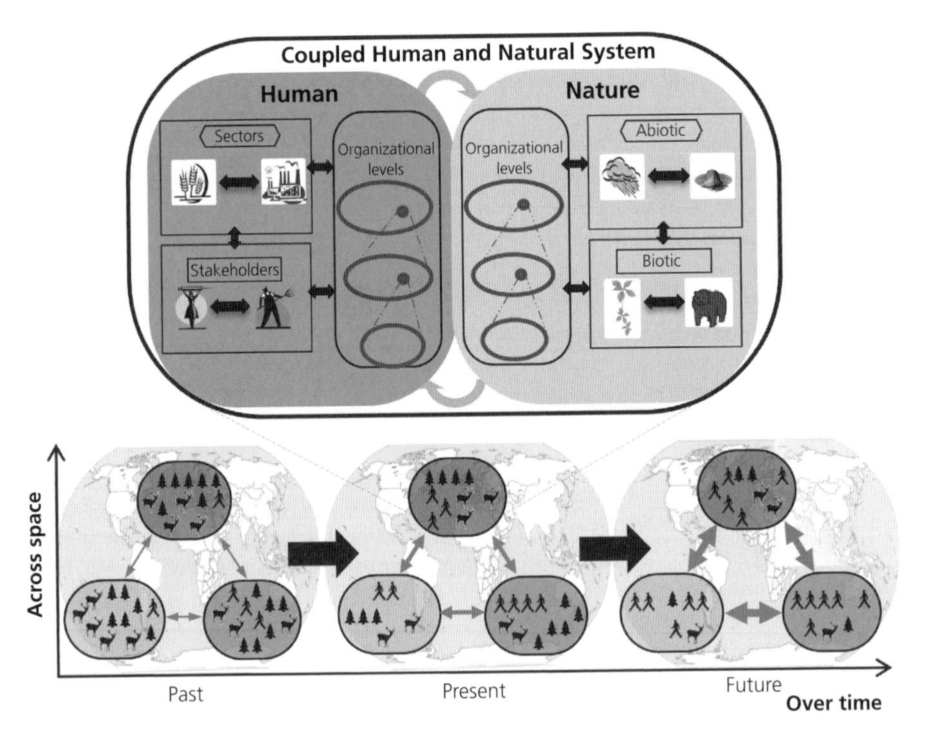

Figure 2.1 Diagram illustrating systems integration of coupled human and natural systems (CHANS) at multiple organizational levels, across space, and over time. The world is a large coupled human and natural system that consists of many smaller ones (exemplified by three here). Each coupled system includes human and natural components at different organizational levels (top inset). Coupled systems are connected through flows of information, matter, and energy as well as feedbacks across space as shown by arrows. Over time, these coupled systems are increasingly human-dominated (represented by more human icons and fewer plant and wildlife icons) and coevolve as their interconnections become stronger from the past and present to the future (represented by increasingly thicker arrows among coupled systems).

the natural subsystem (Figure 2.1). The human subsystem includes diverse stakeholders (e.g., farmers, workers, and traders) working in different sectors (e.g., agriculture, manufacturing, and trade). The natural subsystem consists of various abiotic (e.g., geomorphic and climatic) and biotic (e.g., animal and vegetative) components. As Figure 2.1 shows, a coupled system is often hierarchically structured across multiple levels. It can have multiple tiers (Ostrom, 2009) with many components—each component can divide into subcomponents, and each subcomponent can further divide into subcomponents. The decision to choose which level of variables for analysis is dependent on the questions and objectives of the project, feasibility, spatial and temporal scales, and the types of coupled systems (Ostrom, 2009). Various components interact with each other at the same or across different organizational levels inside a coupled system (i.e., internal or local couplings). For example, different stakeholders or sectors in a human subsystem may cooperate to solve a common problem, or compete for the same resources (Figure 2.1). Biotic components in the natural subsystem interact with each other and rely on abiotic components for growth, survival, and reproduction (Figure 2.1).

Each coupled system also interacts with other coupled systems across spatial boundaries (Figure 2.1; Liu et al., 2013). Such interactions may be carried out by flows or movement of matter, energy, and information. For example, even in many protected areas, people from outside come inside to collect resources (e.g., mineral, timber, and non-timber products), graze, hunt, and visit as tourists. Pollutants and diseases from outside may arrive in the protected areas through water flows (Rogers and Beets, 2001), air circulations (Bytnerowicz et al., 2002), species dispersal (Ervin, 2003), and disposal by people (Saunders et al., 2002). Noises generated outside may also influence organisms and people inside a protected area. On the other hand, wildlife inside a protected area may roam across the boundaries to feed (Douglas-Hamilton et al., 2005), rest (Caro, 1999), and migrate (Gude et al., 2007). Similarly, residents inside a protected area may also use resources elsewhere through trade. Pollinators inside protected areas may fly to pollinate crops outside. The examples here illustrate interactions of individual components in the focal coupled system and other coupled systems. Such interactions lead to interactions at the system level because other components and their relationships may be affected, and ultimately the entire systems may be affected.

Interactions between humans and natural subsystems evolve over time (Figure 2.1). The temporal variations may involve all aspects of human–nature interactions and result from coevolutionary human–nature relationships (Alberti, 2015). Human–nature interactions may go through several phases over time, from emergence to strengthening, weakening, and dissolution. Each phase may last different durations. For example, many people began to use forests in rural China as fuelwood upon their settlement (Liu et al., 1999). The amount of fuelwood collection increased as human population increased. However, as electricity became more available and affordable, fuelwood use decreased. As a result of the new policy banning forest harvesting, forest harvesting was stopped and thus human–nature coupling in the form of forest harvesting was dissolved (Yang et al., 2013a).

The coupled human and natural systems approach also encompasses a number of fundamental concepts that describe complexities in human–nature interactions (Table 2.1). For example, time lags may occur in which the effect of the human subsystem on the natural subsystem (or vice versa) may take a long time to emerge. A new household may not form until approximately two decades after a child is born—when the child gets married (An and Liu, 2010). There are also legacy effects in which past events impact the current conditions and future trajectory of the system. Sometimes these effects appear only after a threshold or tipping point has been reached, when sudden and unexpected changes happen.

Compared to previous notions and practices related to human–nature interactions, the coupled human and natural systems framework explicitly emphasizes the feedbacks between humans and natural subsystems. Feedbacks occur when human activities lead to changes in natural subsystems and changes in natural subsystems in turn affect human subsystems. On one hand, humans rely on natural subsystems (e.g., ecosystem

Table 2.1 Some key concepts of coupled human and natural systems and examples (Liu et al., 2007a, b)

Concept	Definition	Example
Feedback	An effect that reaches a destination and is then fed back to its source (e.g., from natural subsystem to human subsystem and back to natural subsystem again)	Increases in fuelwood collection harm forests, which result in creation of new conservation policies that act as a negative feedback to improve forest cover (Liu et al., 1999)
Time lag	An effect that takes a long time to emerge	Changes in household demographics (e.g., time at marriage, time of first birth) do not substantially impact forest cover until approximately two decades later when children grow up and form new households (An et al., 2001)
Legacy effect	Impact of prior events that affect future conditions	Some systems have evolved over long time frames to be dependent on fossil fuels, a phenomenon called "carbon lock-in" (Unruh, 2000)
Threshold	A transition or tipping point between two alternate states (Speth and Repetto, 2008)	Declines in bird species accelerate at a rapid rate after a certain percentage of suburban housing is reached (Sass et al., 2006)
Surprise	Unexpected outcomes that result from human–nature interactions	People divide into smaller households to take advantage of per-household conservation policy payments, in turn increasing their resource consumption and compromising the policy goals (Liu et al., 2003)
Resilience	Capability of retaining system structure and functioning after disturbance (Gunderson and Holling, 2001)	Off-farm income improves economic stability of households facing risks and uncertainties due to drought, poor soil, or disease (Barrett et al., 2001)
Emergent properties	Properties that arise only after interactions between the human and natural subsystems occur	Some introduced species may not become invasive until after humans have degraded the land to create disturbed habitats that are easier for them to occupy (MacDougall and Turkington, 2005)
Vulnerability	Likelihood of the system to experience harm as a result of exposure to a stressor (Turner et al., 2003)	Investment in high-yield, monoculture cash crops can make both rural households and their surrounding environments more vulnerable to disasters (Eakin and Luers, 2006)
Non-linearity	A disproportionate relationship between two factors (e.g., cause and effect)	Medium-sized groups of people are more effective at participation in conservation programs than larger groups, which are more prone to have "free riders" or cheaters (Yang et al., 2013b)
Heterogeneity	Spatial, temporal, and organizational differences in human–nature couplings	Different rates and patterns of urban sprawl give rise to different magnitudes of threats to wildlife populations facing fragmentation (Hasse and Lathrop, 2003)

services such as clean water, fresh air, food, shelter, bioenergy) for survival and development (5). At the same time, humans alter the natural subsystems, usually in a manner that degrades them. On the other hand, alteration to natural subsystems also affects humans in different ways (Estes et al., 2011, Foley et al., 2005). For example, climate change can have devastating negative effects on agricultural production. While reciprocal human–nature interactions were recognized a long time ago (Diamond, 1997, Ma and Wang, 1984, Marsh, 1864, McDonnell and Pickett, 1993, National Research Council, 1999, Odum, 1971, Thomas Jr, 1956, Turner et al., 1990, Vitousek et al., 1997), their complex patterns and processes are still not well understood (Diamond, 2005, Schneider and Londer, 1984). In the past, ecologists usually focused on natural subsystems and excluded human impacts from their experiments and observations (Redman et al., 2004). Researchers viewed humans as a nuisance—contaminating samples, skewing data, and biasing analyses. In contrast, social scientists often focused on humans but paid less attention to natural subsystems. The coupled system framework provides a platform for natural and social scientists to work together to quantify and integrate human–nature relationships at multiple organizational levels across space and over time (Figure 2.1). Such quantification and integration are essential to a better appreciation of the constraints and consequences of human activities and to developing better practices in socioeconomic development.

Many frameworks of coupled systems have been developed. A major difference between our framework (Figure 2.1) and others is that ours explicitly specifies the interactions and feedbacks between the focal coupled system and other coupled systems while other frameworks do not. In the coupled natural–human system diagram proposed by the US National Science Foundation (NSF), for example, there is no input from other systems or output to other systems (National Science Foundation, 2010, 2014). This framework implicitly treats each coupled system as a closed system, without interactions with the outside. In reality, this is not true for the vast majority of (if not all) systems. Among the ten frameworks reviewed (Binder et al., 2013), none of them explicitly specifies the feedbacks between the focal coupled system and other coupled systems. The framework proposed by Nobel Laureate Elinor Ostrom suggests interactions between the focal coupled system and other "ecosystems" (Ostrom, 2009). This framework also places the focal coupled system under "specific social, economic, and political settings" (Ostrom, 2009). Many studies have treated external factors as drivers of changes in the focal coupled system (DeFries et al., 2010, Lambin and Meyfroidt, 2011, Liu et al., 2012, Simberloff and Von Holle, 1999, Verburg et al., 2002, Stevens et al., 2014). Other separate studies have considered the impact of the focal system on other systems as spatial externalities (Meyfroidt and Lambin, 2009, Wassenaar et al., 2007). Because there are now more interactions and feedbacks across system boundaries than ever before, it is important to consider them (Walker et al., 2009, Folke et al., 2011). They exist not only between the focal system and adjacent coupled systems (Boersma and Parrish, 1999), but also between the focal system and distant coupled systems via "telecoupling" (socioeconomic and environmental interactions over distances; see Chapter 17, Liu et al., 2013).

2.3 Typology of coupled human and natural systems

There is a vast variety of terms used for coupled human and natural systems (see examples in Table 2.2). Some differences among them are trivial. For instance, the order of the words "human" and "natural" varies (e.g., coupled human and natural systems vs coupled natural and human systems). In some cases, the word "and" between "human" and "natural" is replaced by "–" (e.g., coupled human and natural systems vs coupled human–natural systems). In other cases, the word "coupled" is replaced with "interlinked." Others differ in the use of keywords for human and natural subsystems (e.g., "social" instead of "human" and "ecological" instead of "natural"). These differences may be a matter of preferences and traditions. When using different keywords, sometimes they are viewed as either synonyms or specific cases of coupled human and natural systems that focus on specific aspects of human or natural components. Human subsystems encompass many dimensions, such as social, economic, and demographic. Similarly, natural subsystems also encompass many dimensions, including ecological, genetic, and biophysical.

A commonly used concept, social–ecological systems (Brown, 2008, Folke et al., 2005), may be a synonym or a special case of coupled human and natural systems. This concept focuses on social aspects of the human subsystem and ecological aspects of the natural subsystem. However, the definitions of terms such as social and ecological may vary. Sometimes, "ecological" refers to all dimensions of a natural subsystem, while in other situations to only one of the many dimensions of the natural subsystem (e.g., not including the biophysical dimension). Similarly, sometimes "social" refers to all dimensions of humans, while in other situations, it refers to only one of the many dimensions of humans (i.e., social, not economic). The different uses of the same term appear not only in academic articles, but also in other circumstances such as names of institutions. For example, a unit within the NSF is the Directorate for Social, Behavioral, and Economic Sciences; social sciences are treated differently from economic and behavioral sciences. In other situations, social sciences include

Table 2.2 Examples of terms used for coupled human and natural systems

Example terms	Example references
Human and natural system	(Dewulf et al., 2007)
Coupled human–natural system	(Chapman et al., 2009)
Human–natural system	(Ebbin, 2009)
Coupled human–nature system	(Carmenta et al., 2011)
Coupled natural and human system	(Blondel, 2006)
Natural and human system	(Lorenzoni et al., 2000)
Natural–human system	(Bennett and McGinnis, 2008)
Coupled human–environment system	(Turner, 2010)
Human–environment system	(Metzger et al., 2008)
Human and environmental system	(Stroup and Finewood, 2011)
Coupled social and ecological system	(Satake et al., 2008)
Coupled social–ecological system	(Schluter et al., 2009)
Interlinked social–ecological system	(Lacitignola et al., 2010)
Social–ecological system	(Folke et al., 2005)
Socioecological system	(Carpenter et al., 2001)
Coupled socio-ecological system	(Craig, 2010)
Socio-ecological system	(Young, 2010)
Ecological–social system	(Pickett et al., 2004)
Ecosocial system	(Gilioli and Baumgartner, 2007)
Eco-social system	(Waltner-Toews and Kay, 2005)
Socioeconomic–ecological system	(Pedroza and Salas, 2011)
Ecological–economic system	(Costanza, 1996)
Population–environment system	(Miller et al., 2010)
Social–environmental system	(Eakin and Luers, 2006)
Socio-environmental system	(van der Leeuw, 2008)
Socio-natural system	(Aguilar et al., 2009)
Social-economic-natural complex ecosystem	(Ma and Wang, 1984)

economics and other human-related disciplines. For example, the College of Social Sciences at Michigan State University and the Division of Social Science at Harvard University include departments of Anthropology, Economics, Psychology, and Sociology.

Another commonly used concept is human–environment systems (Moran, 2010, Turner et al., 2003). When the word "environment" refers to the natural environment, it would be the same as coupled human and natural systems. However, the term environment sometimes refers to or includes the "built" environment (Ingold, 1995). In this case, human–environment systems seem to emphasize the relationships among different aspects of human subsystems, rather than human–nature relationships. The built environment is constructed by humans, and thus it is considered a part of the human subsystem.

All these and other concepts (e.g., ecological–economic systems and population–environment systems, Table 2.2) are related although there may be some subtle differences. The most important goal seems to be the same across concepts—to emphasize the relationships between human and natural subsystems, either some or all of the major components of human and natural subsystems. Just like other concepts, using different terms to refer to the same thing is common. As time goes by, some of the terms may converge. Of course, sometimes, even the same term may mean different things to different people.

Many coupled human and natural concepts such as those in Table 2.2 are still relatively general. They can be more specific. For example, when wildlife is the main focus of the natural subsystem, coupled human and natural systems can be specified as coupled human and wildlife systems. In Section 2.4, we illustrate the application of the general framework to such a more specific coupled system.

2.4 Application of the general framework to coupled human and wildlife systems

The general conceptual framework (Figure 2.1) can apply to many specific types of coupled human and natural systems. In this section, we use a coupled system in a rural forest setting as an example application (Figure 2.2), which provides the foundation for much of the research discussed throughout the rest of this book. In this coupled system, human populations make their livelihoods largely from direct interactions with local natural resources in forests, and in which conservation policies exist to protect biodiversity, especially endangered wildlife (Figure 2.2). The framework includes five main components that interact with one another, including Local Residents, Policies, Forests, Wildlife, and Contextual Factors. The first two components belong to the human subsystem while Forests and Wildlife are parts of the natural subsystem. Contextual Factors cross human and natural subsystems in the focal coupled system and other coupled systems. For the sake of simplicity, temporal changes and other coupled systems are not illustrated in Figure 2.2. We discuss temporal changes in many other chapters in this book and explicitly discuss other coupled systems in Chapters 16 and 17.

Local Residents include socioeconomic and demographic characteristics (e.g., household size, age and gender structure, family formation through marriage, migration, death, income, expenses, perceptions, and

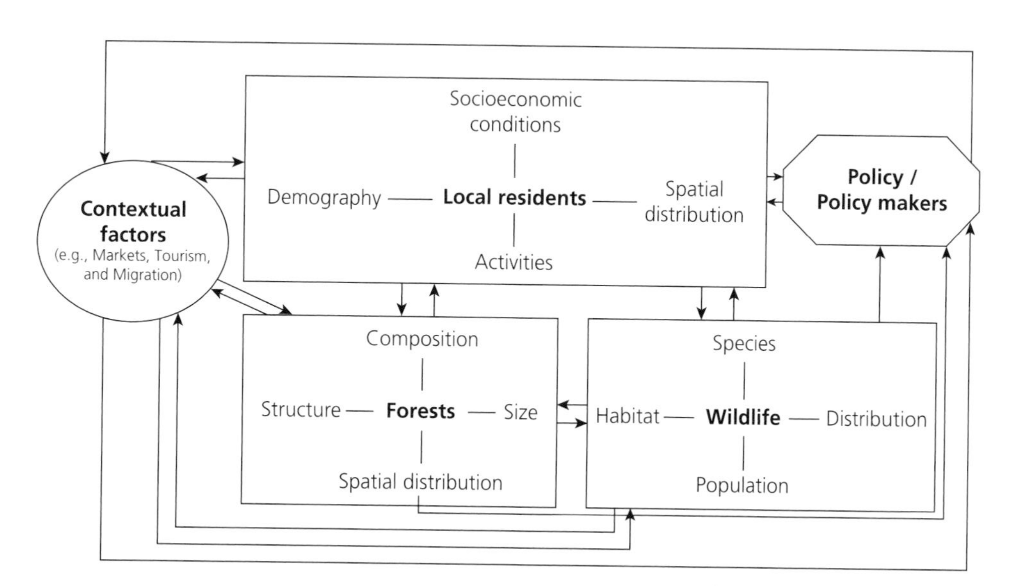

Figure 2.2 Application of the general conceptual framework of coupled human and natural systems (Figure 2.1) to coupled human and wildlife systems. Arrows represent the directions of influence among subsystems, with those going in both directions representing feedbacks. Figure is reprinted with permission from the Missouri Botanical Garden Press (adapted from Liu and Viña, 2014).

attitudes).They also include spatial distribution (e.g., location with respect to forest) and activities (e.g., crop and livestock production, extraction of forest products, or off-farm employment) at multiple organizational levels (e.g., individual, household, village, and township). With regard to Forests, important variables consist of forest type (e.g., deciduous, coniferous, mixed), areal extent, spatial configuration (e.g., continuous or fragmented), and structure (e.g., understory species such as bamboo, stem density, tree canopy cover, and species composition). Forests are important links between Local Residents and Wildlife. For Wildlife, major attributes include species composition, population dynamics, and the distribution and quality of their habitats. The quality of wildlife habitat depends on biophysical conditions (e.g., topography and forest cover) and the degree and nature of human impacts. Policies (and policy makers) serve an important role in coupled systems as they can alter all other components. Policies could originate from the local, provincial, or national organizations or governments. Many of the policies relate to both conservation goals and human livelihoods, such as payments for ecosystem services (PES) programs. Contextual Factors include other components and processes inside a coupled system and those in other coupled systems that interact with the focal coupled system. Examples of Contextual Factors include geographic setting, tourism, migration, and global economic markets.

These five components interact with one another (Figure 2.2). Local Residents rely on ecosystem services (Chapter 5) and participate in a variety of activities that affect both Forests and Wildlife (Chapters 6 and 7). Local Residents may extract resources from Forests (e.g., via timber harvesting, fuelwood collection, or medicinal herb collection) or participate in forest monitoring to prevent such extraction. Local Residents may also influence Forests by raising livestock, which may overgraze the understory or trample tree seedlings. Changes that occur in Forests as a result of such actions in turn indirectly impact Wildlife via changes in the structure and function of their available habitat. Local Residents may also directly impact wildlife via hunting, and Wildlife can affect forests through grazing and reducing regeneration of plants. Policies influence characteristics of Local Residents, Forests, and Wildlife by restricting human activities. These influences can be direct (e.g., preventing timber extraction and fuelwood collection, and spurring tree planting) or indirect (e.g., incentives to use alternatives to fuelwood, such as electricity and natural gas). Contextual Factors influence all other components because they can result in Local Residents leaving the system (e.g., for work in cities), which could benefit both Forests and Wildlife. Alternatively, Contextual Factors such as tourism run by outside companies can increase negative impacts on both Forests and Wildlife by converting Forests and Wildlife habitat into roads and other tourism infrastructures (e.g., hotels and restaurants).

Feedbacks can arise in a number of ways. For example, as Forests near residents shrink, they become more distant from Local Residents, which makes the extraction of timber and non-timber forest products more difficult and time-consuming. Policies can constitute a negative feedback on the conversion of Forests by discouraging deforestation and actively promoting the conversion of farmland to forest. Increases in human–wildlife conflicts (e.g., crop predation) can emerge from successful wildlife conservation. Policies may either further enhance or deter their own success depending on whether Local Residents receive benefits to offset such costs. An increase in available labor after conversion of farmland to forest can encourage out-migration (a Contextual Factor), which could further reduce farming and direct impacts on Forests.

The relationships between different components are complex. For instance, the relationship between changes in Forests and changes in Wildlife habitat may be non-linear because some forested areas are not necessarily suitable wildlife habitat. Local Residents may respond to conservation Policies in a non-linear way, such that increasing benefits could disproportionally increase participation in conservation. The relationship between some components may take a long time to appear, constituting a time lag. For example, a time lag may occur when changes in the fertility rate of Local Residents affect Forests after a delay of years or even decades (An et al., 2011) until children grow up and establish new households.

2.5 Theoretical foundations

Studies of coupled systems can benefit from and contribute to existing systems theories. A foundational theory is general systems theory (von Bertalanffy, 1973). Other related theories focus on specific system attributes. For example, coupled systems can be treated as complex adaptive systems (Holland, 1992, Levin, 1998) with adaptation and interacting feedbacks at multiple scales. Resilience theory addresses another important aspect of systems—the ability to "retain similar structures and functioning after disturbances for continuous development" (Liu et al., 2007b). Hierarchy theory (Gunderson and Holling, 2001, O'Neill et al., 1986) helps frame coupled systems in a hierarchical structure that consists of multiple organizational levels.

Many theories in separate disciplines (e.g., ecology, economics, and sociology) can also be useful for and benefit from coupled systems research. For example, ecological insights are available from niche theory (Chase, 2011), theories of animal behavior (e.g., ideal free distribution theory, Fretwell and Lucas, 1970) and vegetation succession (Connell and Slatyer, 1977). Forest transition theory is informative in understanding forest changes and the underlying mechanisms (Chapter 6). Social norms theory is helpful to elicit the behavior of people participating in conservation programs (Chapter 11, Chen et al., 2009). Theories of the evolution of institutions and social norms can increase our understanding of effects of feedbacks on common-property resource management (Cialdini and Goldstein, 2004, Elster, 1989, Ostrom, 1990). Discrete choice theory is powerful in understanding decisions regarding, for example, fuelwood vs electricity use (Chapter 10, An et al., 2002).

Existing theories can and have provided important insights into coupled systems, but many were not developed in the context of coupled systems. Instead, they were originally developed in the context of natural subsystems (e.g., niche theory, Grinnell, 1917, Hutchinson, 1957, Liu and Ma, 1990) or human subsystems (e.g., theories of household production, Becker, 1981, 1991, Singh et al., 1986). Thus, they may need to be further developed to more effectively address coupled systems. For example, in ecology, a niche is often treated as a hypervolume in an n-dimensional space (Hutchinson, 1957). Each dimension is an environmental condition or resource (biotic and abiotic). The boundaries of the hypervolume define the requirements for a species' growth, reproduction, and dispersal across space and over time. In the context of coupled systems, however, much work is needed to study the effects of human activities (e.g., land use) on niches and species' responses to human activities. The coupled systems framework can advance niche theory in part by contributing an understanding of feedbacks among humans and animals (Hull et al., 2015). For example, human development changes the niches of wildlife pests, which in turn affects human policies for controlling wildlife pests in the future (Morzillo et al., 2014). In economics, household production theory (Becker, 1981, 1991, Singh et al., 1986) can be adapted to understand various household production activities (e.g., cropping, livestock, conservation, and off-farm employment) and their responses to feedbacks between human and natural systems, as households are the basic units of production and consumption (Becker, 1991, Liu et al., 2003). The coupled systems framework can inform household production theory by outlining how complex human–nature feedbacks at fine scales (e.g., households or land parcels) are manifest in changes at broader scales. With more applications and testing of existing theories in coupled systems research, it is possible to revise existing theories and even develop new ones.

2.6 Methods for studying coupled systems

Studying coupled systems requires a portfolio approach (multiple methods, Young et al., 2006) and expertise from a variety of natural and social science disciplines (Alberti et al., 2011, Campbell, 2005). Coupled systems research involves many aspects: the human subsystem, the natural subsystem, and their interactions. Although the research process can be roughly divided into three major steps, these steps are iterative—the information learned in one step would require revisiting an earlier step to make necessary revisions. An important step is to identify the system boundaries, components, and their relationships conceptually, as exemplified in Figures 2.1 and 2.2. Another step is to define research goals, objectives, questions, and hypotheses. The approach to these tasks is similar to those in disciplinary research although the focus often relates to both subsystems and their interactions. The third step is to quantify the components and their relationships. This step includes several procedures: data collection, data cleaning, management, analysis, integration, and visualization. We provide more details on these procedures in the remainder of this section below.

There are many methods for data collection, depending on the types of data. For socioeconomic data, in-person interviews and surveys with households, individuals, and focus groups are common (Chapters 5, 9, 10, and 11; see also Axinn and Ghimire, 2011, Neuman, 2005). Government census data and documents are other major sources of data, especially related to the human population such as individuals, households, and communities (Chapters 12 and 14). For biophysical data, remote sensing data are increasingly available (Chapters 6 and 7). One of the main types of remote sensing data is satellite data, with various spectral, spatial, and temporal resolutions. Field surveys enable researchers to obtain detailed data at fine spatial scales (Chapters 4, 6, and 7). To obtain spatial data such as locations of objects (e.g., field sites and activities), a global positioning system (GPS) receiver is widely used (Tomkiewicz et al., 2010). Such data provide a good basis for ground-truthing. Bio-logging, such as telemetry, GPS collars, and animal-borne imaging, offers essential information on animals and people, their movements, and life histories (Chapter 4; see also Dell et al., 2014).

After or during data collection and entry, it is necessary to clean the data before use. The cleaning process may help fix data errors and ensure data quality for further analyses. As the data for coupled systems are often big and mixed in terms of data types, it is necessary to manage them effectively and efficiently so they can be easily accessible and understandable.

The ultimate purpose of collecting, cleaning, and managing the data is to analyze and integrate them to generate useful information and knowledge. There are numerous methods for data analysis, including specialized statistical methods (Burnham and Anderson, 2002), as well as geographic information systems (GIS). The latter can incorporate multiple layers of data for spatial analysis and mapping, and spatial visualization (e.g., Chapter 4). Systems modeling is a widely used approach for integrating a variety of data and relationships derived from data analysis to understand complex relationships and dynamics. For example, agent-based models are an increasingly popular bottom-up modeling approach that incorporates detailed information about each object to understand emergent properties (Chapters 8 and 14; see also Railsback and Grimm, 2011). Many visualization tools are available to make information visually accessible to the reader, including GIS, and computer software programs such as R (R Development Core Team, 2005).

2.7 Model coupled systems

The world is so large and there are so many coupled systems, it is impossible to study every coupled system in detail for a long time due to constraints of resources (e.g., time, manpower, and money). Thus, innovative approaches are needed to address this challenge. One such approach is the use of model coupled systems, which share important features with many other systems and are studied in depth for a long time. The findings from the model systems can provide insight and enable the design of appropriate research on non-model systems to address fundamental questions and develop effective policies and management strategies. This approach is analogous to successful model systems approaches in other fields such as biology, medicine, and ecology.

In biology and medicine, a large number of model organisms have been used. Model animals include fruit flies (*Drosophila melanogaster*, well known as the subject of genetic experiments by Thomas Morgan). Other examples include mice (*Mus musculus*, classical model vertebrate) and zebrafish (*Danio rerio*, ideal for visual observations of internal anatomy). For plants, *Arabidopsis thaliana*, a member of the mustard (*Brassicaceae*) family (e.g., cabbage and radish) is the most popular model plant. *Arabidopsis thaliana* is also the first plant to have its genome sequenced (TAIR, 2015). Regarding microorganisms, *Escherichia coli* (a common bacterium in the human digestive system) is used for the study of evolution and drug resistance (IECA, 2015). These organisms are ideal models because they are easy to raise, and are inexpensive and convenient to operate with. In addition, these organisms avoid the need for unethical and unfeasible experiments on humans (e.g., human diseases). The model organisms have played huge roles in advancing biological and medical research (Hedges, 2002). For example, many Nobel laureates in physiology or medicine have used model organisms in their research (Marx et al., 2002). While the vast majority of research on model organisms is done in laboratories, some have been researched in the field. For example, checkerspot butterflies have been used to study population biology for more than half a century (Ehrlich and Hanski, 2004). Field studies on Darwin's finches have provided useful insights into the causes of speciation and adaptive radiation (Grant and Grant, 2010).

Model systems also exist at higher levels of organization in ecology, including populations, communities, and ecosystems. Such model systems have been used to study ecological processes, patterns, and dynamics. At the population and community levels, population dynamics of Canadian lynx and snowshoe hares have been used as a classic example of predator–prey interaction cycles appearing in the first well-known ecology textbook (Odum, 1953). This example was chosen in part because long-term data were available that allowed ecologists to analyze their dynamics (Odum, 1953). The Galápagos Islands are excellent sites for studies of bird adaptation and character divergence (Grant and Grant, 2010). Unlike model organisms in the laboratory, model ecosystems are usually associated with a particular place (e.g., a field station, Aigner and Koehler, 2010, Billick and Price, 2010). For example, Hubbard Brook Experimental Forest has been an important source of insights on biogeochemical cycles (Bormann and Likens, 1979). Some model systems research can also integrate across multiple ecosystems, such as pioneering work done on nutrient cycling in the model system of Hawaii (Vitousek, 2004). Some model ecosystems were created by design. Others were developed later after already accumulating a body of knowledge, or after the infrastructure and facilities had been built (Franklin et al., 1990). Many long-term ecological sites have been studied for many years. For example, the Long-Term Ecological Research (LTER) program was initiated in 1980. The program has been supported and expanded by NSF to 26 sites representing diverse ecosystem types across the USA (Collins et al., 2010, ILTER, 2015). Many other countries (e.g., China) adopted the approach, which has led to the formation of the International Long-Term Ecological Research Network (ILTER, 2015). They have produced important ecological insights (Robertson et al., 2012). Although ILTERs are not explicitly described as model ecosystems, they serve the functions of model systems. Many results and insights have been applied to areas beyond the sites (Hobbie et al., 2003, Vihervaara et al., 2013).

Similar to model organisms and model ecosystems, there is a strong need to establish model coupled systems. To some extent, the need for model coupled systems is even greater than model organisms and ecosystems. Coupled systems are more complex and resources for studying coupled systems may be greater. However, the model coupled system approach is far behind that of the model organisms and model ecosystem approaches. Model coupled systems should reflect attributes of many coupled systems and can address many common issues. By doing so, theory, methods, and results from model systems will be useful to many other coupled systems. These model systems can also save time and other resources because limited resources do not allow researchers to study every coupled system in depth and for a long period. As many phenomena in coupled systems play out over multiple organizational, spatial, and temporal scales, it is necessary for deciphering which couplings are important at which scales of interest. Good modeling is about knowing what to ignore. Some properties of coupled systems may be scale-independent and context-independent. In other words, findings in one system may be common to many other different systems at different scales and with different contexts around the world. For example, the effect of group size on collective action has been a theoretical puzzle—the "group size paradox"—and has been debated for over a century (Ostrom, 2005,

Yang et al., 2013b). Previous literature concluded that the group size effect varies across contexts, which is true but lacks theoretical and empirical guidance for further research and practice (Ostrom 2005). But buttressed by years of work in our model coupled system (i.e., Wolong Nature Reserve), our study depicted the whole picture of a non-linear curve of group size effects on both collective action and resource management outcomes, and confirmed the general hypothetical mechanisms at play (Yang et al., 2013b). These findings from our model coupled system provide strong evidence that while there are many scale-dependent and context-dependent phenomena in coupled systems, there indeed are also generalizable properties (e.g., causal mechanisms) across scales and contexts. While it should be cautious to extrapolate findings from one coupled system to another, the study of model coupled systems can play a crucial role in pioneering new thinking, testing new methods, generating empirical evidence, and guiding research under varied contexts. Ultimately, model systems can help to synthesize empirical data, discover generalizable properties and build robust theories to guide decision-making and management practices.

Analogous to the well-known model organisms and model ecosystems, we have used Wolong Nature Reserve in China (Chapter 3) as a model coupled system since 1996. Our research on this model system seeks to understand the complexity of coupled systems and explore the utility of research results in planning, policy making, and governance. The work in Wolong relies on the conceptual framework (Figure 2.2 or earlier versions of this framework) presented above. Like almost all other frameworks, the one guiding our research has evolved over time with modifications to some specific components, but its core remains the same. Our work also builds on the theories and utilizes the methods we have highlighted above, which are constantly evolving. The model system approach has taught us a great deal over the years about our system in Wolong, and in turn about coupled human and natural systems in general. The rest of this book details our findings. We begin with an overview of Wolong (Chapter 3) and then synthesize various research projects we have conducted in Wolong (Chapters 4–14). We also present highlights of the applications of ideas and methods developed in Wolong to some other coupled systems around the world (Chapters 15–18).

2.8 Summary

The ongoing quest to overcome precipitating threats and to strive for global sustainability requires a sound foundation for understanding the complex ways in which humans and nature affect one another. In this chapter, we introduced one promising answer to this call—the coupled human and natural systems framework. This framework provides a blueprint for analyzing reciprocal interactions and feedbacks between humans and nature. The framework goes beyond previous concepts by being more inclusive of diverse human and natural components, by explicitly emphasizing human–nature feedbacks, and by incorporating interactions with other systems. We demonstrated the application of the framework to understanding human–wildlife interactions in a rural setting by characterizing complex interactions between Local Residents, Wildlife, Forests, Policies, and Contextual Factors. We also showed how the framework builds on and contributes to theories such as niche theory and household production theory. We provided an overview of methods for coupled system research, techniques that bring together individuals from diverse disciplines and backgrounds to analyze and integrate diverse data sets. Examples of cutting-edge approaches include systems modeling such as agent-based modeling. We also put forth the need for intensive, long-term study of model coupled systems similar in depth and breadth to classic studies conducted on model organisms (e.g., *Drosophila*) and model ecosystems (e.g., the Galápagos). The rest of this book illustrates our 20-year study of the model coupled system of Wolong Nature Reserve in China and applications to other coupled systems around the world.

References

Aguilar, M.Y., Pacheco, T.R., Tobar, J.M., and Quinonez, J.C. (2009) Vulnerability and adaptation to climate change of rural inhabitants in the central coastal plain of El Salvador. *Climate Research*, **40**, 187–98.

Aigner, P.A. and Koehler, C.E. (2010) The model ecosystem as a paradigm of place-based research. In I. Billick and M.V. Price, eds, *The Ecology of Place: Contributions of Place-Based Research to Ecological Understanding*, pp.359–81. University of Chicago Press, Chicago, IL.

Alberti, M. (2015) Eco-evolutionary dynamics in an urbanizing planet. *Trends in Ecology & Evolution*, **30**, 114–26.

Alberti, M., Asbjornsen, H., Baker, L. A., et al. (2011) Research on coupled human and natural systems (CHANS): approach, challenges and strategies. *Bulletin of the Ecological Society of America*, **92**, 218228.

An, L., Linderman, M., He, G., et al. (2011) Long-term ecological effects of demographic and socioeconomic factors in Wolong Nature Reserve (China). In R.P. Cincotta and L.J. Gorenflo, eds, *Human Population: Its Influences on Biological Diversity*, pp. 179–195. Springer-Verlag, Berlin, Germany.

An, L. and Liu, J. (2010) Long-term effects of family planning and other determinants of fertility on population and environment: agent-based modeling evidence from Wolong Nature Reserve, China. *Population and Environment*, **31**, 427–59.

An, L., Liu, J., Ouyang, Z., et al. (2001) Simulating demographic and socioeconomic processes on household level and implications for giant panda habitats. *Ecological Modelling*, **140**, 31–49.

An, L., Lupi, F., Liu, J., et al. (2002) Modeling the choice to switch from fuelwood to electricity: implications for giant panda habitat conservation. *Ecological Economics*, **42**, 445–57.

Axinn, W.G. and Ghimire, D.J. (2011) Social organization, population, and land use. *American Journal of Sociology*, **117**, 209–58.

Barrett, C.B., Reardon, T., and Webb, P. (2001) Nonfarm income diversification and household livelihood strategies in rural Africa: concepts, dynamics, and policy implications. *Food Policy*, **26**, 315–31.

Becker, G.S. (1981) *A Treatise on the Family*. Harvard University Press, Cambridge, MA.

Becker, G.S. (1991) *A Treatise on the Family: Enlarged Edition*. Harvard University Press, Cambridge, MA.

Bennett, D. and McGinnis, D. (2008) Coupled and complex: human–environment interaction in the Greater Yellowstone Ecosystem, USA. *Geoforum*, **39**, 833–45.

Billick, I. and Price, M.V. (2010) *The Ecology of Place: Contributions of Place-Based Research to Ecological Understanding*. University of Chicago Press, Chicago, IL.

Binder, C.R., Hinkel, J., Bots, P.W., and Pahl-Wostl, C. (2013) Comparison of frameworks for analyzing social–ecological systems. *Ecology and Society*, **18**, 26.

Blondel, J. (2006) The "design" of Mediterranean landscapes: a millennial story of humans and ecological systems during the historic period. *Human Ecology*, **34**, 713–29.

Boersma, P.D. and Parrish, J.K. (1999) Limiting abuse: marine protected areas, a limited solution. *Ecological Economics*, **31**, 287–304.

Bormann, F.H. and Likens, G.E. (1979) *Pattern and Process in a Forested Ecosystem*. Springer-Verlag, New York, NY.

Brown, G. (2008) Social–ecological hotspots mapping: a spatial approach for identifying coupled social–ecological space. *Landscape and Urban Planning*, **85**, 27–39.

Burnham, K.P. and Anderson, D.R. (2002) *Model Selection and Multi-model Inference: a practical information-theoretic approach*. Springer, New York, NY.

Bytnerowicz, A., Tausz, M., Alonso, R., et al. (2002) Summer-time distribution of air pollutants in Sequoia National Park, California. *Environmental Pollution*, **118**, 187–203.

Campbell, L.M. (2005) Overcoming obstacles to interdisciplinary research. *Conservation Biology*, **19**, 574–77.

Carmenta, R., Parry, L., Blackburn, A., et al. (2011) Understanding human–fire interactions in tropical forest regions: a case for interdisciplinary research across the natural and social sciences. *Ecology and Society*, **16**(1), 53.

Caro, T. (1999) Demography and behaviour of African mammals subject to exploitation. *Biological Conservation*, **91**, 91–97.

Carpenter, S., Walker, B., Anderies, J.M., and Abel, N. (2001) From metaphor to measurement: resilience of what to what? *Ecosystems*, **4**, 765–81.

Chapman, D.S., Termansen, M., Quinn, C.H., et al. (2009) Modelling the coupled dynamics of moorland management and upland vegetation. *Journal of Applied Ecology*, **46**, 278–88.

Chase, J.M. (2011) Ecological niche theory. In S.M. Scheiner and M.R. Willig, eds, *The Theory of Ecology*, pp 93–108. The University of Chicago Press, Chicago, IL.

Chen, X., Lupi, F., He, G., and Liu, J. (2009) Linking social norms to efficient conservation investment in payments for ecosystem services. *Proceedings of the National Academy of Sciences of the United States of America*, **106**, 11812–17.

Cialdini, R. and Goldstein, N. (2004) Social influence: compliance and conformity. *Annual Review of Psychology*, **55**, 591–621.

Clark, W.C. and Dickson, N.M. (2003) Sustainability science: the emerging research program. *Proceedings of the National Academy of Sciences of the United States of America*, **100**, 8059–61.

Collins, S.L., Carpenter, S.R., Swinton, S.M., et al. (2010) An integrated conceptual framework for long-term social–ecological research. *Frontiers in Ecology and the Environment*, **9**, 351–57.

Connell, J.H. and Slatyer, R.O. (1977) Mechanisms of succession in natural communities and their role in community stability and organization. *American Naturalist*, **111**, 1119–44.

Costanza, R. (1996) Ecological economics: reintegrating the study of humans and nature. *Ecological Applications*, **6**, 978–90.

Craig, R.K. (2010) "Stationarity is dead"—long live transformation: five principles for climate change adaptation law. *Harvard Environmental Law Review*, **34**, 9–73.

DeFries, R.S., Rudel, T.K., Uriarte, M., and Hansen, M. (2010) Deforestation driven by urban population growth and agricultural trade in the twenty-first century. *Nature Geoscience*, **3**, 178–181.

Dell, A.I., Bender, J.A., Branson, K., et al. (2014) Automated image-based tracking and its application in ecology. *Trends in Ecology & Evolution*, **29**, 417–28.

Dewulf, A., Francois, G., Pahl-Wostl, C., and Taillieu, T. (2007) A framing approach to cross-disciplinary research collaboration: experiences from a large-scale research project on adaptive water management. *Ecology and Society*, **12**(2), 14.

Diamond, J.M. (1997) *Guns, Germs and Steel: The Fates of Human Societies*. Norton, New York, NY.

Diamond, J.M. (2005) *Collapse: How Societies Choose to Fail or Succeed*. Viking, New York, NY.

Douglas-Hamilton, I., Krink, T., and Vollrath, F. (2005) Movements and corridors of African elephants in relation to protected areas. *Naturwissenschaften*, **92**, 158–63.

Eakin, H. and Luers, A.L. (2006) Assessing the vulnerability of social–environmental systems. *Annual Review of Environment and Resources*, **31**, 365.

Ebbin, S.A. (2009) Institutional and ethical dimensions of resilience in fishing systems: perspectives from co-managed fisheries in the Pacific Northwest. *Marine Policy*, **33**, 264–70.

Ehrlich, P.R. and Hanski, I. (2004) *On the Wings of Checkerspots: A Model System for Population Biology*. Oxford University Press, Oxford; New York.

Ehrlich, P.R. and Pringle, R.M. (2008) Where does biodiversity go from here? A grim business-as-usual forecast and a hopeful portfolio of partial solutions. *Proceedings of the National Academy of Sciences of the United States of America*, **105**, 11579–86.

Elster, J. (1989) Social norms and economic theory. *Journal of Economic Perspectives*, **3**, 99–117.

Ervin, J. (2003) Rapid assessment of protected area management effectiveness in four countries. *Bioscience*, **53**, 833–41.

Estes, J.A., Terborgh, J., Brashares, J.S., et al. (2011) Trophic downgrading of planet Earth. *Science*, **333**, 301–06.

Foley, J.A., DeFries, R., Asner, G.P., et al. (2005) Global consequences of land use. *Science*, **309**, 570–74.

Folke, C., Hahn, T., Olsson, P., and Norberg, J. (2005) Adaptive governance of social–ecological systems. *Annual Review of Environment and Resources*, **30**, 441–73.

Folke, C., Jansson, Å., Rockström, J., et al. (2011). Reconnecting to the biosphere. *Ambio*, **40**, 719–738.

Franklin, J.F., Bledsoe, C.S., and Callahan, J.T. (1990) Contributions to the long-term ecological research program. *BioScience*, **40**(7), 509–23.

Fretwell, S.D. and Lucas, H.L. (1970) On territorial behaviour and other factors influencing habitat distribution in birds. I. theoretical development. *Acta Biotheoretica*, **19**, 16–36.

Gilioli, G. and Baumgartner, J. (2007) Adaptive ecosocial system sustainability enhancement in Sub-Saharan Africa. *Ecohealth*, **4**, 428–44.

Grant, P.R. and Grant, B.R. (2010) Ecological insights into the causes of an adaptive radiation from long-term field studies of Darwin's Finches. In I. Billick and M.V. Price, eds, *The Ecology of Place: Contributions of Place-Based Research to Ecological Understanding*, pp. 109–33. University of Chicago Press, Chicago, IL.

Grinnell, J. (1917) The niche-relationships of the California thrasher. *The Auk*, **34**, 427–33.

Gude, P.H., Hansen, A.J., and Jones, D.A. (2007) Biodiversity consequences of alternative future land use scenarios in Greater Yellowstone. *Ecological Applications*, **17**, 1004–18.

Gunderson, L.H. and Holling, C.S. (2001) *Panarchy: Understanding Transformations in Human and Natural Systems*. Island Press, Washington DC.

Hasse, J.E. and Lathrop, R.G. (2003) Land resource impact indicators of urban sprawl. *Applied Geography*, **23**, 159–75.

Hedges, S.B. (2002) The origin and evolution of model organisms. *Nature Reviews Genetics*, **3**, 838–49.

Hird, J.A. (1993) Environmental policy and equity: the case of Superfund. *Journal of Policy Analysis and Management*, **12**, 323–43.

Hobbie, J.E., Carpenter, S.R., Grimm, N.B., et al. (2003) The US long-term ecological research program. *Bioscience*, **53**, 21–32.

Holland, J.H. (1992) Complex adaptive systems. *Daedalus*, **121**, 17–30.

Hull, V., Tuanmu, M.-N., and Liu, J. (2015) Synthesis of human-nature feedbacks. *Ecology and Society* **20**(3), 17.

Hutchinson, G.E. (1957) Concluding remarks. *Cold Spring Harbor Symposia on Quantitative Biology*, **22**, 415–27.

IECA (2015) *International E. coli Alliance E. coli Database Portal*. http://www.uni-giessen.de/ecoli/IECA/index.php.

ILTER (2015) *International Long-term Ecological Research Network* (http://www.ilternet.edu).

Ingold, T. (1995) Building, dwelling, living: how animals and people make themselves at home in the world. In M. Strathern, ed., *Shifting Contexts: Transformations in Anthropological Knowledge*, pp. 57–80. Routledge, London, UK.

IPCC (2008) *Climate Change 2007: Impacts, Adaptation and Vulnerability*. IPCC Secretariat, Geneva, Switzerland.

Kates, R.W., Clark, W.C., Corell, R., et al. (2001) Sustainability science. *Science*, **292**, 641–42.

Lacitignola, D., Petrosillo, I., and Zurlini, G. (2010) Time-dependent regimes of a tourism-based social ecological system: period-doubling route to chaos. *Ecological Complexity*, **7**, 44–54.

Lambin, E.F. and Meyfroidt, P. (2011) Global land use change, economic globalization, and the looming land scarcity. *Proceedings of the National Academy of Sciences of the United States of America*, **108**, 3465–72.

Levin, S.A. (1998) Ecosystems and the biosphere as complex adaptive systems. *Ecosystems*, **1**, 431–36.

Liu, J. (2010) China's road to sustainability. *Science*, **328**, 974–74.

Liu, J., Daily, G.C., Ehrlich, P.R., and Luck, G.W. (2003) Effects of household dynamics on resource consumption and biodiversity. *Nature*, **421**, 530–33.

Liu, J., Dietz, T., Carpenter, S.R., et al. (2007a) Coupled human and natural systems. *Ambio*, **36**, 639–49.

Liu, J., Dietz, T., Carpenter, S.R., et al. (2007b) Complexity of coupled human and natural systems. *Science*, **317**, 1513–16.

Liu, J., Hull, V., Batistella, M., et al. (2013) Framing sustainability in a telecoupled world. *Ecology and Society*, **18**(2), 26.

Liu, J. and Ma, S. (1990) Expanded niche theory. In S. Ma, ed., *Perspectives in Modern Ecology*, pp. 72–90. Science Press, Beijing, China (in Chinese).

Liu, J., Mooney, H., Hull, V., et al. (2015) Systems integration for global sustainability. *Science*, **347**(6225), 1258832.

Liu, J., Ouyang, Z., Taylor, W.W., et al. (1999) A framework for evaluating the effects of human factors on wildlife habitat: the case of giant pandas. *Conservation Biology*, **13**, 1360–70.

Liu, J. and Viña, A. (2014) Pandas, plants, and people. *Annals of the Missouri Botanical Garden*, **100**, 108–25.

Liu, W., Vogt, C.A., Luo, J., et al. (2012) Drivers and socioeconomic impacts of tourism participation in protected areas. *PLoS ONE*, **7**, e35420.

Lorenzoni, I., Jordan, A., O'Riordan, T., et al. (2000) A co-evolutionary approach to climate change impact assessment: Part II. A scenario-based case study in East Anglia (UK). *Global Environmental Change – Human and Policy Dimensions*, **10**, 145–55.

Ma, S. and Wang, R. (1984) The social-economic-natural complex ecosystem. *Acta Ecologica Sinica*, **4**, 1–9 (in Chinese).

Macdougall, A.S. and Turkington, R. (2005) Are invasive species the drivers or passengers of change in degraded ecosystems? *Ecology*, **86**, 42–55.

Marsh, G.P. (1864) *Man and Nature*. Belknap Press of Harvard University Press, Cambridge, MA.

Marx, J., Seife, C., Cho, A., et al. (2002) Nobels run the gamut from cells to the cosmos. *Science*, **298**, 526.

Matson, P.A. (2012) *Seeds of Sustainability: Lessons from the Birthplace of the Green Revolution*. Available online: http://www.islandpress.org/book/seeds-of-sustainability.

McDonnell, M.J. and Pickett, S.T.A. (1993) *Humans as Components of Ecosystems: The Ecology of Subtle Human Effects and Populated Areas*. Springer-Verlag, New York, NY.

Meadows, D.H., Jørgen, R., and Meadows, D.L. (2004) *The Limits to Growth: The 30-Year Update*. Chelsea Green Pub. Co., White River Junction, VT.

Metzger, M.J., Schroter, D., Leemans, R., and Cramer, W. (2008) A spatially explicit and quantitative vulnerability assessment of ecosystem service change in Europe. *Regional Environmental Change*, **8**, 91–107.

Meyfroidt, P. and Lambin, E.F. (2009) Forest transition in Vietnam and displacement of deforestation abroad. *Proceedings of the National Academy of Sciences of the United States of America*, **106**, 16139–44.

Millennium Ecosystem Assessment (2005) *Ecosystems and Human Well-being: Synthesis*. Island Press, Washington, DC.

Miller, B.W., Breckheimer, I., McCleary, A.L., et al. (2010) Using stylized agent-based models for population-environment research: a case study from the Galapagos Islands. *Population and Environment*, **31**, 401–26.

Moran, E.F. (2010) *Environmental Social Science: Human-Environment Interactions and Sustainability*. Wiley-Blackwell, Malden, MA.

Morzillo, A.T., De Beurs, K.M., and Martin-Mikle, C.J. (2014) A conceptual framework to evaluate human–wildlife interactions within coupled human and natural systems. *Ecology and Society*, **19**, 44.

National Research Council (1999) *Our Common Journey*. National Academy Press, Washington, DC.

National Science Foundation (2010) *Dynamics of Coupled Natural and Human Systems (CNH). Program Solicitation*. http://www.nsf.gov/pubs/2010/nsf10612/nsf10612.htm.

National Science Foundation (2014) *Dynamics of Coupled Natural and Human Systems (CNH). Program Solicitation*. http://www.nsf.gov/pubs/2014/nsf14601/nsf14601.pdf.

Neuman, W.L. (2005) *Social Research Methods: Qualitative and Quantitative Approaches*, sixth edition. Allyn and Bacon, Boston, MA.

Odum, E.P. (1953) *Fundamentals of Ecology*. Saunders, Philadelphia, PA.

Odum, H.T. (1971) *Environment, Power and Society*. Wiley Interscience, New York, NY.

O'Neill, R.V., Deangelis, D.L., Waide, J.B., and Allen, G.E. (1986) *A Hierarchical Concept of Ecosystems*. Princeton University Press, Princeton, NJ.

Ostrom, E. (1990) *Governing the Commons: The Evolution of Institutions for Collective Action*. Cambridge University Press, New York, NY.

Ostrom, E. (2005) *Understanding Institutional Diversity*. Princeton University Press, Princeton, NJ.

Ostrom, E. (2007) A diagnostic approach for going beyond panaceas. *Proceedings of the National Academy of Sciences of the United States of America*, **104**, 15181–87.

Ostrom, E. (2009) A general framework for analyzing sustainability of social–ecological systems. *Science*, **325**, 419–22.

Pedroza, C. and Salas, S. (2011) Responses of the fishing sector to transitional constraints: from reactive to proactive change, Yucatan fisheries in Mexico. *Marine Policy*, **35**, 39–49.

Pickett, S.T.A., Cadenasso, M.L., and Grove, J.M. (2004) Resilient cities: meaning, models, and metaphor for integrating the ecological, socio-economic, and planning realms. *Landscape and Urban Planning*, **69**, 369–84.

R Development Core Team (2005) R: A language and environment for statistical computing. Vienna, Austria, R Foundation for Statistical Computing.

Railsback, S.F. and Grimm, V. (2011) *Agent-Based and Individual-Based Modeling: A Practical Introduction*. Princeton University Press, Princeton, NJ.

Redman, C.L., Grove, J.M., and Kuby, L.H. (2004) Integrating social science into the long-term ecological research (LTER) network: social dimensions of ecological change and ecological dimensions of social change. *Ecosystems*, **7**, 161–71.

Robertson, G.P., Collins, S.L., Foster, D.R., et al. (2012) Long-term ecological research in a human-dominated world. *BioScience*, **62**, 342–53.

Rockstrom, J., Steffen, W., Noone, K., et al. (2009) A safe operating space for humanity. *Nature*, **461**, 472–75.

Rogers, C.S. and Beets, J. (2001) Degradation of marine ecosystems and decline of fishery resources in marine protected areas in the US Virgin Islands. *Environmental Conservation*, **28**, 312–22.

Sass, G.G., Kitchell, J.F., Carpenter, S.R., et al. (2006) Fish community and food web responses to a whole-lake removal of coarse woody habitat. *Fisheries*, **31**, 321–30.

Satake, A., Iwasa, Y., and Levin, S.A. (2008) Comparison between perfect information and passive-adaptive social learning models of forest harvesting. *Theoretical Ecology*, **1**, 189–97.

Saunders, D., Meeuwig, J., and Vincent, A. (2002) Freshwater protected areas: strategies for conservation. *Conservation Biology*, **16**, 30–41.

Schluter, M., Leslie, H., and Levin, S. (2009) Managing water-use trade-offs in a semi-arid river delta to sustain multiple ecosystem services: a modeling approach. *Ecological Research*, **24**, 491–503.

Schneider, S.H. and Londer, R. (1984) *The Coevolution of Climate and Life*. Sierra Club Books, San Francisco, CA.

Shapiro, J. (2001) *Mao's War Against Nature: Politics and the Environment in Revolutionary China*. Cambridge University Press, Cambridge, UK.

Simberloff, D. and Von Holle, B. (1999) Positive interactions of nonindigenous species: invasional meltdown? *Biological Invasions*, **1**, 21–32.

Simon, J.L. (1996) *The Ultimate Resource 2*. Princeton University Press, Princeton, NJ.

Singh,I., Squire, L., and Strauss, J. (1986) *Agricultural Household Models: Extensions and Applications*. Johns Hopkins University Press, Baltimore, MD.

Speth, J.G. and Repetto, R. (2008) *Punctuated Equilibrium and the Dynamics of US Environmental Policy*. Yale University Press, New Haven, CT.

Steffen, W., Crutzen, J., and McNeill, J.R. (2007) The Anthropocene: are humans now overwhelming the great forces of Nature? *Ambio*, **36**, 614–21.

Steffen, W., Richardson, K., Rockström, J., et al. (2015) Planetary boundaries: guiding human development on a changing planet. *Science*, **347**(6223), 1259855.

Stevens, K., Irwin, B., Kramer, D., and Urquhart, G. (2014) Impact of increasing market access on a tropical small-scale fishery. *Marine Policy*, **50**, 46–52.

Stroup, L.J. and Finewood, M.H. (2011) The hybrid AMPE approach: towards more effective environmental management. *Society & Natural Resources*, **24**, 85–94.

TAIR (2015) *Arabidopsis Information Resource* (TAIR). https://www.arabidopsis.org/about/index.jsp.

Thomas Jr,W.L. (1956) *Man's Role in Changing the Face of the Earth*. University of Chicago Press, Chicago, IL.

Tomkiewicz, S.M., Fuller, M.R., Kie, J.G., and Bates, K.K. (2010) Global positioning system and associated technologies in animal behaviour and ecological research. *Philosophical Transactions of the Royal Society B: Biological Sciences*, **365**, 2163–76.

Turner, B.L. (2010) Vulnerability and resilience: coalescing or paralleling approaches for sustainability science? *Global Environmental Change – Human and Policy Dimensions*, **20**, 570–76.

Turner, B.L., Clark, W., Kates, R., et al. (1990) *The Earth as Transformed by Human Action: Global and Regional Changes in the Biosphere over the Past 300 Years*. Cambridge University Press, Cambridge, UK.

Turner, B.L., Kasperson, R.E., Matson, P.A., et al. (2003) A framework for vulnerability analysis in sustainability science. *Proceedings of the National Academy of Sciences of the United States of America*, **100**, 8074–79.

Unruh, G.C. (2000) Understanding carbon lock-in. *Energy Policy*, **28**, 817–30.

Van Der Leeuw, S.E. (2008) Climate and society: lessons from the past 10 000 years. *Ambio*, **37**, 476–82.

Verburg, P.H., Soepboer, W., Veldkamp, A., et al. (2002) Modeling the spatial dynamics of regional land use: the CLUE-S model. *Environmental Management*, **30**, 391–405.

Vihervaara, P., D'Amato, D., Forsius, M., et al. (2013) Using long-term ecosystem service and biodiversity data to study the impacts and adaptation options in response to climate change: insights from the global ILTER sites network. *Current Opinion in Environmental Sustainability*, **5**, 53–66.

Vitousek, P.M. (2004) *Nutrient Cycling and Limitation: Hawai'i as a Model System*. Princeton University Press, Princeton, NJ.

Vitousek, P.M., Mooney, H.A., Lubchenco, J., and Melillo, J.M. (1997) Human domination of Earth's ecosystems. *Science*, **277**, 494–99.

Von Bertalanffy, L. (1973) *General System Theory: Foundations, Development, Applications*. G. Braziller, New York, NY.

Walker, B., Barrett, S., Polasky, S., et al. (2009). Looming global-scale failures and missing institutions. *Science*, **325**, 1345–6.

Waltner-Toews, D. and Kay, J. (2005) The evolution of an ecosystem approach: the diamond schematic and an adaptive methodology for ecosystem sustainability and health. *Ecology and Society*, **10**(1), 38.

Wassenaar, T., Gerber, P., Verburg, P., et al. (2007) Projecting land use changes in the Neotropics: the geography of pasture expansion into forest. *Global Environmental Change*, **17**, 86–104.

World Commission on Environment and Development (1987) *Our Common Future*. Oxford University Press, Oxford.

Yang, W., Liu, W., Viña, A., et al. (2013a) Performance and prospects of payments for ecosystem services programs: evidence from China. *Journal of Environmental Management*, **127**, 86–95.

Yang, W., Liu, W., Viña, A., et al. (2013b) Nonlinear effects of group size on collective action and resource outcomes. *Proceedings of the National Academy of Sciences of the United States of America*, **110**, 10916–21.

Young, O.R. (2010) Institutional dynamics: resilience, vulnerability and adaptation in environmental and resource regimes. *Global Environmental Change – Human and Policy Dimensions*, **20**, 378–85.

Young, O.R., Lambin, E.F., Alcock, F., et al. (2006) A portfolio approach to analyzing complex human–environment interactions: institutions and land change. *Ecology and Society*, **11**, 31.

PART II

Model Coupled Human and Natural System

Humans have always been embedded in coupled human and natural systems (CHANS), and never more so than now when human actions dominate many, perhaps most, of earth's ecosystems. The complexity of coupled systems is formidable, with multiple dynamic interactions across space and over time. Perhaps one of the most effective ways to illustrate the dynamics of such CHANS is to immerse oneself fully into one. In this main body of the book, we delve into telling the story of our model CHANS: the Wolong Nature Reserve for giant pandas in southwestern China. We tell the story from multiple perspectives, drawing on diverse disciplines and approaches. Our goal is to take the reader to Wolong's vibrant and dynamic world, providing a glimpse into this remote and majestic place. We seek to weave a tapestry depicting how and why human–nature interactions unfold as a series of intricate patterns and processes related to the long-term sustainability of this globally important system.

In Chapter 3, Hull et al. provide an overview of Wolong Nature Reserve, including a historical and geographic context for understanding Wolong's place in China and the world. They describe the unique geomorphology that has given rise to the complex topography of this reserve, the result of its placement in a hotbed of tectonic activity. They also introduce Wolong as a global biodiversity hotspot, with high species richness across numerous taxa. The chapter then highlights the importance of Wolong for panda conservation, containing a far larger number of wild pandas than any other reserve. They also provide background information on the vibrant human community in the reserve, including the human history, cultural underpinnings, and use of and impact on natural resources. They close the chapter with a discussion of the history of management of human–nature interactions in Wolong and an overview of methods used by our research team for understanding such complex interactions.

In Chapter 4, Hull et al. laid the foundation built in the earlier chapters to investigate an important yet understudied theme in coupled human and natural systems research—interactions between humans and wildlife. Here, they integrate diverse sources of data to analyze patterns of coexistence and competition between pandas and people in Wolong. They bring to light factors that have contributed to long-term coexistence between people and pandas. Examples include the panda's natural avoidance of direct interactions with people and its need for a relatively small amount of space. Zoning regulations in the reserve futher encourage spatial segregation between pandas and people. Hull et al. highlight the mechanisms and consequences of competition between people and pandas that result from activities such as timber harvesting, fuelwood collection, and livestock grazing. They also offer recommendations for policy to promote coexistence between people and pandas, such as promoting natural forest recovery and improving zoning regulations to maximize their efficacy. The complex panda–people interactions they explore here are reminiscent of similar challenges faced in many other coupled systems around the globe where balancing wildlife conservation and human livelihoods is challenging but essential.

In Chapter 5, Yang et al. broaden the scope from pandas–people interactions to examine why nature as a whole is important for humans. Their approach sets a frame that persists throughout the book: appreciating nature as the underpinning for human society. This chapter quantifies human dependence on ecosystem services (benefits that nature provides to humans, such as clean water and air, carbon sequestration, erosion control, and recreation). It presents a new, comprehensive indicator—the index of dependence on ecosystem services—which is an advance over the pioneering framework first offered in the United Nation's Millennium Ecosystem Assessment. The chapter then illustrates the patterns, causes, and effects of changes in the index in Wolong. Analysis of the index brings to light the extent to which households with different amounts of human, financial, and social capital are dependent on ecosystem services. The index can also help to guide priority settings for conservation and development planning. In particular, the findings emphasize the importance of promoting human well-being for poor and marginalized groups who are more vulnerable to declines in ecosystem services.

Long-term and interdisciplinary studies have revealed that Wolong is dynamic with surprises unfolding over time. Chapters 6–10 examine changes in six major components of the Wolong CHANS: landscapes, panda habitat, demography, economy, and energy. In Chapter 6, Viña and Liu document landscape changes using state-of-the-art remote sensing analysis techniques and field data. They found a pattern of decline in cropland in Wolong from 1994 to 2007. The decline occurred in large part due to the implementation of the Grain to Green Program (GTGP), which converts cropland to forest, and a decrease in dependence on agriculture with a diversification of income sources. This pattern is contrasted against a decline in forest cover from 1965 to 1997. What followed was a period of forest recovery by 2007 mainly as a result of the Natural Forest Conservation Program (NFCP), which bans harvesting of natural forests. This recovery was later tempered by the subsequent loss of forest cover in 2008 due to the catastrophic Wenchuan earthquake. These trends are mirrored in giant panda habitat transition documented in Chapter 7. Viña et al. detected such patterns by using multiple satellite sensor systems integrated with field data and novel remotely sensed procedures to analyze the distribution of understory bamboo. Bamboo is a crucial component of panda habitat that previous habitat models were not able to include. They discuss the drivers of these changes and factors that have contributed to the success of the conservation programs that have reversed earlier declines of forests and panda habitat. These include financial incentives provided to local people for their participation in conservation.

The book then turns to examining demographic, economic, and energy transitions, as well as the ripple effects of human behaviors on such transitions. In Chapter 8, An et al. describe increases in population size and decreases in household size over time in Wolong. They use agent-based models to simulate cascading effects of demographic decisions (e.g., marriage and birth) on population size, households, and giant panda habitat. They also describe the complexities of these processes and effects arising from these demographic decisions, such as non-linearities and time lags. For example, when the time interval between marriage and first birth increases from 1 year to 11 years, a lower increase in population size can be seen in about 3–5 years. But it takes nearly 40–50 years to see less panda habitat loss under the longer time intervals. In Chapter 9, Yang et al. document an economic transformation. Local livelihoods shifted from being primarily based on subsistence agriculture to diversifying into varied non-agricultural income sources (e.g., transportation, tourism, and conservation programs) beginning in the early 2000s. They examine the causes of this transformation—shifting social capital and interactions among diverse conservation and development policies. Yang et al. also highlight several examples of interacting impacts of the economic transformation, including changes in wildlife habitat, a shift in labor allocation, and a boom in tourism. In Chapter 10, Liu et al. document an energy switch from fuelwood to electricity around the same period and discuss several economic, demographic, and governance factors potentially driving this shift.

They also discuss the positive effects of this transition for conservation, such as the recovery of forests and panda habitat.

Human–nature dynamics are affected by many factors, including social norms and social capital, natural disasters, and government policies. While these factors are touched upon briefly in some of the previous chapters, they are the prime focus of Chapters 11, 12, and 13. In Chapter 11, Chen et al. explore the role of social norms and social capital in human–nature interactions in Wolong with an emphasis on labor migration and payments for ecosystem services programs. They demonstrate that weak social ties of local residents to people in urban centers have a significant positive effect on labor migration. This effect in turn reduces households' use of fuelwood and thus lessens impacts on forests and panda habitat. Social capital plays a significant role in collective action among groups of households engaged in forest monitoring for NFCP to reduce illegal harvesting. Farmers' decisions about participation in GTGP are also significantly influenced by social norms established by others in the community. While social norms and social capital were rarely explored in coupled systems research in the past, the findings in this chapter demonstrate their importance. These topics help to elucidate the dynamics of coupled systems and illustrate the value of enhancing social capital and shaping social norms among households for management purposes. For instance, stronger social norms and increased social capital can improve collective action in natural resource management and wildlife conservation practices.

Natural disasters have become increasingly common around the world, but their impacts have often been inadequately evaluated due to the lack of long-term data (e.g., data before and after the disasters). Taking advantage of the long-term data collected in Wolong, Yang et al. assess the impacts of a major natural disaster—the Wenchuan earthquake, which occurred in 2008 with its epicenter near Wolong (Chapter 12). They present a summary of the major damage resulting from the earthquake to both human communities and the natural environment. Yang et al. also demonstrate the utility of quantitative indicators (human dependence on ecosystem services and human well-being) and models to understand disaster responses. Results indicate that disadvantaged people who lacked access to social or financial capital suffered most from the disaster. They also examine the efficacy of the reconstruction plan put in place in Wolong. They find that it helped to improve the overall adaptive capacity at the reserve level but did not restore and even harmed adaptive capacity at the household level. The findings suggest that the design of disaster recovery and restoration plans should target capacity building at multiple scales and adapt to local contexts.

Policies are key feedbacks and drivers of changes in coupled systems. Multiple policies are often implemented simultaneously or in sequence in a coupled system, but most studies in the past focused on one policy at a time, and the effects of different policies were often not compared. In Chapter 13, Chen et al. examine three policies in Wolong: a zoning scheme designed to spatially segregate pandas and humans and two payments for ecosystem services programs (NFCP and GTGP). They highlight the advantages and limitations of each policy. For the zoning scheme, human livelihoods have not been well incorporated into the policy design, although construction of infrastructure has been effectively prohibited in the core zone. Although the payments for ecosystem service programs do consider livelihoods by providing economic incentives to local residents, they can be improved in the future. For example, the cost-effectiveness of GTGP is less than that of NFCP but cost-effective targeting could improve GTGP. Studies in Wolong suggest that multiple policies should be implemented together in order to maximize their effectiveness.

While the previous chapters provided important insights into what happened in the past, it is also informative to explore what might occur in the future under different scenarios. Hull et al. close this part of the book by looking into the future of Wolong in Chapter 14. This chapter highlights some of our research group's work on scenario modeling including research on family planning and fuelwood collection, household shifts and bamboo population dynamics, and conservation and development policies. Our group's modeling efforts have revealed surprising results. For example, we found a greater reduction in human impact on the environment through youth-only emigration compared to entire household emigration.

Our models also predicted synergisms between human impacts and bamboo flowering that could threaten the bamboo population over the long run. Our scenarios helped find alternative ways of designing more efficient and more effective policies. Hull et al. also discuss where Wolong might be headed, analyzing emerging socioeconomic and environmental factors that may need to be incorporated into future scenario analyses. Examples include seismic events, market fluctuations, and new forms of conservation policy implementation. Hull et al. conclude with thoughts on how to integrate these factors and lay out sustainability pathways to the future.

CHAPTER 3

Peek into a Home for Pandas and People

Vanessa Hull, Wu Yang, Wei Liu, Yingchun Tan, Jian Yang, Hemin Zhang, Zhiyun Ouyang, and Jianguo Liu

3.1 Introduction

The world is currently experiencing numerous challenges related to sustainability, including climate change, food security, and ecosystem degradation. The coupled human and natural systems (CHANS) approach is an effective means of studying such urgent issues as it accounts for the continually evolving human–nature interactions that may improve or threaten sustainability. Case studies in model coupled systems (Chapter 2) offer in-depth understanding of human–nature interactions and the opportunity to evaluate long-term sustainability in response to various challenges.

In this chapter, we provide an overview of our model coupled system: Wolong Nature Reserve. Wolong was chosen as a model system because it is globally important, spatially diverse, and temporally dynamic with respect to the evolving human–nature interactions. The system also encapsulates the shared challenges experienced worldwide with sustainability issues (e.g., balancing human livelihoods and conservation). It is an area that encompasses a wide range of geophysical variation and biological diversity and is a crucial area with respect to conservation of one of the most famous endangered species, the giant panda. Similar to most nature reserves in developing countries, Wolong is also home to a dynamic human population. This population interacts with the natural environment in a number of complex ways, all mediated by a multifaceted governance system that has evolved over time.

Since 1996, Wolong has served as an excellent laboratory for scientific investigation of coupled systems. We have also applied many ideas and methods developed in Wolong to many other parts of the world (Chapters 15–18; see also Liu et al., 2003a, Liu and Diamond, 2005, Liu et al., 2007). For example, our findings on household dynamics in Wolong have led to a paper that documented similar patterns in 141 countries (Liu et al., 2003a). Our observations on ecological degradation in Wolong, even after establishment of the protected area, were a catalyst to rethink the efficacy of protected areas worldwide (Liu et al., 2001). Insights from Wolong provided the foundation for the development of the coupled system framework (Liu et al., 1999, Liu et al., 2007). Lessons and experiences gained from Wolong helped improve the management of other protected areas in China (e.g., Liu et al., 2003b) and beyond (e.g., Liu et al., 2003a).

3.2 Historical and geographic context

Wolong Nature Reserve was founded in 1963 in Sichuan Province of China (31°04′30″ N, 103°13′41″ E, Figure 3.1). Wolong was initially 200 km^2 in area but expanded to 2000 km^2 in 1975 (Ministry of Forestry, 1998). The name Wolong translates as "crouching dragon" and originates from local folklore. The story goes that a dragon once flew over the area and was inspired to make a home there after being enraptured by the beautiful mountains, where it still rests (Schaller, 1994). Wolong is one of the first nature reserves established in China, in response to a

Pandas and People. Edited by Jianguo Liu, Vanessa Hull, Wu Yang, Andrés Viña, Xiaodong Chen, Zhiyun Ouyang, and Hemin Zhang. Published 2016 by Oxford University Press.

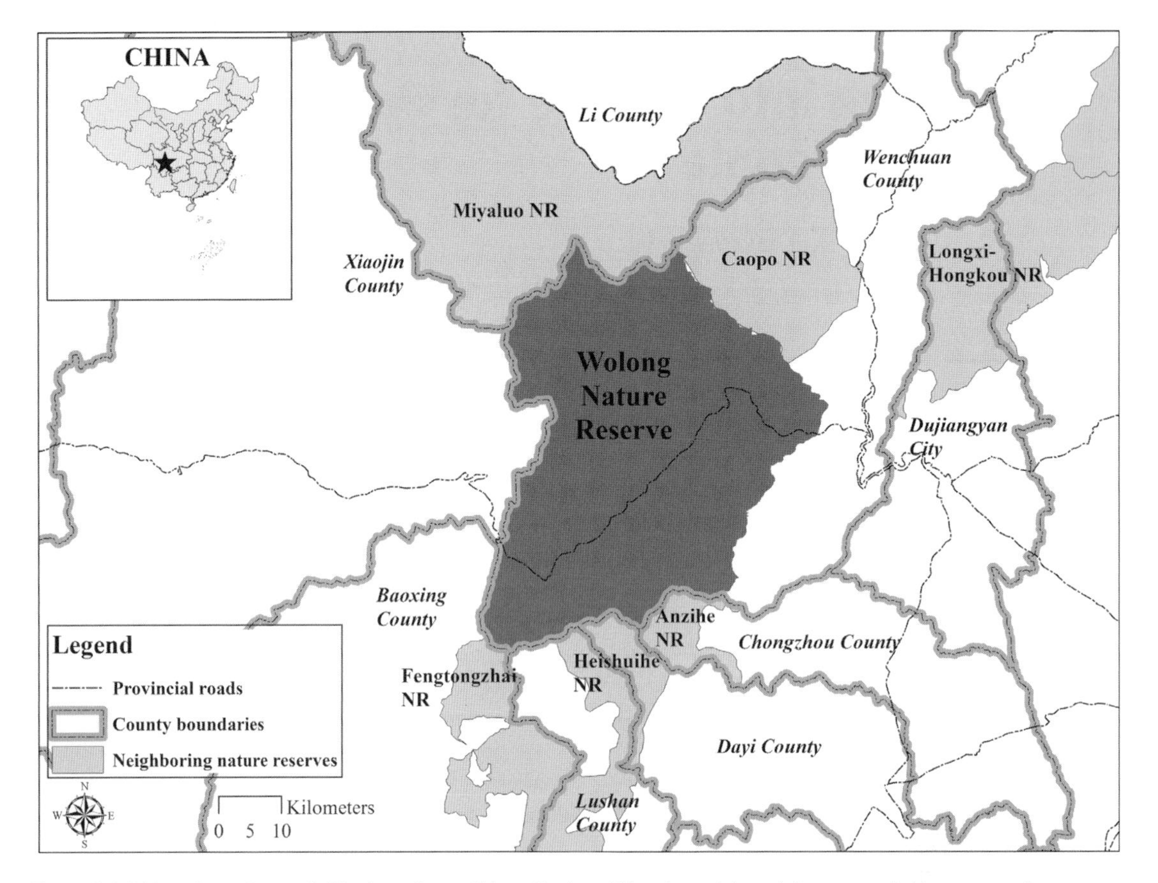

Figure 3.1 Wolong Nature Reserve in Wenchuan County, Sichuan Province, China, the model coupled system studied by our research group.

call from the Third National People's Congress in 1956 to set aside areas for conservation (Harkness, 1998). The main objective of Wolong is to ensure "the protection of [the] giant panda, other valuable rare animals and plants, and the typical natural ecosystem" (Ministry of Forestry, 1998: X).

Sichuan Province is located in the southwestern region of China. It is the fifth-largest province in terms of land area (485 000 km^2), supporting the fourth-largest human population (80.5 million people; People's Government of Sichuan Province, 2012). Sichuan contains a wide range of topographical features, with one of the four largest basins of China found in the eastern portion of the province and mountain peaks reaching above 6000 m in elevation on the western side. Wolong is in the southeastern part of the Qionglai Mountain range, which extends from north to south in the central part of the province. It lies approximately 130 km northwest of Chengdu, the capital city of Sichuan. Chengdu is a historical landmark and economic center of the country, and is the fourth-largest Chinese city in permanent human population (14 million residents; National Bureau of Statistics of China, 2010). Provincial Road 303 runs through the middle of Wolong, which connects with other roads that link Chengdu to other rural communities located to the west, including Xiaojin and Baoxing counties. Wolong is also connected to five other nature reserves: Caopo and Miyaluo reserves to the north and Anzihe, Heishuihe, and Fengtongzhai reserves to the south (Figure 3.1).

3.3 A dramatic geomorphological template

Wolong is situated in a unique geomorphological position in that it lies within the transition zone between the Sichuan Basin to the east and the Tibetan highlands to the west. The transition zone is evident even within the boundaries of Wolong itself through the wide elevation range, spanning from 1150 m to 6250 m (Wolong Administration Bureau, 2004; Figure 3.2A). The tallest peak in Wolong is part of the famous Siguniang ("Four Sisters") Mountains, which include the tallest peak in the Qionglai Mountain range.

The region as a whole has undergone significant geomorphic transformations since India's landmass collided with Eurasia's landmass over 50 million years ago (Hubbard and Shaw, 2009). The area with the highest relief within the region can be found in the Longmen Mountains (Hubbard and Shaw, 2009), the base of which contains the Longmen Mountain fault zone that encompasses Wolong. Shortening of the earth's crust is believed to have produced this thrust fault zone (Hubbard and Shaw, 2009). The fault zone has caused unstable tectonic activity throughout the region, most notably with the magnitude 8.0 (surface wave magnitude [M_s] 8.0 or moment magnitude [M_w] 7.9) Wenchuan earthquake in 2008 that had an epicenter along Wolong's boundary. Tectonic activity has also caused Wolong to experience violent uplifts that have resulted in a high topographical relief, producing steep cliffs and dramatic overlapping peaks (Wolong Administration Bureau, 2004).

Rock strata present in Wolong are dominated by limestone and phyllite from the Silurian to Triassic periods (Schaller et al., 1985). Soils in Wolong are diverse, ranging from yellow mountain soils at the lowest elevations, gradually changing to dark brown with increasing elevation (Ministry of Forestry, 1998). Alpine and tundra soils occur at the highest elevations (Ministry of Forestry, 1998). The main

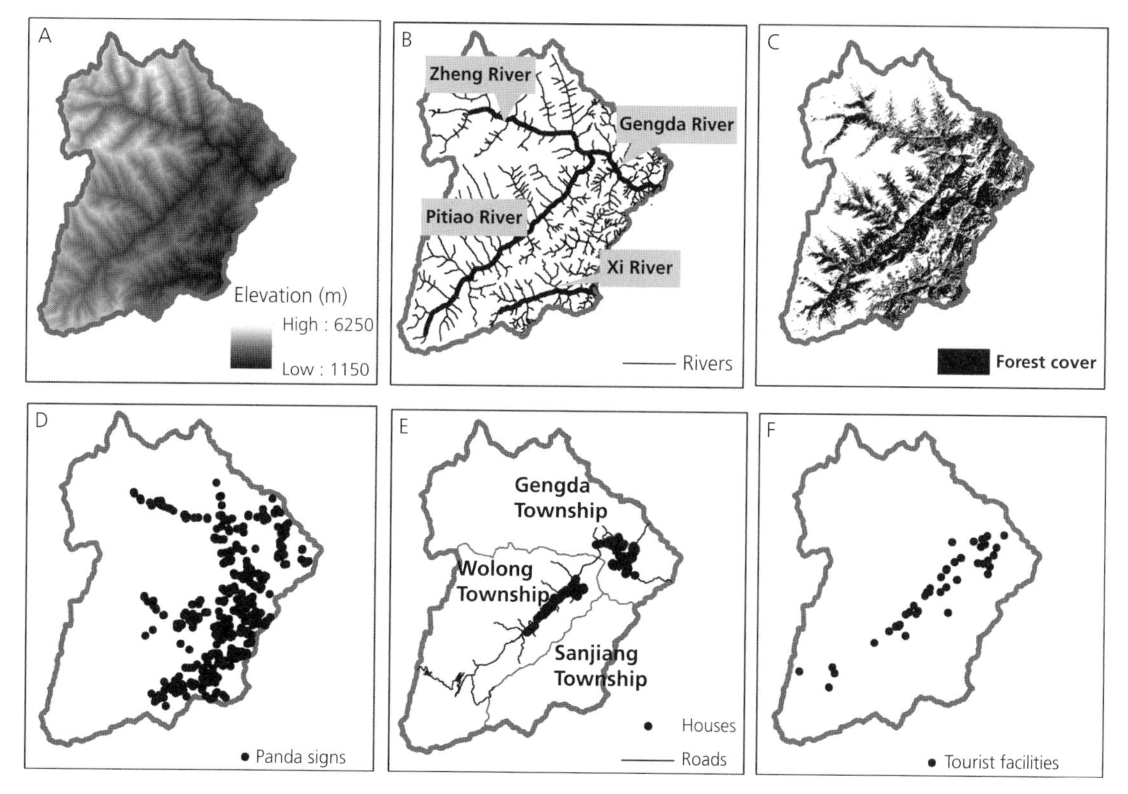

Figure 3.2 Key attributes of the model system in Wolong Nature Reserve. These include (A) elevation, (B) rivers, (C) forest cover, (D) giant panda signs, (E) roads, townships, and houses, and (F) tourism facilities. Elevations depicted in (A) were estimated using a digital elevation model acquired by the US National Aeronautics and Space Administration's advanced spaceborne thermal emission and reflection radiometer. Forest cover in (C) was estimated from supervised classification of remote sensing data acquired in 2007 from Viña et al. (2011). Giant panda signs in (D) were obtained from the third national giant panda survey conducted in Wolong in 2001 (State Forestry Administration, 2006).

rivers in Wolong include the Pitiao (60 km), Zheng (45 km), and Xi (37 km) (Li et al., 1992a; Figure 3.2B). The Pitiao River originates in the Balang Mountains and extends along the valley of the reserve (Ministry of Forestry, 1998). It serves as an important headwater area for the Minjiang River, a major branch of the Yangtze River. Rivers in Wolong usually flow abundantly due to the short flow lengths relative to the steep drop in elevation. The fastest rate of stream flow occurs in May, with the lowest water levels observed in November of each year. River flooding is frequent in the summer months of July and August (Ministry of Forestry, 1998).

The climate in Wolong is typical of the Qinghai-Tibetan Highland belt, with a temperature range of around −12 °C to 30 °C, averaging around 9 °C throughout the year (Wolong Administration Bureau, 2004). The warmest months are in the summer (July–August) and the coldest months are in the winter (November–March; Schaller et al., 1985). The area is known for its high humidity, with a yearly average humidity of 80% (Wolong Administration Bureau, 2004). There are clear dry and wet seasons. The average annual rainfall of around 888 mm mostly occurs in April–November, when it often rains at least 15 days per month (Schaller et al., 1985, Wolong Administration Bureau, 2004). Snowfall occurs between October and April, with annual snowfall around 930 mm (Reid and Hu, 1991).

3.4 Biodiversity hotspot

Wolong is within one of the global biodiversity hotspots (Liu et al., 2003b, Myers et al., 2000). The wide variations in elevation, topography, climate, and soils give rise to remarkably diverse plant and animal communities. The Qionglai mountain range as a whole is among the most biodiverse places on Earth. Its wide elevation range, humid climate, complex topography, and extensive watersheds allow it to support thousands of plant and animal species (Mackinnon, 2008). Wolong alone "contain(s) more plants, butterflies, amphibians, and birds than in most European countries" (Mackinnon, 2008: 96). There are about 4000 species of plants in Wolong, which make up 40% of all plant species present in Sichuan Province (Li et al., 1992a). Exact numbers of each type of plant are unknown. The only reliable survey, from the 1970s, found 174 species of moss, 191 species of ferns, 20 gymnosperms, and 1604 angiosperms (Li et al., 1992a). The survey also found about 870 plants that have medicinal value for humans (Li et al., 1992a). In addition, 24 plant species are labeled as rare and receive legal protection within China (Wolong Administration Bureau, 2004). Examples of such plants include the dove tree (*Davidia involucrata*), katsura tree (*Cercidiphyllum japonicum*), and hardy rubber tree (*Eucommia ulmoides*) (Li et al., 1992a). There are about 450 species of higher animals, including 96 species of mammals (nearly half of all mammal species occurring throughout Sichuan Province). In addition, there are 283 species of birds, 20 species of reptiles, 15 species of amphibians, and 6 species of fish (according to a 1970s survey, Li et al., 1992a). Insects alone are believed to include 1700 species (Li et al., 1992a). There are also 57 animal species that are rare and receive legal protection within China (Li et al., 1992a). Examples include snub-nosed monkeys (*Rhinopithecus* spp.), snow leopard (*Panthera uncia*), and takin (*Budorcas taxicolor*) (Li et al., 1992a).

Regarding the spatial distribution of plants and animals in Wolong, the most notable pattern is a distinct vertical zonation pattern, which can be attributed to the elevation (and thus soil) gradient. From low to high elevation, habitat types gradually change from evergreen broadleaf forest, to evergreen and deciduous broadleaf forest, to mixed coniferous and deciduous broadleaf forest, to subalpine coniferous forest, to alpine meadow and thicket, and finally to rock (Schaller et al., 1985). Total percent of the reserve covered with forest was estimated to be around 38% as of 2007 (Viña et al., 2007, Yang et al., 2013a; Figure 3.2C). The forest type that takes up the largest area of the reserve is the subalpine coniferous habitat, which covers around 16% of the total area (Schaller et al., 1985).

3.5 Stronghold for giant pandas

Wolong is currently the third-largest reserve for giant pandas in terms of area. The reserve was estimated to support almost 10% (154 individuals) of the total panda population as of the year 2000 (State Forestry Administration, 2006). The panda population in Wolong is larger than in any other panda reserve and has twice the number of individuals in the nature reserve with the second-largest population (State Forestry Administration, 2006). The population in Wolong in 2000 was similar to that estimated in a survey conducted in the 1970s (145 pandas; Schaller et al., 1985). Nonetheless, there is some uncertainty in the population estimates for this elusive species. Wolong's pandas live in three distinct river drainage areas (Figure 3.2D). These include the Zheng–Gengda drainage to the north, the Zhong–Xi drainage to the southeast, and the Pitiao drainage to the southwest (Schaller et al., 1985). The northern subpopulation is believed to be largely isolated from the other two as a result of the highway running through the reserve and associated human establishments. In fact, the panda population in Wolong is recognized as being at risk of future extinction due to genetic isolation among subpopulations (Hu, 2001). However, the northern subpopulation is potentially well connected to other panda populations outside the reserve. This northern area was identified in a broad-scale habitat suitability model as being a key linkage that helps connect panda habitats across the greater Qionglai mountain range (Xu et al., 2006).

Pandas in Wolong can be found in the mid-elevations, mainly from 1500 m to 3250 m but optimally from 2500 m to 3000 m (Liu et al., 1999). The habitat in this zone consists of evergreen and deciduous broadleaf, coniferous and deciduous broadleaf, and subalpine coniferous forests, all of which support bamboo in the understory (Schaller et al., 1985). The dominant tree species found in panda habitat in Wolong include birch, beech, maple, hemlock, and larch. The main species of bamboo that are consumed by pandas in Wolong include umbrella bamboo (*Fargesia robusta*) and arrow bamboo (*Bashania faberi*). Each covers roughly 40% (together 80%) of all of the bamboo distribution in Wolong (Li et al., 1992a). Other less common bamboo species found in localized patches throughout Wolong include *Yushania brevipaniculata, Fargesia angustissima, Phyllostachys nidularia, Fargesia nitida, Phyllostachys mannii*, and *Chimonobambusa pachystachys* (Shiqiang Zhou, pers. comm.). Pandas cannot occupy the lowest elevations due to loss of habitat as a result of human establishments. Pandas do not occupy the highest elevations because bamboo does not grow well in the alpine region, which covers approximately half of the reserve's area.

3.6 A vibrant human community

The area that is now contained in Wolong was inhabited by people long before the reserve was established. Tibetan people migrated to the area from the Tibetan Plateau during the late seventeenth century (Ghimire, 1997). One trail in the reserve was once part of the historic "Tea-Horse Road." Spanning more than 2000 km from Sichuan to Tibet, this road served as an important trade route connecting the Tibetans and Han people (Wolong Administration Bureau, 2009). The Han people (the current ethnic majority in China), who occupied the region at that time, avoided creating settlements within Wolong because they preferred the lowlands (Ghimire, 1997). Tibetans, on the other hand, were well adapted to the high altitudes and mountain living. This minority group successfully set up a pastoral way of life, mainly relying on high-altitude yak grazing for subsistence living (Ghimire, 1997). The Tibetans are believed to have later moved down to the lower elevations where they set up permanent homesteads and crops after the emergence of agriculture in the early eighteenth century (Ghimire, 1997). Small numbers of Han people migrated to the area later, around the beginning of the twentieth century. Until 2012, the reserve was dominated by members of Tibetan ethnicity (72%), with smaller numbers of Han (23%), Qiang (4%), and other ethnic minorities (<1%). The Han influence, however, can be strongly felt with the dominance of agriculture in Wolong. Their contribution included new and diverse crops and non-highland livestock such as pigs and cows (Ghimire, 1997). The majority of residents use the dominant (Han-derived) Chinese language and have lost their native tongues (Ghimire, 1997).

Dramatic changes took place in the community during the twentieth century. The human population in Wolong increased significantly from 1582 people in 1949 (Li et al., 1992a) to 2560 in 1975 and 4933 people in 2012 (Figure 3.3). The number of households increased at an even faster rate, by a factor of 3.4 between 1975 and 2012, to reach approximately 1436 households. Today, the human population in the reserve is distributed in two townships—Gengda and Wolong (Figure 3.2E)—each containing three villages further divided into a number of smaller units consisting of 26 groups of

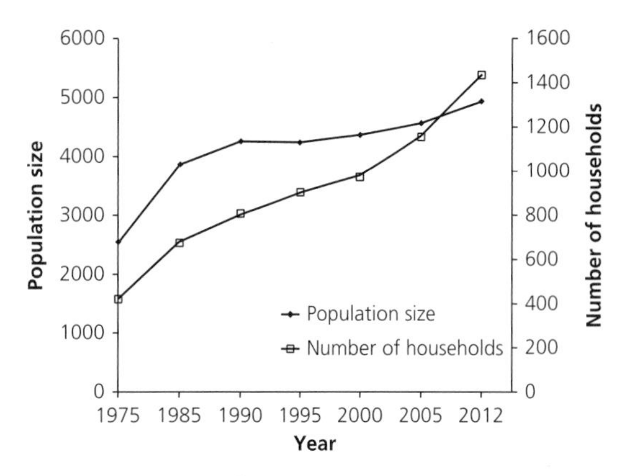

Figure 3.3 Dynamics of human population size and household number in Wolong Nature Reserve.

households. The boundary of Sanjiang Township is also inside the reserve (Figure 3.2E), but all households lie outside Wolong's boundaries. The trend in increased human population is due to a high fertility rate within local communities (Chapter 8). Of note is the fact that the majority of Wolong residents are exempt from China's one-child policy because they are ethnic minorities. The exemption allowed for a fertility rate of roughly 2.5 in the 1990s (Liu et al., 1999). The increase in number of households was due to a number of factors, including changing economic markets that promote smaller households and changing social norms and cultural values that discourage multigenerational family living (Liu et al., 1999, 2003a).

Infrastructure and facilities to support the growing human community also have evolved over time. Roads have shown marked improvements in the last several decades, beginning with the construction of a paved mud–rock road in 1960. The provincial highway was later completed in 1999 (Figure 3.2E). The road was then upgraded and widened in 2006 after the Sichuan Giant Panda Sanctuary was elected as a World Heritage Site (Liu et al., 2012, Wolong Administration Bureau, 2004). After the approval of the reserve's first ecotourism development master plan in the early 2000s, the Wolong Administrative Bureau implemented a series of upgrades. Using outside investments, the government initiated tourism infrastructure development projects including a new four-star hotel, a plank road in the Panda Valley, and several other scenic sites (He et al., 2008, Liu et al., 2012, Liu et al., 2015). Local residents also built a few dozen private hotels and restaurants for the anticipated increase of tourists.

Access to education and health care have historically been low and continue to be challenges for the local communities. A small number of rudimentary health care centers were established in the reserve in the 1950s (He et al., 2008, Liu et al., 2012, Wolong Administration Bureau, 2004). But even today local residents need to travel at least a couple hours to the nearest city to receive care for serious health conditions. While most residents of Wolong are literate with a primary school education, access to high-quality education is still limited. An elementary school and a middle school were opened in the 1950s (He et al., 2008, Liu et al., 2012). The facilities have since expanded to include multiple elementary schools and one combined middle and high school. However, the quality of education received in these schools is much lower than in nearby cities, making many local families choose to send their children to boarding schools outside Wolong.

The infrastructure (e.g., roads, hotels, residential houses, schools, hospitals, and tourism facilities) was largely destroyed by the 2008 Wenchuan earthquake and subsequent landslides, mud–rock flows, and mountain torrents (Yang et al., 2013b). Compared to other areas affected by the Wenchuan earthquake, where reconstruction was largely completed by May 2010, the post-earthquake reconstruction in Wolong lags far behind. This discrepancy is due to frequent associated disasters occurring in Wolong after the earthquake and inappropriate management (Yang, 2013). Most residential houses in Wolong had been repaired or rebuilt by 2012. Schools and the rudimentary hospital in Wolong were not rebuilt until 2011 and 2012, respectively. Completion of the reconstruction of Wolong was postponed to 2016.

3.7 Close ties between humans and nature

Despite changes that have taken place throughout the last century in Wolong, people have continued to mainly adopt an agricultural way of life. This lifestyle puts them in close contact with the natural world. By the 1990s, the main livelihood activities consisted of crop production, raising livestock, and collection of non-timber forest products (Ghimire, 1997). The per capita net income of Wolong households was around 1624 yuan in 2008 (Wolong Nature Reserve, 2008). This amount is small, considering the average across rural China was 4760 yuan in 2008 (National Bureau of Statistics of China, 2010). Crop production on average contributed 6% to total household income and occurred on approximately 3750 mu of land (2.5 km^2) throughout the reserve as of 2007 (Yang et al., 2013b). Main crops include cabbages, carrots, corn, and potatoes. In recent years, local people have adopted agricultural technologies to improve crop yields, including chemical fertilizers, e.g., carbamide, phosphate, and potash (Wolong Nature Reserve, 2008). A number of livestock species also occupy homesteads, grazing areas, and occasionally forests. They include yak (3010), goats and sheep (1668), cattle (1396), pigs (795), and horses (344)—estimates reflect an agricultural survey conducted in 2008 (Wolong Nature Reserve, 2008). Regarding non-timber forest products (e.g., medicinal herbs, wild vegetables, raw lacquer, and bamboo shoots), approximately 18% of households were engaged in this activity as of 2007. This activity comprises an average of 1.4% of household incomes (Yang et al., 2013b). Some main plants of interest include tianma (*Gastrodia elata*) and beimu (*Fritillaria cirrhosa*).

In recent years, local people have begun to diversify by exploring new livelihood strategies, including transportation of goods, road construction, tourism, and migrant work. These human activities impact nature in numerous ways. More roads and transportation cause more habitat fragmentation and noise. Tourism can create more waste, expansion of human infrastructure (Figure 3.2F), and direct human footprints on hiking trails (Liu, 2012, State Forestry Administration, 2006). Facing a lack of opportunities to gain income to meet livelihood needs, some residents have also turned to hunting and illegal logging, both of which impact local wildlife and their habitats (Schaller, 1994).

Another important component of human–nature interactions in the reserve involves timber harvesting (for timber sale and house construction) and fuelwood collection (for heating and cooking human food and pig fodder). Official records of timber harvesting go as far back as 1916 (Li et al., 1992b). In the mid-1990s, residents consumed approximately 10 000 m^3 of wood for cooking and heating, and ~1000 m^3 for house construction each year (Liu et al., 1999). As a result, in the last several decades, significant losses of forests and giant panda habitat have occurred (Liu et al., 2001). Timber harvesting has lessened in recent years due to the implementation of the Natural Forest Conservation Program (NFCP; see Section 3.8). Fuelwood collection has decreased due to improved availability and reduced cost of electricity after the construction of a hydropower station in 2002 (Yang et al., 2013a). More information about timber harvesting and fuelwood collection is available in Chapters 6, 7 and 10.

Human communities in Wolong are also profoundly affected by natural disturbances. The most salient recent example was the Wenchuan earthquake as discussed in Section 3.6 and Chapter 12. This profound disturbance has threatened the livelihoods of local people over the long term. They have lost opportunities to engage in tourism and the sale and transport of goods, and educational opportunities for their children have been compromised.

It is also important to appreciate that human–nature interactions in Wolong are affected by numerous contextual factors originating from outside the reserve (Chapter 17). Some examples include educational opportunities, demand for labor in cities, and agricultural market fluctuations in cities. All factors impact both the out-migration rates of locals and their choices regarding engagement in income-generating activities (Chen et al., 2012, Liu et al., 2005). National-level factors have also influenced local human–nature interactions in Wolong. Examples include national conservation policies that provide incentives for locals to participate in conservation and national policies that promote tourism development (Liu et al., 2008, 2012). Global factors whose effects are manifested locally include growth in the international tourism market, international aid (e.g., for earthquake recovery and giant panda conservation), and global market forces (Liu et al., 2012, Wolong Administration Bureau, 2009).

3.8 A storied management history

At the time Wolong was established as a nature reserve, there was no formal regulatory framework governing the operations of nature reserves. However,

Wolong grew to become a model reserve that served to guide the development of such frameworks elsewhere across China. A turning point occurred when the reserve joined the UNESCO Man and Biosphere Programme (MAB) network in 1980, an event that attracted newfound national and international attention (Ministry of Forestry, 1998). The Chinese government officially designated the reserve as the country's first "special district" for conservation in 1983. This decision meant that the administration governing the area would concurrently manage both conservation and development (Wolong Administration Bureau, 2004). The reserve is now managed by the Wolong Administration Bureau, which is overseen by both the central government's State Forestry Administration and the Forestry Department of Sichuan Province. The reserve was recognized as a National Key Nature Reserve by China's State Council in 1985, signifying it had achieved national-level recognition and support (Ministry of Forestry, 1998). The central government began to invest in Wolong by building six conservation stations (Wolong Administration Bureau, 2004). Around the same time, the management of the reserve also began to collaborate with WWF in scientific research, particularly on the giant panda. This collaboration led to the creation of what would later become the largest captive breeding facility in the world (Ministry of Forestry, 1998). Such recognition extended to the global level over time, later prompting the reserve to be included as a central part of the Sichuan Giant Panda Sanctuary, a World Heritage Site designated by UNESCO in 2006.

A key event in the history of Wolong's development was the drafting of the Wolong Nature Reserve Master Plan in 1998 (Ministry of Forestry, 1998). This document was the first complete management plan for a nature reserve managed by the then Ministry of Forestry (now State Forestry Administration) in China (Ministry of Forestry, 1998). The plan was designed to conserve the ecological system while also allowing for human development. One of the central components was a zoning scheme that divided the reserve into three spatial designations: an experimental zone for human development, a core zone for biodiversity protection, and a buffer zone to soften the border between the two (Ministry of Forestry, 1998). The zoning scheme has been maintained to spatially separate the reserve's two main goals, but it also has limitations, such as challenges with enforcement (Chapter 13). Another component of the management plan included a community development plan, which outlined ways to improve the livelihoods of the local communities. Examples include improvement of infrastructure, expanded opportunities to participate in tourism, and better access to education and health care (Ministry of Forestry, 1998). As alluded to in Section 3.7, to date successes on these fronts have been mixed. While the establishment of the nature reserve has created new income-earning opportunities, education and health care are still undersupplied, and tourism benefits are unequally distributed (Ghimire, 1997, Liu et al., 2012).

A number of policies have been implemented to help achieve reserve goals. In the early 1980s, the government sought to relocate approximately 100 families in Gengda Township from areas considered to be panda habitat (at higher elevations) down to the roadside (Ghimire, 1997). Although the World Food Programme supported the construction of a large apartment complex, the policy failed. The families did not move because the policy makers did not foresee farmers' need for agricultural land (Ghimire, 1997). A more successful policy also implemented in the 1980s dealt with forest replanting. Through subsidies provided by the World Food Programme, local farmers received small amounts of compensation for converting approximately 113 ha of their farmland and other degraded areas to forest plantations (Wolong Administration Bureau, 2004).

The legacy of reforestation policies continued with the launching of two national-level conservation programs in China in the late 1990s: NFCP and the Grain to Green Program (GTGP). The former is a comprehensive forest initiative that includes a ban on timber harvesting in all natural forests. Locally, NFCP incorporates a payments for ecosystem services (PES) component that provides monetary rewards for locals to participate in forest monitoring (Liu et al., 2008). The latter is a forest replanting program that also provides PES to locals in the form of subsidies (grain, money, and seedlings) for converting steeply sloped farmlands into forest plantations (Liu et al., 2008). These ambitious conservation initiatives had measurable results for nature, as from the mid-1980s to the early 2000s around 70% of local croplands were converted to tree plantations (Liu et al., 2012). Forests and panda habitats are recovering (Chapters 6 and 7; see also Viña et al., 2007). Both policies also have generated both positive (e.g., release of labor from farming and fuelwood collection) and negative (e.g., increase of wildlife-induced losses) effects for local residents (Liu et al., 2012, Yang et al., 2013a).

Another recent management activity that has shaped the reserve is the development of tourism. In the early 2000s, Wolong Nature Reserve developed its first ecotourism development master plan, the first for any national nature reserve in China. This plan provided a framework to accommodate the growing number of domestic and international tourists visiting Wolong for various reasons: viewing captive giant pandas, hiking, bird watching, and enjoying the cooler climate (Liu et al., 2012). Growth in this sector has been significant. Between 1998 and 2006, the revenue generated from tourism increased from 2 million yuan to over 42 million yuan (Liu et al., 2012). However, the impact on the local human community has also been mixed. It is a continuing challenge to figure out how best to transfer benefits to local communities that have received few direct benefits from tourism growth (He et al., 2008, Liu et al., 2012). Negative impacts of tourism on nature are numerous (e.g., trash accumulation, noise, and infrastructure), but remain understudied.

3.9 Toolbox of research approaches

Our long-term interdisciplinary study in Wolong is guided by an integrated conceptual framework (Chapter 2). To untangle complex human–nature interactions and address fundamental questions, we integrate multiple methods from a variety of social and natural science disciplines. Data integration is the hallmark of the coupled system approach. The types of data collection methods we have used in Wolong include in-person interviews with local residents and government officials, archival and government document collection, field surveys, and GPS measurements. After we collect the data, we manage them in relational databases and GIS. To analyze and integrate various data, we use remote sensing software, GIS, and systems modeling and simulation. Detailed descriptions of these methods are available in many of the subsequent chapters and in published papers. Below is a brief introduction.

Our team has conducted household surveys in Wolong since 1996. Since 1999, approximately 220 households selected using a stratified random sampling method based on administrative units have been surveyed almost every year through face-to-face interviews with household heads or their spouses. Information obtained in these surveys includes household demographics, income, expenditures, energy consumption characteristics, attitudes, intentions, and behaviors in response to policy changes, and participation in and attitudes toward conservation programs and tourism activities. We conducted interviews with relevant government officials on a variety of topics, including reserve policy implementation, reserve planning, and perceptions of long-term changes in human–nature interactions in the reserve. Government documents utilized in our research include agricultural and population censuses, household registration lists, management plans, and policy papers, and paper and digital maps of key locations of interest in the reserve.

We also have conducted field surveys since 1996 at a variety of locations throughout the reserve, including in both natural forests and human-altered habitat types (e.g., agricultural fields, forest plantations, and GTGP plots). Information obtained in these surveys includes topographic features (e.g., slope, aspect, and elevation), vegetation features (e.g., forest type and tree and bamboo species, density, and cover), animal data (e.g., presence and timing of animal signs, and food consumption by animals), and human disturbances (e.g., timber harvesting, fuelwood collection, and the presence of trash). Typically, these data are collected in fixed area plots of varying sizes (e.g., 10 × 10 m or 30 × 30 m plots). Plots were located along transects selected to capture maximum variation in habitats across the landscape. GPS measurements are taken at each plot to record the location, a practice applied to other locations of interest throughout the reserve, including households, tourist destinations, and other infrastructure. GPS collars were used to monitor the behavior of wild giant pandas by recording their locations every 4 hours for 1–2 years from 2010–2013 (Chapter 4).

Remote sensing and geospatial analyses have been ongoing since the inception of the project. We have collected and analyzed remotely sensed data at different resolutions (from 0.6 m to 250 m) acquired between 1965 and 2010 by several satellite systems such as Corona, Landsat TM, ETM +, MODIS, IKONOS, and Quickbird. Analyses have relied on several techniques, including unsupervised classification, supervised classification, generation of vegetation indices, analysis of phenological signatures, and artificial neural networks (Liu et al., 2001, Tuanmu et al., 2010, Viña et al., 2007, 2008, 2010). Such analyses have sought to characterize changes in quantity and quality of forests and giant panda habitat over time in Wolong. This objective has involved prediction of forest cover and type and prediction of bamboo presence using cutting-edge procedures developed by our team (Tuanmu et al., 2010, Viña et al., 2008). We have in turn integrated such dynamics in forest cover and giant panda habitat with information about changes in human population and policies to identify linkages between different components of the coupled system.

Modeling and simulation have played central roles in our study. We have integrated our various sources of data to create a variety of different models, including land-cover change models (Chapters 6 and 7; see also Linderman et al., 2006, Liu et al., 2001), animal resource selection models (Chapter 4; see also Bearer et al., 2008), agent-based models of human behavior (Chapter 8; see also An et al., 2005), econometric models (Chapter 9; see also Chen et al., 2009a), censored regression models (Chapter 11; see also Yang et al., 2013c, Chen et al., 2009b), causal inference models (Chapters 11 and 13; see also Yang et al., 2013c, Chen et al., 2012), and ecosystem service and human well-being models (Chapters 5 and 12; see also Yang et al., 2013b, Yang et al., 2015). Many of these models were spatially explicit, which was accomplished by integrating remote sensing data and GPS data on various natural and human components into a GIS prior to model construction. Simulation also has allowed us to test hypotheses about how the system may change in the future under different policy or management scenarios (Chapter 14). They include conditions under which local residents would switch their energy source from fuelwood to electricity (An et al., 2002), re-enroll in a conservation program (Chen et al., 2009b), and plan for maximizing the effectiveness of a conservation program (Viña et al., 2013). Other scenarios pertain to the effects of various potential household growth patterns and resident behavior patterns on panda habitat in the future (Linderman et al., 2006). Such analyses have been made possible by the availability of long-term data and aid in understanding system functioning and guiding policy and management of this system in the future.

3.10 Summary

Research on human–nature interactions benefits from in-depth and long-term case studies in model coupled human and natural systems. In this chapter, we introduced the model coupled system we have studied since 1996—Wolong Nature Reserve. Wolong is a high-profile protected area that covers 2000 km^2 in Sichuan Province of southwestern China. It is globally important as a UNESCO MAB Reserve and a World Heritage Site. The reserve contains diverse ecosystems along an elevational range of 1150 m to 6250 m, including thousands of plant and animal species, most notably over 100 endangered giant pandas. Wolong also is home to a growing human community of nearly 5000 people (mostly farmers). There are complex interactions and feedbacks (e.g., crop production, timber harvesting, raising livestock, and road construction) among local residents, forests, and wildlife. These interactions are also influenced by government policies, contextual factors originating from other systems (e.g., tourism), and natural disasters (e.g., the 2008 Wenchuan earthquake and associated subsequent disasters that destroyed or damaged the majority of the infrastructure). To understand our model coupled system in an integrated manner, we have drawn on diverse natural and social sciences to design field and social surveys, build systems models that integrate various sources of data, and run simulations to understand and predict complex processes and patterns across space and over time. Ultimately, we aim to make our work useful in understanding and helping govern sustainably not only the model coupled system, but also many other coupled systems around the world.

References

An, L., Linderman, M., Qi, J., et al. (2005) Exploring complexity in a human-environment system: an agent-based spatial model for multidisciplinary and multiscale integration. *Annals of the Association of American Geographers*, **95**, 54–79.

An, L., Lupi, F., Liu, J., et al. (2002) Modeling the choice to switch from fuelwood to electricity: implications for giant panda habitat conservation. *Ecological Economics*, **42**, 445–57.

Bearer, S., Linderman, M., Huang, J., et al. (2008) Effects of fuelwood collection and timber harvesting on giant panda habitat use. *Biological Conservation*, **141**, 385–93.

Chen, X., Frank, K.A., Dietz, T., and Liu, J. (2012) Weak ties, labor migration, and environmental impacts toward a sociology of sustainability. *Organization & Environment*, **25**, 3–24.

Chen, X., Lupi, F., He, G., and Liu, J. (2009a) Linking social norms to efficient conservation investment in payments for ecosystem services. *Proceedings of the National Academy of Sciences of the United States of America*, **106**, 11812–17.

Chen, X., Lupi, F., He, G., et al. (2009b) Factors affecting land reconversion plans following a payment for ecosystem service program. *Biological Conservation*, **142**, 1740–47.

Ghimire, K.B. (1997) Conservation and social development: an assessment of Wolong and other panda reserves in China. In K.B. Ghimire and M.P. Pimbert, eds, *Environmental Politics and Impacts of National Parks and Protected Areas*, pp. 187–213. Earthscan Publications, London, UK.

Harkness, J. (1998) Recent trends in forestry and conservation of biodiversity in China. *The China Quarterly*, **156**, 911–34.

He, G., Chen, X., Liu, W., et al. (2008) Distribution of economic benefits from ecotourism: a case study of Wolong Nature Reserve for giant pandas in China. *Environmental Management*, **42**, 1017–25.

Hu, J. (2001) *Research on the Giant Panda*. Shanghai Science and Education Publishing House, Shanghai, China (in Chinese).

Hubbard, J. and Shaw, J.H. (2009) Uplift of the Longmen Shan and Tibetan plateau, and the 2008 Wenchuan (M = 7.9) earthquake. *Nature*, **458**, 194–97.

Li, C., Zhou, S., Xiao, D., et al. (1992a) A general description of Wolong Nature Reserve. In Wolong Nature Reserve, Sichuan Normal College, eds, *The Animal and Plant Resources and Protection of Wolong Nature Reserve*, pp. 313–25. Sichuan Publishing House of Science and Technology, Chengdu, China (in Chinese).

Li, C., Zhou, S., Xiao, D., et al. (1992b) The history and status of Wolong Nature Reserve. Wolong Nature Reserve, Sichuan Normal College, eds, *The Animal and Plant Resources and Protection of Wolong Nature Reserve*, pp. 326–42. Sichuan Publishing House of Science and Technology, Chengdu, China (in Chinese).

Linderman, M.A., An, L., Bearer, S., et al. (2006) Interactive effects of natural and human disturbances on vegetation dynamics across landscapes. *Ecological Applications*, **16**, 452–63.

Liu, J., An, L., Batie, S.S., et al. (2005) Beyond population size: examining intricate interactions among population structure, land use, and environment in Wolong Nature Reserve, China. In B. Entwisle and P.C. Stern, eds, *Population, Land Use, and Environment: Research Directions*, pp. 217–37. National Academies Press, Washington, DC.

Liu, J., Daily, G.C., Ehrlich, P.R., and Luck, G.W. (2003a) Effects of household dynamics on resource consumption and biodiversity. *Nature*, **421**, 530–33.

Liu, J. and Diamond, J. (2005) China's environment in a globalizing world. *Nature*, **435**, 1179–86.

Liu, J., Dietz, T., Carpenter, S.R., et al. (2007) Complexity of coupled human and natural systems. *Science*, **317**, 1513–16.

Liu, J., Li, S., Ouyang, Z., et al. (2008) Ecological and socioeconomic effects of China's policies for ecosystem services. *Proceedings of the National Academy of Sciences of the United States of America*, **105**, 9477–82.

Liu, J., Linderman, M., Ouyang, Z., et al. (2001) Ecological degradation in protected areas: the case of Wolong Nature Reserve for giant pandas. *Science*, **292**, 98–101.

Liu, J., Ouyang, Z., Pimm, S.L., et al. (2003b) Protecting China's biodiversity. *Science*, **300**, 1240–41.

Liu, J., Ouyang, Z., Taylor, W.W., et al. (1999) A framework for evaluating the effects of human factors on wildlife habitat: the case of giant pandas. *Conservation Biology*, **13**, 1360–70.

Liu, W. (2012) *Patterns and Impacts of Tourism Development in a Coupled Human and Natural System*. Doctoral Dissertation, Michigan State University, East Lansing, MI.

Liu, W., Vogt, C.A., Luo, J., et al. (2012) Drivers and socioeconomic impacts of tourism participation in protected areas. *PLoS ONE*, **7**, e35420.

Liu, W., Vogt, C.A., Lupi, F., et al. (2015) Evolution of tourism in a flagship protected area of China. *Journal of Sustainable Tourism*. DOI: 10.1080/09669582.2015.1071380.

Mackinnon, J. (2008) Species richness and adaptive capacity in animal communities: lessons from China. *Integrative Zoology*, **3**, 95–100.

Ministry of Forestry (1998) *Wolong Nature Reserve Master Plan*. Ministry of Forestry, Beijing, China (in Chinese).

Myers, N., Mittermeier, R.A., Mittermeier, C.G., et al. (2000) Biodiversity hotspots for conservation priorities. *Nature*, **403**, 853–58.

National Bureau of Statistics of China (2010) *Sixth National Population Census of the People's Republic of China*. National Bureau of Statistics of China, Beijing (in Chinese).

People's Government of Sichuan Province (2012) *Sichuan Overview*. http://www.sc.gov.cn/10462/10758/11799/11800/2012/7/25/10219536.shtml (in Chinese).

Reid, D.G. and Hu, J. (1991) Giant panda selection between *Bashania fangiana* bamboo habitats in Wolong Reserve, Sichuan, China. *Journal of Applied Ecology*, **28**, 228–43.

Schaller, G.B. (1994) *The Last Panda*. University of Chicago Press, Chicago, IL.

Schaller, G.B., Hu, J., Pan, W., and Zhu, J. (1985) *The Giant Pandas of Wolong*. University of Chicago Press, Chicago, IL.

State Forestry Administration (2006) *The Third National Survey Report on the Giant Panda in China*. Science Publishing House, Beijing, China (in Chinese).

Tuanmu, M.-N., Viña, A., Bearer, S., et al. (2010) Mapping understory vegetation using phenological characteristics derived from remotely sensed data. *Remote Sensing of Environment*, **114**, 1833–44.

Viña, A., Bearer, S., Chen, X., et al. (2007) Temporal changes in giant panda habitat connectivity across boundaries of Wolong Nature Reserve, China. *Ecological Applications*, **17**, 1019–30.

Viña, A., Bearer, S., Zhang, H., et al. (2008) Evaluating MODIS data for mapping wildlife habitat distribution. *Remote Sensing of Environment*, **112**, 2160–69.

Viña, A., Chen, X.D., McConnell, W.J., et al. (2011) Effects of natural disasters on conservation policies: the case of the 2008 Wenchuan Earthquake, China. *Ambio*, **40**, 274–84.

Viña, A., Chen, X., Yang, W., et al. (2013) Improving the efficiency of conservation policies with the use of surrogates derived from remotely sensed and ancillary data. *Ecological Indicators*, **26**, 103–11.

Viña, A., Tuanmu, M.–N., Xu, W., et al. (2010) Range-wide analysis of wildlife habitat: implications for conservation. *Biological Conservation*, **143**, 1960–69.

Wolong Administration Bureau (2004) *The Chronicle of Wolong Nature Reserve*. Sichuan Science and Technology Press, Chengdu, China (in Chinese).

Wolong Administration Bureau (2009) *Wolong National Nature Reserve Post-earthquake Reconstruction Master Plan*. Wolong Administration Bureau, Wolong, China (in Chinese).

Wolong Nature Reserve (2008) *Wolong Nature Reserve Annual Agricultural Report 2008*. Wolong Administration Bureau, Wolong, China (in Chinese).

Xu, W., Ouyang, Z., Viña, A., et al. (2006) Designing a conservation plan for protecting the habitat for giant pandas in the Qionglai mountain range, China. *Diversity and Distributions*, **12**, 610–19.

Yang, W. (2013) *Ecosystem Services, Human Well-being, and Policies in Coupled Human and Natural Systems*. Doctoral Dissertation, Michigan State University, East Lansing, MI.

Yang, W., Dietz, T., Kramer, et al. (2015) An integrated approach to understanding the linkages between ecosystem services and human well-being. *Ecosystem Health and Sustainability*, **1**, 19.

Yang, W., Dietz, T., Liu, W., et al. (2013b) Going beyond the Millennium Ecosystem Assessment: an index system of human dependence on ecosystem services. *PLoS ONE*, **8**, e64581.

Yang, W., Liu, W., Viña, A., et al. (2013a) Performance and prospects of payments for ecosystem services programs: evidence from China. *Journal of Environmental Management*, **127**, 86–95.

Yang, W., Liu, W., Viña, A., et al. (2013c) Nonlinear effects of group size on collective action and resource outcomes. *Proceedings of the National Academy of Sciences of the United States of America*, **110**, 10916–21.

CHAPTER 4

Pandas–People Coexistence and Competition

Vanessa Hull, Jindong Zhang, Wei Liu, Jinyan Huang, Shiqiang Zhou, Scott Bearer, Weihua Xu, Mao-Ning Tuanmu, Andrés Viña, Hemin Zhang, Zhiyun Ouyang, and Jianguo Liu

4.1 Introduction

With the expansion of the human population in the last few centuries, human–wildlife interactions are increasing like never before (Carter et al., 2014, Woodroffe et al., 2005). Nearly all wildlife populations and associated habitats have been altered by humans in a multitude of ways. Examples include poaching, livestock grazing, resource extraction, pollution, and climate change (Pimm and Raven, 2000). Wildlife in turn affect humans in numerous ways, including destruction of crops, killing livestock and people, providing economic income through wildlife tourism, and controlling pests (Carter et al., 2014). Human–wildlife interactions can be framed from many different perspectives, but two of the most common contrasting paradigms are human–wildlife competition and human–wildlife coexistence (Woodroffe et al., 2005). Competition occurs when humans and wildlife engage in antagonistic interactions with one another due to limited available resources (Treves and Karanth, 2003, Woodroffe et al., 2005). Coexistence can be achieved when one or both sides have adapted to the other's presence so that both can be sustained while sharing the same space (Madden, 2004, Woodroffe et al., 2005). Understanding factors that trigger competition and those that foster coexistence can help guide conservation policy and management planning into the future to promote sustainability of coupled human–wildlife systems.

The giant panda is one endangered species that has been profoundly shaped by human–wildlife interactions. As outlined in Chapter 1, giant pandas and humans have a long and complex history of interacting with one another. The story of human–panda interactions is one that can be told from many different perspectives, and themes of competition and coexistence are interwoven even in the earliest records of ancient China (Chapter 1). The portrayal of pandas as peaceful mountain-dwelling neighbors was juxtaposed against the harvest of pandas by elites for sport and fur (Schaller, 1994). The dominant pattern for centuries has been coexistence, considering that pandas have survived in the face of considerable human development pressures when many other large mammals have not. For example, elephants are almost extinct in China and rhinos, gibbons, and snub-nosed monkeys have nearly disappeared due to the combined effects of hunting and habitat loss (Corlett, 2007). A survey of over 50 reserves in South China found "very few signs of large and medium-sized mammals" (Fellowes et al., 2004, cited in Corlett, 2007). The subspecies of tiger once living in South China has recently been extirpated (Tilson et al., 2004). So have the leopard and gray wolf, with several other species at regional risk (Lau et al., 2010). Pandas are also found to persist at densities higher than many other bears around the world, which highlights differences in their biology (Garshelis, 2004), and a higher capacity to coexist with people (Hull et al.,

Pandas and People. Edited by Jianguo Liu, Vanessa Hull, Wu Yang, Andrés Viña, Xiaodong Chen, Zhiyun Ouyang, and Hemin Zhang. Published 2016 by Oxford University Press.

2014a). A recent synthesis suggests that pandas may be more flexible and adaptable in their habitat selection choices than previously realized (Hull et al., 2014a).

However, in the last hundred years, human communities have increasingly expanded and competed with pandas for space and resources (Hu, 2001, Pan et al., 2001, Viña et al., 2010). Humans have had a decisive competitive edge over pandas, shrinking the panda habitat that once covered nearly all of China to small and fragmented pockets in the southwestern region (Chapter 7, see also Viña et al., 2010). This drastic change occurred due to various human activities such as timber harvesting, farming, and road construction. Humans have scaled back these impacts in recent years, buoyed by conservation policies aimed at restoring panda habitat (Liu et al., 2008, Viña et al., 2011). But there is still a long way to go to level the playing field for pandas. In this era of rapid industrialization and human expansion in China and across the world, pandas need to be able to continue to coexist with people in order to survive over the long term.

In this chapter, we explore coexistence and competition between pandas and humans in our model coupled system in Wolong Nature Reserve. A stronghold for the wild giant panda population, Wolong is an ideal laboratory to investigate this topic. In addition to being a flagship nature reserve that drives panda policy, it is also a relatively well-studied system. Researchers have done extensive long-term research on wild panda behavior and population dynamics dating back to the 1980s (e.g., Schaller et al., 1985). A variety of socioeconomic research has also been conducted on the local human population (Chapters 5, 8, 9, 10, and 11). The reality is that nearly 5000 people live inside Wolong and relocation is ethically questionable and immensely difficult. Thus the livelihood and well-being of rural residents need to be considered in any panda conservation program in order for sustainability to be achieved (Liu et al., 2001). We use our unique and interdisciplinary data set to tackle important questions relating to coexistence and competition. We hope that our findings can help inform policy making to allow for long-term sustainability of the species alongside human communities.

4.2 Hermits in the forest—coexistence between pandas and people

There is a tendency in the modern wildlife research community to portray humans and wildlife as adversaries. However, some scholars have begun to recognize that the paradigm of human–wildlife coexistence is valuable (Peterson et al., 2010). This alternate view is reflective of the long history of coexistence and is also more constructive for finding future solutions to pressing conservation issues (Peterson et al., 2010). This research investigates how animals and people can adapt their behaviors in response to one another in ways that foster coexistence. Coexistence can be achieved by using different resources such as spaces (within the same overall habitat; i.e., patch segregation) or by using the same resources but at different times (i.e., temporal segregation). For example, tigers in Nepal have been found to adapt their behaviors to be more active at night when humans are not out traversing their habitat (Carter et al., 2012). This form of temporal separation may allow them to coexist alongside humans in the same space at fine scales (Carter et al., 2012).

For giant pandas, it is instructive to consider the factors that have contributed to their long history of surviving alongside humans. Research in the field shows that pandas and humans do not coexist side by side in the same locations at the same times (Pan et al., 2001, Schaller et al., 1985). Pandas are endearingly known as hermits who secretively enshroud themselves in the dense forests, rarely making their existence known to people living nearby. The species is so shy that even researchers studying pandas in the wild for several years may never see one there. Pandas are also well known for having a strong sense of smell and sensitivity to noise (Schaller et al., 1985, Swaisgood et al., 1999). These acuities allow them to live in close proximity to human communities while still avoiding costly direct confrontations with people (Schaller et al., 1985). Since pandas largely subsist on a bamboo diet, they rarely raid farmers' crops (a common problem with deer and wild boar in panda habitat). They also do not prey on livestock in the way that large carnivores such as leopards and tigers do in other regions (Berger, 2006, Carter et al., 2014). This biological attribute allows them to coexist and have minimal direct tension with local residents.

Another attribute of giant pandas that has allowed them to coexist with people relates to their space use patterns. Space is a fundamental requirement for any animal, and pandas are no exception. Some radio collar studies were conducted in the 1980s that offered glimpses into the world of wild pandas (Pan et al., 2001, Schaller et al., 1985, Yong et al., 2004). But the Chinese government later banned telemetry on pandas for over a decade due to concerns over animal safety. In 2006, our research group obtained permission to conduct a GPS (Global Positioning System) collar study on five wild giant pandas in Wolong. We were able to learn new information about pandas' space use that can help shed light on competition between pandas and humans (Hull et al., 2015). Our findings corroborated earlier studies (including another recent study by Zhang et al., 2014) suggesting that pandas have lower space requirements than many other bear species because they have small home ranges (2.7–6 km^2 in our study). Their space use patterns also overlap with neighboring pandas (up to 35% overlap in our study; Figure 4.1). This aspect of their biology allows them to persist at higher densities than other bears and helps to lessen competition. They do not require the large expanses such as several hundred square kilometers for large carnivores, an area difficult to conserve under human pressures (Gittleman, 2001, Carter et al., 2014). We also found surprising evidence that the GPS-collared pandas were somewhat flexible in the habitat they used, more so than had been believed in the past. Pandas used areas of non-forest, making up 18–42% of the home ranges (Hull, 2014). The spatial pattern of non-forest was important. Pandas could tolerate and in some instances even preferred some small patches of non-forest within their home range (Hull, 2014).

The issue of coexistence across different seasons is also interesting to consider. Certain times of the year may bring pandas and humans into closer contact. For example, pandas have long been observed to move to lower elevations closer to humans to forage on umbrella bamboo during the bamboo growing season in Wolong (Schaller et al., 1985). This seasonal foraging pattern puts them in closer

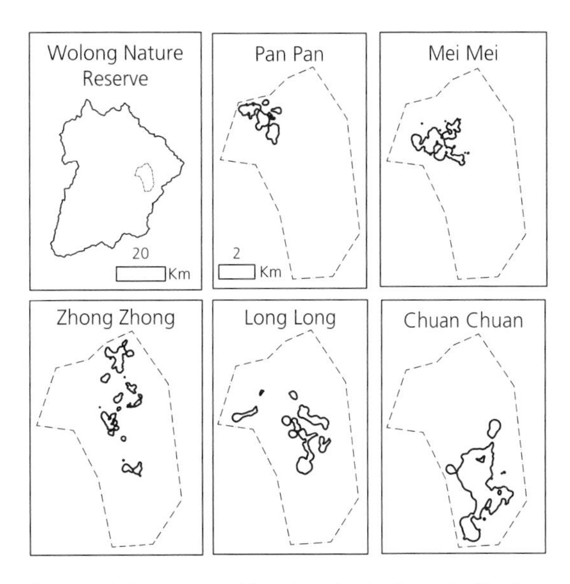

Figure 4.1 Home ranges of five GPS-collared wild pandas (Pan Pan, Mei Mei, Zhong Zhong, Long Long, and Chuan Chuan) monitored over 6–12 month periods in Wolong Nature Reserve. Home ranges were estimated using biased random bridge movement models. Figure is adapted from Hull et al. (2015). Reproduced with permissions from Oxford University Press.

proximity to humans and may make them more vulnerable to disturbance. But the payoff is likely worth it for pandas. The energetic return from foraging on succulent, low-elevation bamboo shoots is higher than other times of year (Nie et al., 2015, Schaller et al., 1985). Further research is needed to understand how coexistence is being achieved in these periods and whether panda migration patterns are affected by the degree of disturbance in a given area.

Because pandas avoid direct interactions with people, it is important to consider how to maintain adequate space for pandas to live as undisturbed as possible. Our research to date stands to contribute to understanding how pandas and humans can coexist in a system by means of spatial segregation. Nature reserves across China address the issue of thousands of people living inside nature reserves by mandating a zoning scheme (Liu and Li, 2008). This regulation requires delineation of a core zone (set aside for conservation of biodiversity with very limited human activities allowed) and an experimental zone (for human communities to use and develop). Also required is a buffer zone (an area where some human activities are allowed, designed to soften the division between the two other zones; Liu and Li, 2008). In Wolong, the zoning scheme is designed around the existing human establishments. The experimental zone is largely limited to the main road and the core zone to higher elevations. The buffer zone is distributed throughout some low- to mid-elevation areas (Hull et al., 2011; Figure 4.2). The scheme is designed to allow coexistence between humans and pandas, with both afforded their own spaces to subsist. However, at the start of our work, there was little information on the efficacy of the zoning policy on the ground.

We set out to fill this information gap by conducting an interdisciplinary analysis of the efficacy of the zoning scheme for panda conservation in Wolong. We conducted a spatial overlay of the zoning

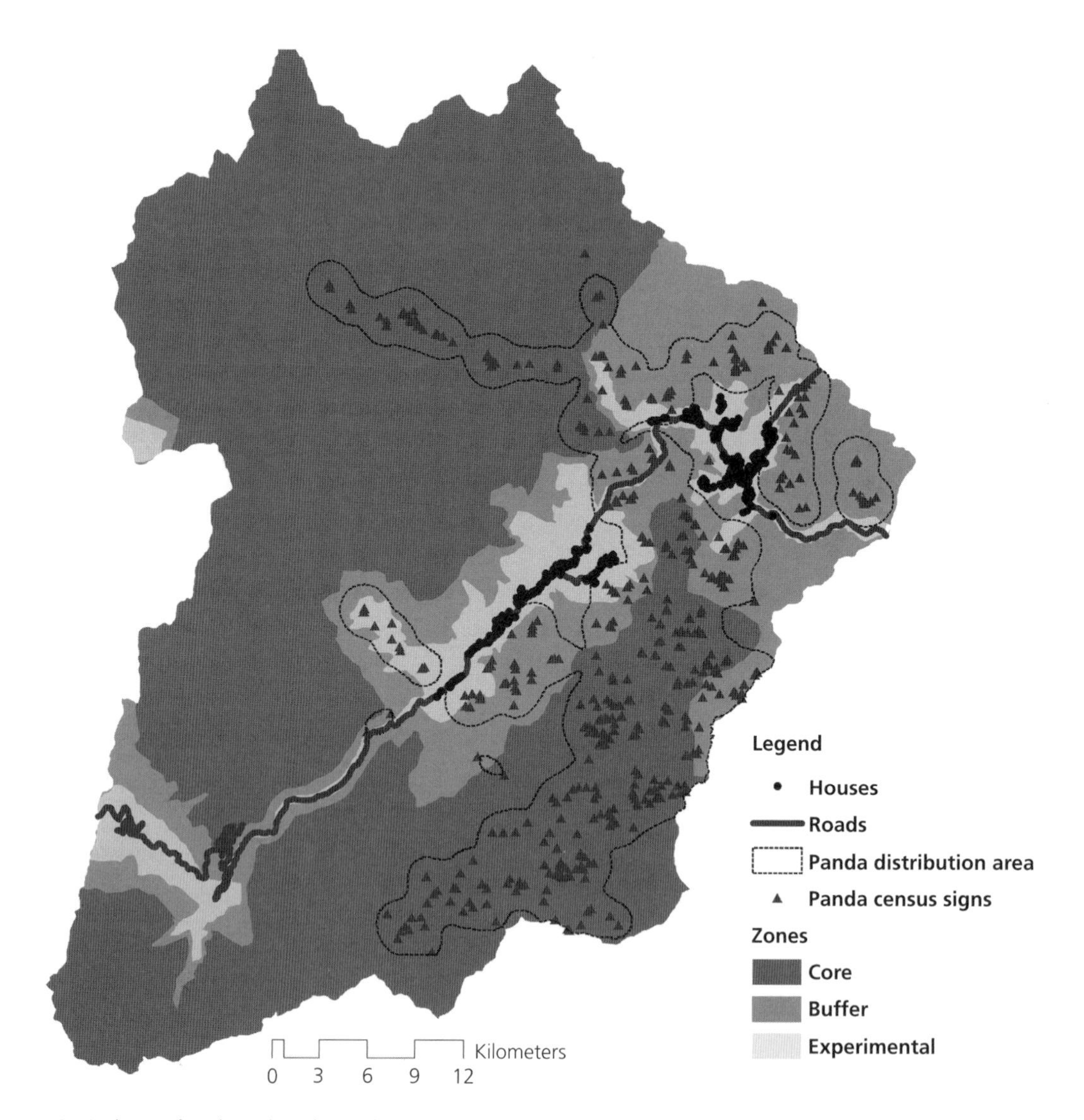

Figure 4.2 Distribution of pandas and people in Wolong Nature Reserve. Panda locations were obtained from the third national survey of wild giant pandas (State Forestry Administration, 2006) and are surrounded by an estimated distribution area predicted using a 95% kernel (bandwidth = 1000 m). Also shown are zoning designations for the reserve. Figure is adapted from Hull et al. (2011). Reprinted with permission from Elsevier.

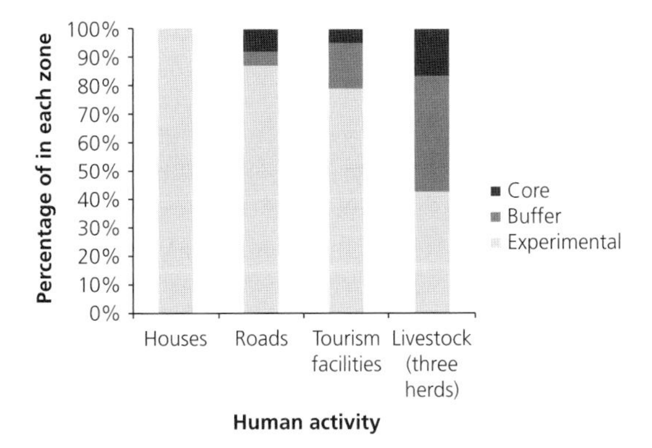

Figure 4.3 Distribution of houses, roads, tourism facilities, and livestock in each management zone (core, buffer, and experimental) in Wolong Nature Reserve. House locations (n = 1060) were measured with GPS units in 2002, roads were obtained from government documents, tourism facilities (n = 19) were recorded with GPS units in 2006, and livestock (three herds of horses only) were monitored using GPS collars and field sampling. Reprinted with permission from Elsevier (originally printed in Hull et al., 2011).

scheme with the distribution of pandas in the reserve (as estimated by the national giant panda survey) along with various human activities. We found mixed evidence regarding the idea that the zoning scheme helped foster coexistence between humans and pandas (Hull et al., 2011). We noticed that pandas were distributed nearly equally in the core and buffer zones, with little presence in the experimental zone. Human development in the form of roads and infrastructure was largely well contained in the experimental zone, with little spilling over into the area set aside for panda conservation. However, human activities such as livestock grazing were not well contained in the experimental zone. We found livestock grazing in all three zones in the reserve (Hull et al., 2011; Figure 4.3). In particular, our analysis suggested that pandas and humans have a fair amount of spatial overlap in the buffer zone. Because this area is not well regulated with clear guidelines, pandas are potentially placed in jeopardy (Hull et al., 2011).

4.3 Dueling adversaries—conflicts between pandas and people

Competition between humans and wildlife has been well documented among diverse species around the world (Morzillo et al., 2014, Treves and Karanth, 2003, Woodroffe et al., 2005). From insects to fish to mammals, competition with humans shapes animal populations in diverse habitats (e.g., deserts, oceans, and forests; Distefano, 2005, Pomeroy et al., 2007, Woodroffe et al., 2005). As Earth's finite resources become more limited with the expansion of the human population, the shared need for both humans and wildlife to procure fundamental resources for survival becomes a contest unfolding as a series of complex interactions and feedbacks between the two sides. Such interactions play out in sometimes unpredictable ways over space and time (Carter et al., 2014, Morzillo et al., 2014). Complex human–wildlife conflicts often arise and are as varied as the species involved. For example, wildlife pests invade native habitats (Morzillo et al., 2014), wildlife increasingly crash into vehicles (Ng et al., 2008), and large predators kill livestock (Berger, 2006, Carter et al., 2014). In response (as a feedback), farmers take retaliation tactics such as poisoning or killing wildlife (Carter et al., 2014).

Despite the fact that pandas have largely been peaceful neighbors sharing the mountains with people for centuries, the recent expansion of the human population throughout the pandas' range has put them in closer contact and potential conflict with native panda populations. This increase in pandas–people interactions is especially true in the decades since the industrial revolution and rapid expansion of the human population in China (Pan et al., 2001). The mechanisms driving competition include conflicts between pandas' needs for survival and human desire for maintaining and expanding communities (Liu et al., 1999). The three main resource needs that overlap between both parties and are thus sources of contention include space, trees, and food.

Pandas have relatively small home ranges compared to many large wildlife species, yet competition for space between pandas and humans is still an issue. Space available for pandas has been severely decreased and fragmented by people over the last several decades (Chapter 7; see also Viña et al., 2010). The reality of the present era is that there is not a huge amount of space left as adequate habitat for pandas—only an estimated 21 300 km^2 across the entire panda range (Viña et al., 2010). Panda habitat is defined by areas that support bamboo, the food source that makes up 99% of the pandas' diet (Chapter 7; see also Schaller et al., 1985). Bamboo requires adequate climate and soils and some degree of tree cover to grow and thrive, limiting pandas to forested areas at particular elevational ranges (Schaller et al., 1985, Taylor et al., 2004). The competition over space between pandas and humans thus unfolds along an elevational gradient. Both people and pandas cannot occupy the alpine zone at the highest elevations (i.e., above 3600 m). This area is above the treeline and does not support bamboo. It also is largely inaccessible or inhospitable for human occupancy, aside from occasional resource collection such as livestock grazing or medicinal herb collection (Hull et al., 2011, Schaller et al., 1985). Fossil records suggest that pandas once occupied the lowest elevations available in panda habitat (Peng et al., 2001), but humans have since deforested and occupied these areas, thus relegating pandas to the mid-elevation belt between the human communities and the alpine zone. In Wolong, we see this pattern as well, given the absence of pandas along the major roadway and adjacent to human establishments (Figure 4.2).

Aside from space itself, trees are also important for pandas and are a source of competition with humans. Historically, one of the most dominant sources of competition between humans and pandas is over forest resources. Trees are important for pandas as they provide the necessary conditions for bamboo growth. They also support important activities such as sunning, denning, and scent marking for communication purposes (Nie et al., 2012, Pan et al., 2001, Schaller et al., 1985, Zhang et al., 2007). Timber harvesting was the single greatest threat to giant pandas and their habitat as of the third national giant panda survey in the early 2000s (State Forestry Administration, 2006). In recent decades, fuelwood collection also became a threat to panda habitat. Locals needed a fuel source for cooking for themselves and for their pigs, in addition to heating their homes (Chapter 10; see also Liu et al., 1999). To understand competition between humans and pandas over these resources, we conducted the first comprehensive analysis of giant panda habitat use across different ages of forests (Bearer et al., 2008). We took extensive forest sampling throughout Wolong in areas used and unused by giant pandas (as detected by presence/absence of giant panda feces). Sampling was distributed across different forest ages (0–10 years, 11–30 years, 31–100 years, and old-growth (unharvested) forests) and different sizes of harvested areas (<10 ha equated to fuelwood collection sites and >10 ha timber harvesting sites). We found that panda use was significantly lower in newly harvested forests (0–30 years) than in both late successional forests harvested at least 31–100 years prior and old-growth forests (Figure 4.4; Bearer et al., 2008). However, panda use in the 31–100 year-old forests and old-growth forests did not differ significantly from one another. Our results suggest that a forest needs roughly four decades of natural recovery time after harvesting to become giant panda habitat again. This finding shows the extent of competition between pandas and humans for the same resource. On the other hand, this result also suggests that competition is dynamic and the competitive advantage can shift over time, allowing pandas to make up for earlier losses.

Aside from space and trees, pandas and humans also compete for bamboo—the panda's main food source. In some areas of panda habitat, humans

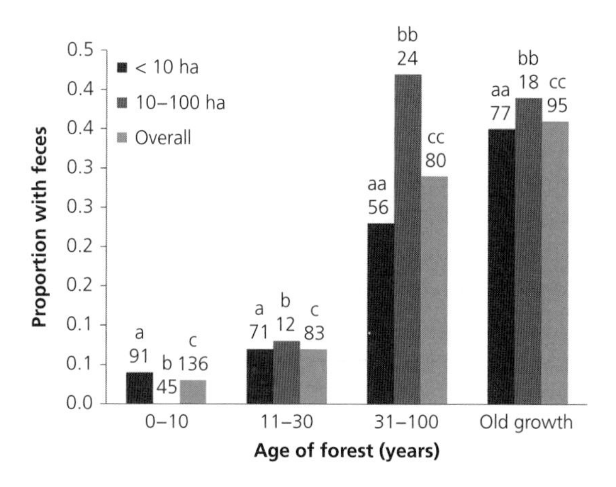

Figure 4.4 Proportion of plots with giant panda feces in areas with varying forest ages and sizes. Italicized numbers above bars represent plot sample size. Symbols (a, aa, b, bb, c, cc) above bars represent significance groupings in the proportion of survey plots with panda feces by age class for each size using binomial proportion test. The same number and type of letter-symbols above bars indicate no significant difference. Reprinted with permission from Elsevier (originally printed in Bearer et al., 2008).

harvest bamboo for their own uses (Wei et al., 2004). In Wolong, people search for bamboo shoots during the bamboo shooting season to sell as food. But more research is needed to better understand the extent and impacts of this practice. Recently, livestock raised by humans have begun to threaten the panda's food source. Livestock grazing was identified as the most significant threat to giant pandas in the most recent fourth national giant panda survey (State Forestry Administration of China, 2015). It was the most commonly observed disturbance to giant panda habitat across their entire range (34% of plots, 14% of all disturbances; State Forestry Administration of China, 2015). However, there are few in-depth field studies that document the nature of the impacts of livestock grazing on giant pandas.

We set out to investigate livestock grazing in Wolong, where we observed a new livestock sector (domestic horse raising) taking shape and having unexpected effects on panda habitat. Local residents in Wolong have raised pigs, sheep, goats, cattle, and yaks for centuries, but horse raising is a more recent phenomenon. There were only 25 horses in 1998, but there were nearly 350 in 2008 (Wolong Nature Reserve, 2008). Thus, we were interested in investigating horse encroachment in panda habitat, which could be potentially more severe than impacts of other types of livestock in Wolong.

To understand this emerging issue, we selected four local herds involved in forest encroachment to monitor in Wolong using GPS collars and field surveys. We found that around 50% of three of the horse herd home ranges were distributed in highly suitable or suitable giant panda habitat (Hull et al., 2014b; Figure 4.5). Several long-term field plots experienced extensive horse foraging, some over 75%. Horses also appear to have affected panda habitat use. Panda signs observed in pre-horse occupancy surveys at one horse-affected location were five to ten times higher than observed in the post-horse occupancy surveys (Hull et al., 2014b; Figure 4.6).

Foraging by horses could potentially threaten the availability of bamboo for giant pandas, if horse populations are unchecked. Horses need to consume up to 10 kg of plant biomass a day (Menard et al., 2002). In our study areas, horses consumed arrow bamboo (*Bashania faberi*; a species important to pandas), which has around 0.4–0.9 kg of biomass per m^2 (Taylor and Qin, 1987). However, horses only grazed the top 10% of bamboo stems and leaves. Therefore, a herd of 20 horses could forage up to 20% of bamboo culms in a 1 km^2 area in a single year (Hull et al., 2014b).

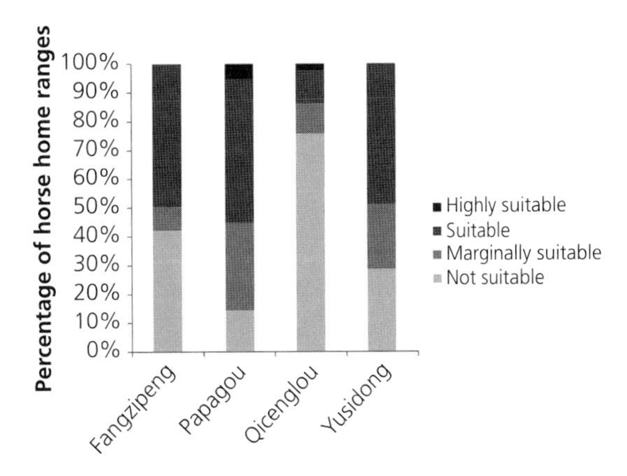

Figure 4.5 Proportion of horse home ranges in different giant panda habitat suitability classes (classes derived from Liu et al., 1999). The four horse herds (signified by Chinese place names) were monitored by placing a GPS collar on one member of the herd for 8–11 month periods. Figure is adapted from Hull et al. (2014b). Reprinted with permission from Elsevier.

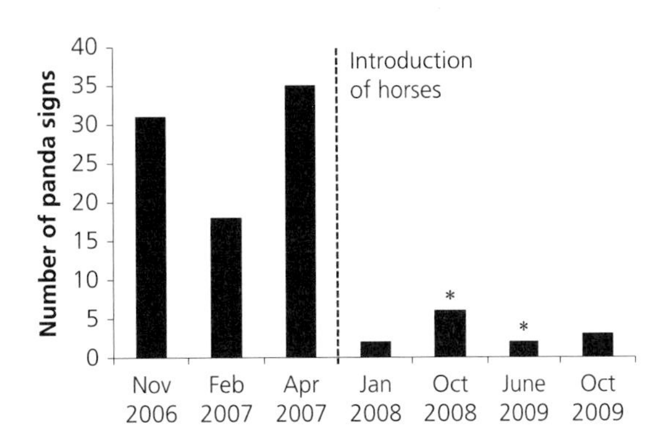

Figure 4.6 Number of panda signs observed during repeated sampling in transects over time in a roughly 2 km^2 area of panda habitat in Wolong called Fangzipeng that was inhabited by a group of 16 horses. Sampling periods denoted with a star (*) represent twice as many signs as were actually found during the survey (i.e., estimates were doubled to account for a one-half search effort performed during these periods). Reprinted with permission from Elsevier (originally printed in Hull et al., 2014b).

4.4 Ways forward—promoting coexistence and minimizing competition for sustainability

As our earlier discussion shows, coexistence between pandas and humans has been the dominant pattern throughout history. However, pandas have more recently been put into a competitive situation due to humans encroaching on their habitat, threatening their once isolated cocoons enshrouded from the outside world. Pandas cannot directly compete with humans in the way that a large carnivore or wildlife pest could by directly compromising the economy and health of human communities (Distefano, 2005). But pandas have pushed back indirectly in recent years by inspiring conservation efforts. Their charismatic and endearing persona has served as a sort of secret weapon that has inspired people to make their own efforts to scale back or self-handicap against the advantage of humans. This change in human behavior has given pandas the chance to thrive under careful management.

Our research on coexistence and competition between people and pandas provides some direction for the future management of Wolong and other coupled systems throughout the habitat of this endangered species. The reality is that people cannot be removed from Wolong (and many other areas throughout panda habitat). From both a logistical and ethical standpoint, human livelihoods need to be adequately considered in any conservation initiative. Therefore, managers need to find ways to promote pandas–people coexistence and minimize competitive interactions that could be harmful to one or both sides. In this section, we put forth some ideas based on our research findings.

With respect to the issue of timber harvesting and fuelwood collection, our findings give hope for further panda habitat recovery in protected areas for pandas in the future via natural successional processes. Our results also highlight the value of secondary (previously logged) forests for panda conservation, and indicate that efforts to protect late-successional and old-growth forests from harvesting are worthwhile. Also, programs of reforestation of native tree species can provide an alternate fuelwood source and may be important mitigation tools in regions where local people require fuelwood. Managers should realize that pandas may be flexible enough to use the areas from which they were previously driven by humans and invest in the protection of such areas. Managers should also consider the spatial distribution of panda habitat when drawing up management plans. Areas with disturbed or lower-quality habitat may still serve as important parts of panda home ranges if they are patchily distributed among high-quality habitats.

Our livestock study was unique in being able to document a human disturbance near the start of its emergence, an endeavor that has value in potentially driving future management measures. In fact, after we brought our concerns about the impact of horses to the managers of Wolong Nature Reserve, the administration implemented a successful horse removal plan from the Reserve in late 2012. However, our recent observations in the field suggest that some villages have increased other livestock species (e.g., sheep and goats) since the horse ban. Thus, the complex interplay between pandas and livestock in competing for food needs to continue to be examined over the long term. In the future, we recommend that the reserve managers undertake a comprehensive spatial planning approach to livestock grazing management to ensure that people maintain livestock in locations that do not compromise giant panda habitat. This provision would allow for coexistence of livestock and pandas. In addition, proper monitoring and enforcement need to be conducted to prevent illegal grazing in forests (as well as other illegal activities such as poaching which we have personally observed to be increasing in Wolong after the 2008 earthquake).

Furthermore, zoning is one of the most important tools that can help foster coexistence because it allows spatial segregation between pandas and people. Our research suggests that the government should maintain zoning as an effective means of controlling the spatial pattern of development in the reserve. However, the zoning scheme could also be improved. First, several key areas used by pandas in the experimental and buffer zones are not well protected by the current zoning scheme. By conducting a spatial overlay to find key locations within the existing panda distribution area not currently developed by humans, we identified these

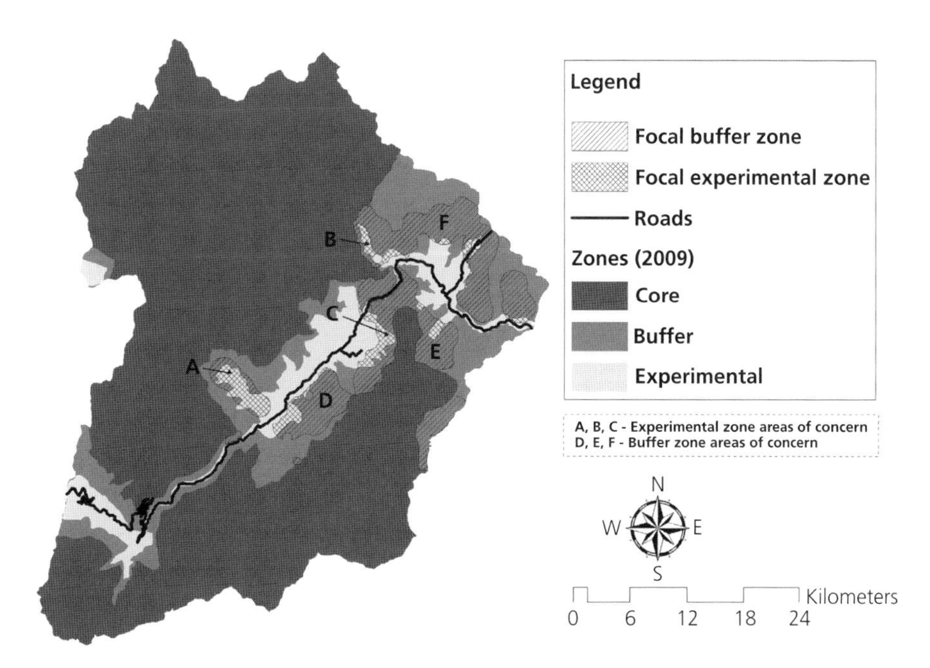

Figure 4.7 Proposed focal areas for zoning revisions in Wolong Nature Reserve for improved giant panda conservation. Zoning designations (core, buffer, and experimental) are presented along with focal experimental zones (areas of experimental zone that should be considered for conversion to buffer and/or core zone) and focal buffer zones (areas of buffer zone that should be considered for partial or full conversion to core zone). Both focal zones represent areas that support giant pandas and are also outside existing human establishments. Letters represent focal experimental (A–C) and buffer (D–F) zones of particular importance that are recommended for revision to better protect the panda population. Reprinted with permission from Elsevier (originally printed in Hull et al., 2011).

areas as targets for zoning revisions toward a higher level of protection (Figure 4.7; Hull et al., 2011). In addition, clearer guidelines should be put in place by the central government dictating rules about how the government may revise the zoning scheme over time as management needs change. For example, recent revisions to Wolong's zoning scheme allowed for tourism development in a previously protected region. This change could be harmful to pandas because this area overlaps with a key part of the panda distribution area (Hull et al., 2011). In addition, managers should recognize that zoning does not do as well at managing human activities such as livestock grazing and should be combined with other policies that promote coexistence. Programs involving payments for ecosystem services, discussed at length in other chapters of this book (e.g., Chapters 5 and 13), provide good examples of ways to reward local people for their help with conservation. Such an approach is necessary and worthwhile for fostering coexistence and serves as a good complement to exclusionary policies like zoning.

4.5 Summary

Humans have interacted with wildlife in diverse ways, ranging from coexistence to competition. In this chapter, we explored the ways in which our interdisciplinary research has revealed novel insights into the relationship between the pandas and humans in the coupled human and natural system of Wolong Nature Reserve. We brought to light factors that have contributed to long-term coexistence between people and pandas. These factors include the panda's natural avoidance of direct interactions with people, plant-based diet (so they do not prey on livestock), and relatively small space requirements (overlapping home ranges of 2.7–6 km^2). We also described how the reserve's zoning regulations foster coexistence by encouraging spatial segregation between pandas and people, with human development limited to specified experimental zones in the reserve. We then highlighted the mechanisms and consequences of competition between people and pandas via activities such as timber harvesting, fuelwood collection, and livestock grazing. Our research also demonstrates the effectiveness of existing governmental policies and helped us to set forth recommendations for future policy making to promote coexistence between people and pandas. Other examples include improving zoning regulations to protect a greater proportion of panda habitat and developing a comprehensive livestock policy to address ever-evolving grazing issues. The complex panda–human interactions we explored here are reminiscent of the same challenges faced with balancing wildlife conservation and human livelihoods in many other coupled systems around the globe. Such complexities require a renewed effort toward analysis of wildlife–human relationships for better management of endangered wildlife and human well-being.

References

Bearer, S., Linderman, M., Huang, J., et al. (2008) Effects of fuelwood collection and timber harvesting on giant panda habitat use. *Biological Conservation*, **141**, 385–93.

Berger, K.M. (2006) Carnivore–livestock conflicts: effects of subsidized predator control and economic correlates on the sheep industry. *Conservation Biology*, **20**, 751–61.

Carter, N.H., Shrestha, B.K., Karki, J.B., et al. (2012) Coexistence between wildlife and humans at fine spatial scales. *Proceedings of the National Academy of Sciences of the United States of America*, **109**, 15360–65.

Carter, N.H., Viña, A., Hull, V., et al. (2014) Coupled human and natural systems approach to wildlife research and conservation. *Ecology and Society*, **19**, 43.

Corlett, R.T. (2007) The impact of hunting on the mammalian fauna of tropical Asian forests. *Biotropica*, **39**, 292–303.

Distefano, E. (2005) Human–wildlife conflict worldwide: collection of case studies, analysis of management strategies and good practices. *SARD. Initiative Report*, FAO, Rome, Italy.

Fellowes, J., Lau, M., Chan, B., et al. (2004) Nature reserves in South China: Observations on their role and problems in conserving biodiversity. In Y. Xie, S. Wang, and P. Schei, eds, *China's Protected Areas*, pp. 341–55. Tsinghua University Press, Beijing, China.

Garshelis, D.L. (2004) Variation in ursid life histories: is there an outlier? In D. Lindburg and K. Baragona, eds, *Giant Pandas: Biology and Conservation*, pp. 53–73. University of California Press, Berkeley, CA.

Gittleman, J.L. (2001) *Carnivore Conservation* (**vol. 5**). Cambridge University Press, Cambridge, UK.

Hu, J. (2001) *Research on the Giant Panda*. Shanghai Science and Education Publishing House, Shanghai, China (in Chinese).

Hull, V. (2014) *Giant Panda Behavior across a Coupled Human and Natural System*. Doctoral Dissertation, Michigan State University, East Lansing, MI.

Hull, V., Roloff, G., Zhang, J., et al. (2014a) A synthesis of giant panda habitat selection. *Ursus*, **25**, 148–62.

Hull, V., Xu, W., Liu, W., et al. (2011) Evaluating the efficacy of zoning designations for protected area management. *Biological Conservation*, **144**, 3028–37.

Hull, V., Zhang, J., Zhou, S., et al. (2014b) Impact of livestock on giant pandas and their habitat. *Journal for Nature Conservation*, **22**, 256–64.

Hull, V., Zhang, J., Zhou, S., et al. (2015) Space use by endangered giant pandas. *Journal of Mammalogy*, **96**, 230–36.

Lau, M.W.-N., Fellowes, J.R., and Chan, B.P.L. (2010) Carnivores (Mammalia: Carnivora) in South China: a status review with notes on the commercial trade. *Mammal Review*, **40**, 247–92.

Liu, J., Li, S., Ouyang, Z., et al. (2008) Ecological and socioeconomic effects of China's policies for ecosystem services. *Proceedings of the National Academy of Sciences of the United States of America*, **105**, 9477–82.

Liu, J., Linderman, M., Ouyang, Z., et al. (2001) Ecological degradation in protected areas: the case of Wolong Nature Reserve for giant pandas. *Science*, **292**, 98–101.

Liu, J., Ouyang, Z., Taylor, W.W., et al. (1999) A framework for evaluating the effects of human factors on wildlife habitat: the case of giant pandas. *Conservation Biology*, **13**, 1360–70.

Liu, X. and Li, J. (2008) Scientific solutions for the functional zoning of nature reserves in China. *Ecological Modelling*, **215**, 237–46.

Madden, F. (2004) Creating coexistence between humans and wildlife: global perspectives on local efforts to address human–wildlife conflict. *Human Dimensions of Wildlife*, **9**, 247–57.

Menard, C., Duncan, P., Fleurance, G., et al. (2002) Comparative foraging and nutrition of horses and cattle in European wetlands. *Journal of Applied Ecology*, **39**, 120–33.

Morzillo, A.T., De Beurs, K.M., and Martin-Mikle, C.J. (2014) A conceptual framework to evaluate human–wildlife interactions within coupled human and natural systems. *Ecology and Society*, **19**, 44.

Ng, J.W., Nielson, C., and St Clair, C.C. (2008) Landscape and traffic factors influencing deer–vehicle collisions in an urban environment. *Human–Wildlife Interactions*, Paper 79. http://digitalcommons.unl.edu/hwi/79/.

Nie, Y., Swaisgood, R.R., Zhang, Z., et al. (2012) Giant panda scent-marking strategies in the wild: role of season, sex and marking surface. *Animal Behaviour*, **84**, 39–44.

Nie, Y., Zhang, Z., Raubenheimer, D., et al. (2015) Obligate herbivory in an ancestrally carnivorous lineage: the giant panda and bamboo from the perspective of nutritional geometry. *Functional Ecology*, **29**, 26–34.

Pan, W., Lü, Z., Zhu, X.J., et al. (2001) *A Chance for Lasting Survival*. Beijing University Press, Beijing, China (in Chinese).

Peng, J., Jiang, Z., and Hu, J. (2001) Status and conservation of giant panda (*Ailuropoda melanoleuca*): a review. *Folia Zoologica*, **50**, 81–88.

Peterson, M.N., Birckhead, J.L., Leong, K., et al. (2010) Rearticulating the myth of human–wildlife conflict. *Conservation Letters*, **3**, 74–82.

Pimm, S.L. and Raven, P. (2000) Biodiversity: extinction by numbers. *Nature*, **403**, 843–45.

Pomeroy, R., Parks, J., Pollnac, R., et al. (2007) Fish wars: conflict and collaboration in fisheries management in Southeast Asia. *Marine Policy*, **31**, 645–56.

Schaller, G.B. (1994) *The Last Panda*. University of Chicago Press, Chicago, IL.

Schaller, G.B., Hu, J., Pan, W., and Zhu, J. (1985) *The Giant Pandas of Wolong*. University of Chicago Press, Chicago, IL.

State Forestry Administration (2006) *The Third National Survey Report on the Giant Panda in China*. Science Publishing House, Beijing, China (in Chinese).

State Forestry Administration of China (2015) *The Fourth National Giant Panda Survey*. http://www.forestry.gov.cn/main/4462/content-743596.html (in Chinese).

Swaisgood, R.R., Lindburg, D.G., and Zhou, X.P. (1999) Giant pandas discriminate individual differences in conspecific scent. *Animal Behaviour*, **57**, 1045–53.

Taylor, A.H., Huang, J., and Zhou, S. (2004) Canopy tree development and undergrowth bamboo dynamics in old-growth *Abies–Betula* forests in southwestern China: a 12-year study. *Forest Ecology and Management*, **200**, 347–60.

Taylor, A.H. and Qin, Z. (1987) Culm dynamics and dry matter production of bamboos in the Wolong and Tangjiahe giant panda reserves, Sichuan, China. *Journal of Applied Ecology*, **24**, 419–33.

Tilson, R., Defu, H., Muntifering, J., and Nyhus, P.J. (2004) Dramatic decline of wild South China tigers *Panthera tigris amoyensis*: field survey of priority tiger reserves. *Oryx*, **38**, 40–47.

Treves, A. and Karanth, K.U. (2003) Human–carnivore conflict and perspectives on carnivore management worldwide. *Conservation Biology*, **17**, 1491–99.

Viña, A., Chen, X.D., McConnell, W.J., et al. (2011) Effects of natural disasters on conservation policies: the case of the 2008 Wenchuan Earthquake, China. *Ambio*, **40**, 274–84.

Viña, A., Tuanmu, M.-N., Xu, W., et al. (2010) Range-wide analysis of wildlife habitat: implications for conservation. *Biological Conservation*, **143**, 1960–69.

Wei, F., Yang, G., Hu, J., and Stringham, S. (2004) Balancing panda and human needs for bamboo shoots in Mabian Nature Reserve, China: predictions from a logistic-like model. In D. Lindburg and K. Baragona, eds, *Giant Pandas: Biology and Conservation*, pp. 201–09. University of California Press, Berkeley, CA.

Wolong Nature Reserve (2008) *Wolong Nature Reserve Annual Agricultural Report, 2008*.

Woodroffe, R., Thirgood, S., and Rabinowitz, A. (2005) *People and Wildlife, Conflict or Coexistence?* Cambridge University Press, Cambridge, UK.

Yong, Y., Liu, X., Wang, T., et al. (2004) Giant panda migration and habitat utilization in Foping Nature Reserve, China. In D. Lindburg and K. Baragona, eds, *Giant Pandas: Biology and Conservation*, pp. 159–69. University of California Press, Berkeley, CA.

Zhang, Z., Sheppard, J.K., Swaisgood, R.R., et al. (2014) Ecological scale and seasonal heterogeneity in the spatial behaviors of giant pandas. *Integrative Zoology*, **9**, 46–60.

Zhang, Z., Swaisgood, R.R., Wu, H., et al. (2007) Factors predicting den use by maternal giant pandas. *The Journal of Wildlife Management*, **71**, 2694–98.

CHAPTER 5

Quantifying Human Dependence on Ecosystem Services

Wu Yang, Thomas Dietz, Wei Liu, and Jianguo Liu

5.1 Introduction

5.1.1 General background

Human society directly and indirectly benefits from various goods and services provided by nature—ecosystem services (ES; Daily, 1997, Millennium Ecosystem Assessment, 2005). These benefits range from things such as clean water, timber, food, air purification, and carbon sequestration to aesthetic enjoyment, spiritual encouragement, nature-based tourism (e.g., ecotourism), and recreation. Such substantial dependence on ES dates back to the dawn of human history when human subsistence was based on hunting and foraging for plants. Over human history, and especially during the past few decades, the human use of ES has been escalating due to the unprecedented scale of population growth, household proliferation, affluence, and technology advances (Dietz et al., 2007, Liu et al., 2003, Millennium Ecosystem Assessment, 2005). Conceptually, the scientific community has recognized the associated consequences of these changes on ecosystems. But systematic assessments of the temporal and spatial patterns, causes, and associated environmental and socioeconomic impacts at global, national, regional, and local scales were largely missing until the completion of the Millennium Ecosystem Assessment (MA).

Launched in 2001, the MA has issued a series of reports that assess the consequences of ecosystem change. The methods and results of the MA also provide guidance for sustainable management of ecosystems for human well-being (Millennium Ecosystem Assessment, 2005). Perhaps the key finding of the MA is about the relationship between ES and human well-being. The MA revealed that over the past five decades, the overuse of ES substantially contributed to net gains in human well-being. Nevertheless, this trend also increased risks of abrupt changes in ecosystems and further marginalized some groups of the population. Unless significant changes in institutions, policies, and practices can be made, the degradation of ES and exacerbation of poverty for marginalized groups may get even worse. This prospect poses a huge barrier to achieving environmental and socioeconomic sustainability (Millennium Ecosystem Assessment, 2005).

The influence of the MA has been substantial, but it is only an initial effort toward systematic understanding of how ecosystem change affects human well-being (Carpenter et al., 2006). A robust theory linking biodiversity to ecosystem dynamics, the provision of ES, and changes in human well-being is lacking (Carpenter et al., 2006, Yang et al., 2013a, b). When attempting to formulate such a theory, many important questions remain unanswered. For instance, how does human dependence on ES change across time, space and among different population groups such as households at various income levels? To address this question, a quantitative, longitudinal approach is urgently needed. This chapter primarily is based on a published article describing our work (Yang et al., 2013b) and intends to adopt a quantitative approach to understanding human dependence on ES. Specific objectives include (1)

Pandas and People. Edited by Jianguo Liu, Vanessa Hull, Wu Yang, Andrés Viña, Xiaodong Chen, Zhiyun Ouyang, and Hemin Zhang. Published 2016 by Oxford University Press.

proposing a quantitative method to measure the degree of human dependence on ES—an index system of human dependence on ES (IDES); (2) empirically demonstrating the value of the quantitative index system at Wolong Nature Reserve; and (3) using the proposed quantitative index system to understand the patterns, causes, and effects of changes in human dependence on ES.

5.1.2 Human dependence on ES at Wolong Nature Reserve

Chapter 3 provided a general description of the natural and human contexts at Wolong Nature Reserve. Here we provide more detailed illustrations to qualitatively demonstrate local households' dependence on a large set of ES. These illustrations lay a foundation for quantification of such dependence later on.

The local ecosystem in Wolong Nature Reserve supports a wide range of provisioning services (i.e., products generated by ecosystems). Examples include crops such as maize, potatoes, carrots, and cabbages, livestock such as yaks, cattle, and pigs, and natural resources such as clean water, mushrooms, wild vegetables, honey, fuelwood, and traditional medicinal plants. There are also many types of regulating services (i.e., benefits obtained from the regulation of ecosystem functions and processes) such as water retention, flood control, erosion control, air purification, and carbon sequestration, although many of these services go far beyond the reserve boundary. For example, eight small hydropower stations with a total capacity of 34 million watts were established in the reserve before 2001 and the majority of generated electricity was transported outside (Yang et al., 2013c). The giant pandas, beautiful scenery, and cool weather in summer also attracted thousands of tourists worldwide every year before the 2008 Wenchuan earthquake (Liu et al., 2015). Many primary schools, high schools, and universities in Sichuan Province and around the world also choose the reserve as their field site for environmental and scientific education.

The majority of local residents are farmers and the local economy was primarily subsistence-based before the early 2000s, largely relying on various types of ES mentioned above. Currently, local households' revenue primarily comes from sales of agricultural and forest products and non-agricultural jobs such as local and migrant labor work. Small businesses related to tourism and recreation also contribute considerably. Examples include stores, temporary market stalls, restaurants, taxis, and farmers' inns. Government programs such as low-income subsidies and payments for ecosystem services (PES) are also economically important.

There are three major PES programs that have been implemented in the reserve. One is the Natural Forest Conservation Program (NFCP). This program uses payment as an incentive to motivate conservation behavior and aims to protect and restore natural forests via logging bans, afforestation, and monitoring (Liu et al., 2008, Yang et al., 2013c, d). The other two are the Grain to Green Program (GTGP) and the Grain to Bamboo Program (GTBP). Both provide farmers with payments to convert cropland on steep slopes to forest/grassland (GTGP) or bamboo (GTBP; Liu et al., 2008, Wolong Nature Reserve, 2005, Yang et al., 2013c). NFCP and GTGP are national policies developed in response to China's massive droughts in 1997 and floods in 1998. They were designed and implemented mainly to improve regulating services such as soil erosion control, water conservation, carbon sequestration, and air purification. GTBP is a local policy mimicking GTGP to increase bamboo for captive pandas and improve aesthetic appeal. At Wolong Nature Reserve, the implementation of GTGP, NFCP, and GTBP started in 2000, 2001, and 2002, respectively (Yang et al., 2013c).

5.2 An index system of human dependence on ES

5.2.1 Conceptual basis of the index system

We used the existing MA framework as a conceptual basis for developing a new index system of human dependence on ES (IDES). We chose to use the MA framework as a starting point because, though imperfect, it is the most commonly used approach and thus facilitates analyses at multiple scales and units. In addition, the MA framework defines the components of both ES and human well-being as well as their linkages to one another. Thus, it serves as a common platform for integration of indicators for ES and human well-being.

We designed the IDES system to include an overall index and three subindices (i.e., subindices for provisioning, regulating services, and cultural services). The overall index is defined as the ratio of net benefits acquired from ecosystems to the absolute value of total net benefits. The latter includes both net benefits that are obtained from ecosystems and other socioeconomic activities not obtained from ecosystems (e.g., small businesses and migrant work that are not directly related to ES; see Table 5.1). Each subindex is similar to the overall index, but the numerator only includes the corresponding category of ES. So the overall index equals the sum of three subindices. The general form of equations for the IDES system is below (Yang et al., 2013b).

$$IDES_i = ENB_i / \left| \sum_{i-1}^{3} ENB_i + SNB \right| \quad \text{(Equation 5.1)}$$

$$IDES = \sum_{i-1}^{3} IDES_i \quad \text{(Equation 5.2)}$$

where i is the category of three types of ES (i.e., provisioning, regulating, and cultural services). $IDES_i$ is the subindex for category i. ENB_i is the total net benefit obtained from ES in category i. SNB is the total net benefit acquired from other socioeconomic activities. IDES is the overall index. A higher value of overall index or subindex represents a higher dependence on the corresponding ES and a higher vulnerability to its decline or degradation.

The IDES system reflects tradeoffs and synergies among different types or categories of ES because it uses the net benefits. It also incorporates ES and ecosystem disservices through the use of net benefits. Therefore, both the overall index and subindices can be negative since net benefits are not necessarily positive. When an index has a negative value, it means that the gross benefit obtained from ES is smaller than the sum of costs for generating the corresponding ES and costs of ecosystem disservices. Taking the provision of agricultural products as an example, the gross benefits from sales of agricultural products may be less than the total costs of seeds, fertilizers, and pesticides.

5.2.2 Methods for the index system construction

The construction of an IDES system relies on assessing gross benefits and costs of a chosen unit of analysis (e.g., household). Thus the methodology is similar to cost-benefit analyses (CBA; Boardman et al., 2006, Hanley et al., 2001) and ecosystem service assessments (ESA; Chang et al., 2011, Millennium Ecosystem Assessment, 2005, Nelson et al., 2009, Yang et al., 2008). First, the unit of analysis must be identified. Then, the aggregation of data from a variety of sources can be done according to the specified unit of analysis.

The gross benefits and costs can be assessed with a set of economic valuation methods that are commonly used for CBA and ESA. Existing literature has reviewed these economic valuation methods and provided detailed guidelines (e.g., Barbier, 2011, Boardman et al., 2006, Hanley et al., 2001, Richard et al., 2001). Bateman et al. (2011) also summarized advantages and critiques of using them and thus this chapter does not expand on such issues. The general view is that, when markets for the gross benefits and costs exist, it is straightforward to assess them with market-based valuation methods. Examples include the market price method, the appraisal method, and the replacement cost method (Barbier, 2011, Chee, 2004, Scott et al., 1998, Yang et al., 2008). When no market data are available, one can employ non-market valuation methods. Examples include the contingent valuation method, the travel cost method, the stated preference method, and the hedonic price method (Barbier, 2011, Bateman et al., 2011, Scott et al., 1998, Yang et al., 2008). In some cases one may combine both market-based and non-market methods such as the benefit transfer method and unit-day value method (Ready and Navrud, 2006, Shrestha et al., 2007, Wilson and Hoehn, 2006). In addition, one may also use integrated tools such as the Integrated Valuation for Ecosystem Services and Tradeoffs (InVEST). This tool first assesses ecological production and then applies economic valuation methods (Kareiva et al., 2011, Nelson et al., 2009). Below we demonstrate how we used various

Table 5.1 Detailed classification of household net income and avoided costs by type of related ES at Wolong Nature Reserve. Source: adapted from Yang et al. (2013b).

Category	Subcategory	Item	Type of related ES*
Operating income	Crop income	Cabbage	P0
		Radish	P0
		Potato	P0
		Corn	P0
		Other crops	P0
	Animal husbandry income	Bacon	P1
		Pig	P0
		Goat	P0
		Cattle	P0
		Yak	P0
		Horse	P0
		Poultry and eggs	P0
		Honey bee	P0
		Other husbandry	P0
	NTFP income	Non-timber forest products	P0
	Other agricultural operating income	Other agricultural operating income	P0
	Non-agricultural operating income	Restaurants and hotels	C1 or NA†
		Ecotourism	C1 or NA†
		Transportation	C1 or NA†
		Contract work	NA
		Other small businesses	C1 or NA†
Wage income		Wage and bonus	NA
		Local labor income	NA
		Migrant labor income	NA
Property income	Land and housing rents	Land and housing rents	C1 or NA†
	Other property income	Interest income	NA
		Land acquisition compensation	NA
		Other rents	NA
Transfer income	Gift income from relatives and friends	Gift income from relatives and friends	NA
	Payments for ecosystem services (PES) income	Natural forest conservation program (NFCP)	R0
		Grain to Green program (GTGP)	R0
		Grain to Bamboo program (GTBP)	R0
	Social security benefits	Low income subsidy	NA
		Pension	NA
		Other subsidies	NA
Other income		Remaining other socioeconomic income	NA
Avoided costs		Use of fuelwood for energy	P0
		Subsidized electricity fees due to watershed conservation	R1

Notes:
*Capitalized letters P, R, C, and NA refer to provisioning, regulating, cultural services, and other benefits unrelated to ES, respectively. The digits "0" and "1" following "P, R or C" refer to direct and first-order indirect benefits, respectively. At Wolong Nature Reserve, payments from PES programs mainly compensate the forgone benefits for improving regulating services (e.g., water conservation, soil erosion control, carbon sequestration, and air purification). Therefore, PES payments were classified as benefits related to regulating services. Disservices and costs of ES provision have already been incorporated in the net income data.
†For each household, the benefit is included as a benefit related to cultural services if it is related to ecotourism and recreation; otherwise, it is classified as a benefit unrelated to ES.

economic valuation methods to empirically assess the gross benefits and costs in order to construct the IDES system at Wolong Nature Reserve.

5.2.3 Empirical application at Wolong Nature Reserve

We used household survey data to assess net benefits or equivalently gross benefits and costs. Our research team has collected household survey data in Wolong almost every year since 1999. Here we used the data from the summer of 1999 (to measure activities in 1998, An et al., 2001) and the end of 2007 (to measure 2007), before the 2008 Wenchuan Earthquake. We randomly sampled 180 households in 1999 (An et al., 2001) and revisited them in 2007 (Yang et al., 2013b). Since household heads or their spouses are often the decision-makers for household affairs, they were chosen as interviewees (An et al., 2001). Our household survey used the same categories for household income and expenditures as those of the National Bureau of Statistics of China (National Bureau of Statistics of China, 2011). This design feature allowed for consistency with standard economic survey methods for cross-context comparisons. For goods and services that have market prices but do not involve market transactions (e.g., self-consumed agricultural products), their monetary values were calculated using corresponding market prices. These estimates were then incorporated into the income and expenditure data for assessments of gross benefits and costs.

Household income and expenditure data only reflect benefits and costs involving market transactions. There are also benefits and costs that do not involve market transactions, such as the avoided costs. It is impractical, if not impossible, to assess all of the ES in any given location. This study attempted to take into account as many major ES as possible with the best available data. We were able to include two major items of avoided costs. One is the avoided costs of electricity fees due to the discount on unit electricity price as mentioned below. Forest conservation largely improves water yield for hydropower stations. As a result, local households received an electricity price reduced by 0.07 yuan per kilowatt-hour in both 1998 and 2007. Therefore, the avoided electricity fees can be easily calculated by multiplying the consumed amount of electricity by the discount price. Another avoided cost came from the use of fuelwood. Without fuelwood, local households would have to pay for alternative energy such as electricity and coal. Previously, if one household lacked laborers during the fuelwood collection season, one would exchange laborers by working for other households in different seasons or hire laborers from other households. Therefore, in this case, one can estimate the economic value of collected fuelwood by the equivalent market value of labor spent on fuelwood collection. Our survey assessed the amount of fuelwood collected and the amount of labor spent collecting it. The calculated shadow price of fuelwood was approximately 0.10 yuan per kilogram in 1998 and 0.20 yuan per kilogram in 2007. All the monetary values here are adjusted for inflation into the values in 1998 based on the consumer price index.

After obtaining the net benefits (or gross benefits and associated costs), the next step is to link different net benefits with the corresponding ES. Such an approach allows one to use Equations 5.1 and 5.2 to calculate the overall IDES index and subindices. In doing so, one needs to define the scope of accounting for direct and indirect ecosystem benefits. Direct ecosystem benefits refer to ES that directly generate human well-being. For instance, agricultural products (e.g., food) are provisioning services that generate direct benefits to humans. Some other services may indirectly contribute to human well-being: sometimes such services are one step away from direct benefits (i.e., first-order indirect benefits) while other times they are more distant (i.e., secondary or more distant indirect benefits). For example, incomes from providing services to tourists (e.g., transportation, food, and accommodation services) are first-order indirect benefits associated with cultural services of ecotourism and recreation. But the development of ecotourism and recreation often encourages infrastructure development (e.g., road construction), creates more job opportunities, and thus generates indirect benefits that are several steps away from the cultural services. Often, to avoid double counting, the CBA literature suggests including only direct benefits. Exceptions to this occur if there is a strong rationale for the inclusion of some indirect benefits that do not lead to double counting (Boardman et al., 2006). Here, as a first approximation, both direct benefits and first-order indirect benefits were incorporated into the analysis since together they captured the major benefits at Wolong and did not involve double counting (Table 5.1).

5.3 Patterns of change in human dependence on ES

Figures 5.1 and 5.2 show the temporal patterns of changes in human dependence on ES at Wolong Nature Reserve from 1998 to 2007. Results indicated that the overall households' dependence on ES increased by an average of 37% from 1998 to 2007. The average overall indices were 0.45 in 1998 and 0.61 in 2007, suggesting that households obtained approximately 45% and 61% of their total net benefits from ecosystems in 1998 and 2007, respectively. The results also showed that the majority of households were highly dependent on ES, of those some were fully dependent on ES. Specifically, there were approximately 54% and 63% of households with overall indices higher than 0.50, and 9% and 16% of households with overall indices of 1.00 in 1998 and 2007, respectively.

Table 5.2 suggests that there was also a significant spatial pattern of changes in human dependence on ES at Wolong Nature Reserve. In 1998, there was no significant difference in terms of the overall index between the two townships (i.e., Gengda and Wolong) within Wolong Nature Reserve. But in 2007, the overall index for Wolong Township was significantly lower than that of Gengda Township. Compared to Wolong Township, Gengda Township was more dependent on ES. This difference at the township level might be because on average local residents at Gengda were better informed due to closer access to outside markets and thus strategically adjusted their production of ES to maximize the benefits they obtained. For instance, residents in Gengda switched from traditional subsistence crops (e.g., potatoes and maize) to cash crops (e.g., cabbages and traditional Chinese medicinal plants) and participated

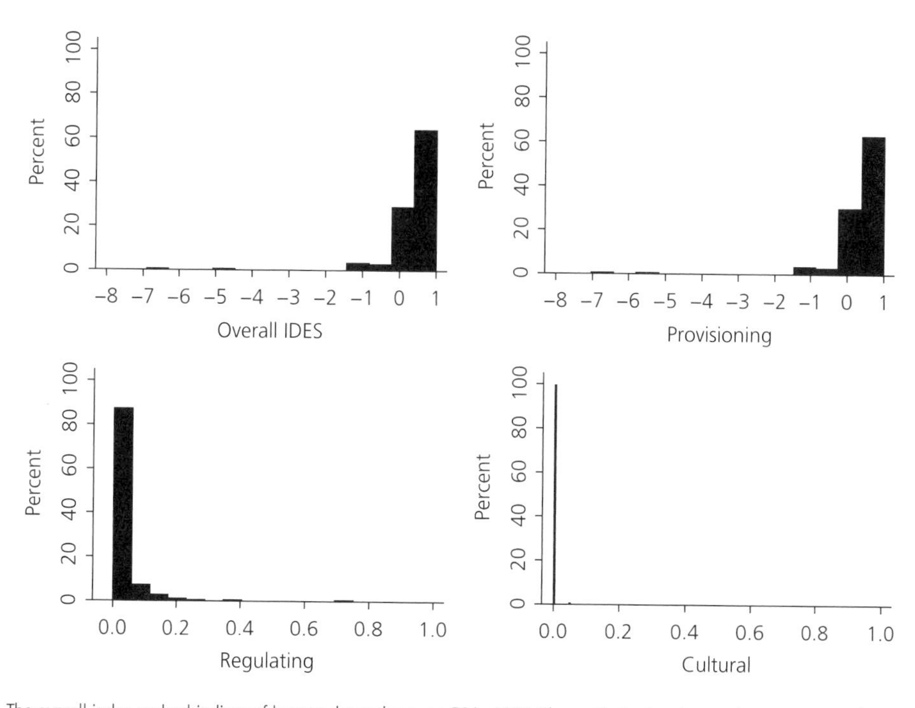

Figure 5.1 The overall index and subindices of human dependence on ES in 1998. The vertical axis refers to the percentage of surveyed households. A negative value means that the gross benefit obtained from ES is smaller than the total costs for generating the corresponding ES and costs of ecosystem disservices. $N = 180$. IDES originally calculated in Yang et al. (2013b).

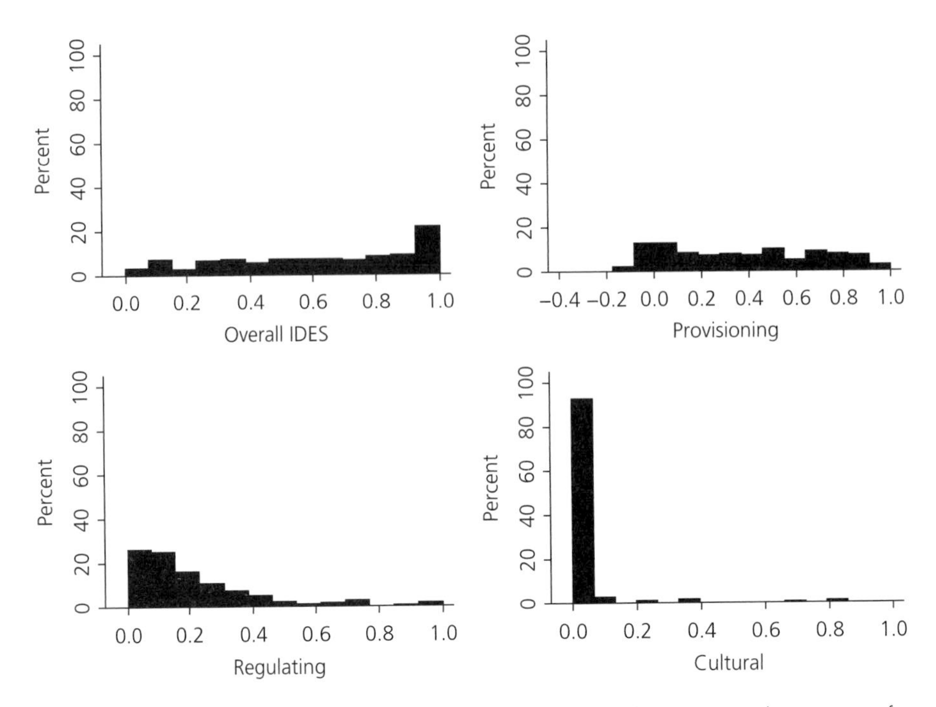

Figure 5.2 The overall index and subindices of human dependence on ES in 2007. The vertical axis represents the percentage of surveyed households. A negative value means that the gross benefit obtained from ES is smaller than the total costs for generating the corresponding ES and costs of ecosystem disservices. $N = 180$. IDES originally calculated in Yang et al. (2013b).

in ecotourism businesses more quickly. Meanwhile, at the household level, households' dependence on ES also changed with their household locations. In 1998, there was no significant difference between the overall index and the distance from each household to the main road. In 2007, those households located further from the main road had a significantly higher dependence than those closer to the main road. This is because households located further away from the main road often had larger areas of cropland (Spearman's rho coefficients were 0.247, $p < 0.001$ in 1998 and 0.433, $p < 0.001$ in 2007, respectively). As the reserve's economy became more connected to outside markets over time, access to the road also became important for alternative non-agricultural income sources. Similar to the pattern at the township level, households closer to the road were better informed about other employment opportunities, reacted more quickly to market changes, and thus were more likely to diversify their income sources from solely agricultural activities to a mixture of agricultural and non-agricultural sources. Such diversification decoupled those households closer to the road from ecosystems and thus reduced their dependence on ES. This example illustrates how the effect of the road on dependence on ES differed at two different scales.

Results also show an uneven distribution of human dependence on ES among households across different income levels. In both 1998 and 2007, the overall indices were negatively correlated with household income (Table 5.3), providing support for the common view that low-income households are more dependent on ES. Agricultural income share (i.e., the ratio of agricultural income to total income) is often used as an indicator to approximate a rural household's dependence on ES. Considering this practice, we compared the performance of IDES with agricultural income share. Results confirm that the overall IDES is better than the agricultural income share in terms of reflecting households' dependence on ES. While overall IDES indices were negatively associated with household income in both 1998 and 2007, the association

Table 5.2 Factors associated with the overall IDES. Dependent variables are overall IDES in 1998 and 2007, respectively. Numbers outside and inside parentheses are coefficients and robust standard errors of bivariate regressions, respectively. Total number of observations is 180. Unit of analysis is the household. Source: adapted from Yang et al. (2013b).

	Indicator	IDES 1998	IDES 2007
Natural capital	Area of cropland (mu, 1 mu = 1/15 ha)	0.020 (0.014)	0.042*** (0.007)
Human capital	Household size	−0.077 (0.049)	−0.037* (0.016)
	Number of laborers	−0.070 (0.056)	−0.080*** (0.018)
	Average education of adults (year)	−0.032† (0.017)	−0.032*** (0.009)
	Average age of adults (year)	0.011* (0.005)	0.008*** (0.002)
Financial capital	Household income (yuan, log)	0.071 (0.075)	−0.152*** (0.025)
	Per capita income (yuan, log)	0.143 (0.095)	−0.126*** (0.028)
Manufactured capital	Type of house (0 for low-quality non-concrete sheds and 1 for high-quality concrete house)	−0.022 (0.139)	−0.189*** (0.048)
	Distance to the main road (meter, log)	−0.029 (0.027)	0.042*** (0.010)
Social capital	Social ties to local township and reserve-level officials (0: low; 1: high)	0.065 (0.129)	−0.188** (0.065)
Township	0: Gengda; 1: Wolong	0.098 (0.133)	−0.101* (0.047)

†$p < 0.10$; *$p < 0.05$; **$p < 0.01$; ***$p < 0.001$.

Table 5.3 Comparison of agricultural income share and overall IDES for their associations with gross household income. Numbers are Spearman's rho coefficients. Total number of observations is 180. Source: adapted from Yang et al. (2013b).

	Agricultural income share	Overall IDES
Household income in 1998	−0.355***	−0.194**
Household income in 2007	−0.012	−0.405***

**$p < 0.01$; ***$p < 0.001$.

between agricultural income share and household income was negatively significant in 1998, but not in 2007. The temporal changes of IDES subindices from 1998 (Figure 5.1) to 2007 (Figure 5.2) suggest the reason. In 1998, as the majority of households' dependence on ES was from agriculture, the agricultural income share was almost equivalent to the overall IDES. However, in 2007, households' dependence on ES expanded to a variety of sources, including both agriculture and non-agricultural sources such as ecotourism and recreation, and PES programs (see also Chapter 9).

Results further suggest that human dependence on ES varies not only across groups having different levels of financial capital but also across groups at different levels of natural, human, manufactured, and social capital (Table 5.2). It is reasonable to expect that households with more access to natural capital should also be more dependent on ES because ES flow from natural capital. Our results confirmed this expectation with the positive associations between the overall IDES and the area of cropland in 2007 (Table 5.2). Results also show that the overall IDES was negatively associated with indicators of human, manufactured, and social capital.

5.4 Causes behind the changes in human dependence on ES

The dramatic changes in households' dependence on ES across time, space, and different forms of capital result from economic development at Wolong Nature Reserve under the macro-background of China's rapid economic growth and urbanization. In addition, part of the changes can be attributed to the implementation of conservation and development policies such as NFCP, GTGP, GTBP, and tourism development. Before 2000, the local economy of

Wolong Nature Reserve primarily relied on agriculture and had limited connections with the outside world. Since 2000, the implementation of NFCP, GTGP, GTBP, and tourism development has led to intensive infrastructure development (e.g., construction of roads, restaurants, hotels, and tourism sites). This trend in turn has contributed to a dramatic increase in exchanges of information (e.g., about new types of crops and market prices), materials (e.g., vehicles and construction materials), and people (e.g., tourists and migrant workers) with the outside world (see also Chapter 17). For instance, the number of tourists visiting Wolong Nature Reserve increased by 11 times (from 20 000 to 235 500) from 1996 to the peak year of 2006 (Liu et al., 2012). Some farmers enrolled their cropland in GTGP and GTBP and went to cities for migrant work; other farmers started to grow cash crops to sell to outside markets. On average, from 1998 to 2007, per household net benefits from provisioning services increased 270% (from 2309 to 8545 yuan) and other socioeconomic activities increased 403% (from 2456 to 12 350 yuan), respectively (Yang et al., 2013b). There were even higher increases in regulating services (3619%, from 78 to 2901 yuan) and cultural services (50 767%, from 3 to 1526 yuan) over the same period. These increases also explain why some indicators of financial and social resources were statistically significantly associated with overall IDES in 2007 but not in 1998 (Table 5.2). Examples included the type of house, distance to the main road, area of cropland, and social ties to local leaders.

5.5 Effects of changes in human dependence on ES

The effects of changes in human dependence on ES are multifaceted. On one hand, changes in human dependence on ES could lead to changes in the impacts on ecosystems. But a distinction should be made between dependence on ES and the impacts on ecosystems since there are ways to sustainably extract ES without having a negative impact on the ecosystem. For instance, the uses of regulating and cultural services such as air purification and ecotourism are usually non-consumptive and may have little detrimental impact on ecosystems when managed well. But the uses of many provisioning services such as timber and food are often consumptive and lead to ecosystem degradation if not appropriately managed. These findings suggest that reducing the impacts on ecosystems can be done through reducing the overall dependence on, or moving toward more sustainable extraction of, ES. Alternatively, impacts can be reduced by shifting the structure of dependence such as from high dependence on provisioning services to high dependence on regulating and cultural services. Specifically, at Wolong Nature Reserve, we estimated that approximately 11 000 ha of forest (5.5% of the total land area) were recovered between 2001 and 2007 (Yang et al., 2013c). This recovery might at least partly be attributed to the shift in dependence structure from the decrease in dependence on provisioning services to increase in dependence on regulating and cultural services.

On the other hand, changes in human dependence on ES might also affect human well-being. As identified earlier in Section 5.2.1, the indices of dependence on ES reflect the vulnerability of the system to the degradation or decline of corresponding ES. For instance, households' well-being can be dramatically affected once there are major disturbances to ES (e.g., natural disasters or pollution incidents) on which the households are highly dependent. Chapter 12 will investigate this point in more detail by demonstrating the response of Wolong Nature Reserve to a major earthquake.

We note that the causes and effects of changes in human dependence on ES are not static and often form feedback loops. For instance, the PES programs led to the shift in dependence structure on ES at Wolong Nature Reserve and reduced forest degradation (Liu et al., 2013, Yang et al., 2013c). This shift in turn encouraged the central government to renew and extend the implementation period of those PES programs (Liu et al., 2013, Yang et al., 2013c).

5.6 Discussion and concluding remarks

This chapter presented an index system to quantify human dependence on ES and demonstrated its validity and utility at Wolong Nature Reserve. Results confirmed that the overall index and subindices effectively captured the general patterns of households' dependencies on ES. The index also captured changes in ES dependencies across time, space, and different access levels of natural, human, manufactured, and social capital. With the index system, we confirmed the proposition that the poor are more dependent on ES. We further generalized this proposition by demonstrating that those disadvantaged groups who have lower levels of access to other forms of capital (i.e., human, financial, manufactured, and social capital) are more dependent on ES. Thus, these groups are more vulnerable to degradation or decline of corresponding ES.

The conceptual basis and methods of the IDES system were designed, based on the MA framework, to be generalizable across different sites and units of analysis. Due to the different contexts at various sites, it is possible that the classification of various net benefits into different categories of ES may be different and should be adjusted according to the specific context. But the overall index and subindices, based on the MA framework, should be comparable across contexts. Given that it is impractical to assess all benefits and costs, in practice the most important benefits and costs in each context should be estimated to reflect the general pattern of dependence on ES. We believe that further applications and elaborations would improve the estimates of human dependence on ES and its linkages with human well-being.

We also note a distinction between dependence on ES and dependence on PES programs. The PES programs often compensate for only one or a few but not all the forgone benefits of local households. In the case of Wolong Nature Reserve, NFCP payment is mainly intended to buy an expected increase in regulating services (e.g., soil erosion control, carbon sequestration, and water conservation) while offsetting the forgone provisioning services (e.g., timber). The payment therefore did not cover benefits from cultural services (e.g., recreation and ecotourism); thus we took benefits from ecotourism and recreation into account separately.

Besides methodological innovation, the IDES system also has many theoretical contributions and management implications for understanding and managing ES for human well-being. First, the quantification of human dependence on ES hopefully can be integrated into decision-making. Such a step can rectify the current tendency of overlooking the linkages between poor people and ES in statistical reports (e.g., statistics yearbooks), poverty assessments, and natural resource management. Thus, the IDES system may avoid strategies that ignore or overlook such dependence and lead to further marginalization of disadvantaged groups. As a result, it could reduce the pressure and negative impacts that disadvantaged groups place on ecosystems. Second, this index system may help with establishing priority settings in conservation and development planning (e.g., PES program enrollment and poverty alleviation). This goal can be achieved by targeting priority population groups such as those having low access to capital and high dependence on provisioning services. Third, this index system may help to draw stakeholders' attention to and manage the previously unmanaged risks and unrealized opportunities associated with dramatic changes in ecosystems and their provision of services. Rapid global changes such as climate change and land-use change have dramatically altered the Earth's ecosystem structure and functions and have threatened the sustainable provision of ES. People and organizations that heavily depend on ES are also those who are most vulnerable to the degradation or decline of ES. Finally, future research may involve the construction of integrated models that combine IDES with indicators of indirect drivers, direct drivers, and human well-being. Hopefully such advances can improve the theoretical understanding and management of feedback loops in coupled human and natural systems. Consider the poverty alleviation and biodiversity conservation traps as an example. Poverty forces disadvantaged people to extract ES in an unsustainable way and leads to biodiversity losses, which reduce the provision of ES, further aggravating poverty, and creating vicious circles. Later in this book, we make an initial attempt to construct such integrated models (Chapter 12).

5.7 Summary

Human society substantially depends on a variety of ecosystem services (ES). To improve the management of risks and opportunities related to ES, a quantitative understanding of human dependence on ES, as well as the patterns, causes, and effects of its changes is essential. This chapter presented an index of dependence on ecosystem services (IDES) system to quantify human dependence on ES. The index is defined as the ratio of net benefits acquired from ecosystems to the absolute value of total net benefits, an amount calculated using economic valuation methods. We also demonstrated the construction of the IDES system and illustrated the patterns, causes, and effects of changes in IDES at Wolong Nature Reserve. Empirical analyses confirmed the validity of the index system in reflecting the general patterns of households' dependencies on ES. Households obtained approximately 45% and 61% of their total benefits from ecosystems in 1998 and 2007, respectively. Dependence on ES also shifted away from provisioning services and toward more regulating services and cultural services. This shift may have helped improve forests in Wolong and had mixed effects on human communities. Findings supported the proposition that disadvantaged groups who have less access to other forms of capital (i.e., human, financial, manufactured, and social capital) than those with greater control are more dependent on ES. It is promising that further applications and elaborations of the IDES system may be able to improve theoretical understanding and management of ES for human well-being in a rapidly changing global environment.

References

An, L., Liu, J., Ouyang, Z., et al. (2001) Simulating demographic and socioeconomic processes on household level and implications for giant panda habitats. *Ecological Modelling*, **140**, 31–49.

Barbier, E.B. (2011) Pricing nature. *Annual Review of Resource Economics*, **3**, 337–53.

Bateman, I.J., Mace, G.M., Fezzi, C., et al. (2011) Economic analysis for ecosystem service assessments. *Environmental & Resource Economics*, **48**, 177–218.

Boardman, A.E., Greenberg, D.H., Vining, A.R., and Weimer, D.L. (2006) *Cost-Benefit Analysis: Concepts and Practice* (third edition). Prentice Hall, Upper Saddle River, NJ.

Carpenter, S.R., DeFries, R., Dietz, T., et al. (2006) Millennium ecosystem assessment: research needs. *Science*, **314**, 257–58.

Chang, J., Wu, X., Liu, A.Q., et al. (2011) Assessment of net ecosystem services of plastic greenhouse vegetable cultivation in China. *Ecological Economics*, **70**, 740–48.

Chee, Y.E. (2004) An ecological perspective on the valuation of ecosystem services. *Biological Conservation*, **120**, 549–65.

Daily, G.C. (1997) *Nature's Services: Societal Dependence on Natural Ecosystems*. Island Press, Washington, DC.

Dietz, T., Rosa, E.A., and York, R. (2007) Driving the human ecological footprint. *Frontiers in Ecology and the Environment*, **5**, 13–18.

Hanley, N., Shogren, J.F., and White, B. (2001) *Introduction to Environmental Economics*. Oxford University Press, New York, NY.

Kareiva, P., Tallis, H., Ricketts, T.H., et al. (2011) *Natural Capital: Theory and Practice of Mapping Ecosystem Services*. Oxford University Press, Oxford, UK.

Liu, J., Daily, G.C., Ehrlich, P.R., and Luck, G.W. (2003) Effects of household dynamics on resource consumption and biodiversity. *Nature*, **421**, 530–33.

Liu, J., Li, S., Ouyang, Z., et al. (2008) Ecological and socioeconomic effects of China's policies for ecosystem services. *Proceedings of the National Academy of Sciences of the United States of America*, **105**, 9477–82.

Liu, J., Ouyang, Z., Yang, W., et al. (2013) Evaluation of ecosystem service policies from biophysical and social perspectives: the case of China. In S.A. Levin, ed., *Encyclopedia of Biodiversity* (second edition), vol. **3**, pp. 372–84. Academic Press, Waltham, MA.

Liu, W., Vogt, C.A., Luo, J., et al. (2012) Drivers and socioeconomic impacts of tourism participation in protected areas. *PLoS ONE*, **7**, e35420.

Liu, W., Vogt, C.A., Lupi, F., et al. (2015) Evolution of tourism in a flagship protected area of China. *Journal of Sustainable Tourism*. DOI: 10.1080/09669582.2015.1071380.

Millennium Ecosystem Assessment (2005) *Ecosystems and Human Well-being: Synthesis*. Island Press, Washington, DC.

National Bureau of Statistics of China (2011) *Rural Household Survey Instrument*. Report for National Bureau of Statistics of China (Beijing) (in Chinese).

Nelson, E., Mendoza, G., Regetz, J., et al. (2009) Modeling multiple ecosystem services, biodiversity conservation, commodity production, and tradeoffs at landscape scales. *Frontiers in Ecology and the Environment*, **7**, 4–11.

Ready, R. and Navrud, S. (2006) International benefit transfer: methods and validity tests. *Ecological Economics*, **60**, 429–34.

Richard, T.C., Nicholas, E.F., and Norman, F.M. (2001) Contingent valuation: controversies and evidence. *Environmental and Resource Economics*, **19**, 173–210.

Scott, M.J., Bilyard, G.R., Link, S.O., et al. (1998) Valuation of ecological resources and functions. *Environmental Management*, **22**, 49–68.

Shrestha, R., Rosenberger, R., and Loomis, J. (2007) Benefit transfer using meta-analysis in recreation economic valuation. *Environmental Value Transfer: Issues and Methods*, **9**, 161–77.

Wilson, M.A. and Hoehn, J.P. (2006) Valuing environmental goods and services using benefit transfer: the state-of-the art and science. *Ecological Economics*, **60**, 335–42.

Wolong Nature Reserve (2005) *Development History of Wolong Nature Reserve*. Sichuan Science and Technology Press, Chengdu, China (in Chinese).

Yang, W., Chang, J., Xu, B., et al. (2008) Ecosystem service value assessment for constructed wetlands: a case study in Hangzhou, China. *Ecological Economics*, **68**, 116–25.

Yang, W., Dietz, T., Kramer, D.B., et al. (2013a) Going beyond the Millennium Ecosystem Assessment: an index system of human well-being. *PLoS ONE*, **8**, e64582.

Yang, W., Dietz, T., Liu, W., et al. (2013b) Going beyond the Millennium Ecosystem Assessment: an index system of human dependence on ecosystem services. *PLoS ONE*, **8**, e64581.

Yang, W., Liu, W., Viña, A., et al. (2013c) Performance and prospects on payments for ecosystem services programs: evidence from China. *Journal of Environmental Management*, **127**, 86–95.

Yang, W., Liu, W., Viña, A., et al. (2013d) Nonlinear effects of group size on collective action and resource outcomes. *Proceedings of the National Academy of Sciences of the United States of America*, **110**, 10916–21.

CHAPTER 6

Landscape Changes in Space and Time

Andrés Viña and Jianguo Liu

6.1 Introduction

The anthropogenic conversion of natural landscapes around the world has created international concern due to greenhouse gas emissions and species extinctions (Foley et al., 2005, Seppelt et al., 2014). This conversion also contributes to habitat reduction, population decline, and fragmentation of many endangered species populations (Carr, 2004, 2005, Liu et al., 2015, Myers, 1990, Pahari and Murai, 1999, Pimm et al., 2014). These negative impacts are not limited to human-dominated areas (Alberti et al., 2003, Collins et al., 2000). Such impacts are also common in many of the more than 170 000 protected areas worldwide (e.g., national parks, nature reserves, and natural heritage sites; Le Saout et al., 2013). This conversion is mainly driven by land use such as agricultural expansion and urbanization. Land use has changed drastically in the past several decades. It is common to observe land-use transition, which refers to "change in land use systems from one state to another one" (Lambin and Meyfroidt, 2010). For example, much land has changed from forest to agricultural and residential use worldwide.

On the other hand, in many places around the world, there are also many positive changes for sustainability. Examples include recovery of habitat for wildlife species (Chapron et al., 2014) and forest transition (i.e., forest-cover change from net loss to net gain: Mather, 1992, 2004, Mather et al., 1998, 1999, Mather and Fairbairn, 2000). Forest transition started to occur in the eighteenth century in Western Europe (Mather, 1992). This phenomenon has since spread to some other developed countries around the world (Foster et al., 1998, Totman, 1986). More recently, forest transition has also been occurring in some developing countries. In the case of Asia, four countries (China, India, Vietnam, and Bangladesh) have been experiencing it since the 1980s (Mather, 2007, Rudel, 2005). Much has been written about the many factors that play important roles as determinants of forest transition (Foster and Rosenzweig, 2003, Kaimowitz, 1997, Klooster, 2003, Lambin and Meyfroidt, 2010, Nagendra et al., 2005, Pan and Bilsborrow, 2005, Perz and Skole, 2003). Two general mechanisms are considered the most important (Rudel, 1998, Rudel et al., 2005): forest scarcity and economic development. In the first case, deforestation encourages planting trees to compensate for losses (Hardie and Parks, 1996, Hart, 1968, Liu, 2014, Walters, 1997). In the second case, industrialization creates many off-farm job opportunities that attract laborers to shift from on- to off-farm activities. Such opportunities lead to the abandonment of marginal farmland and its reconversion to forests (Aide et al., 1995, Bentley, 1989, Bowen et al., 2007, Hart, 1968). While these two mechanisms are compelling, they do not explain all forest-transition phenomena. Neither development nor forest plantation alone can guarantee the emergence of a forest transition (Klooster, 2003, Perz and Skole, 2003, Perz, 2007). Thus, a third mechanism has also been proposed: government intervention. For instance, it has been suggested that the forest transition that took place over a century or more in Western Europe and North America may occur in other regions in just decades. This prediction is largely attributed

Pandas and People. Edited by Jianguo Liu, Vanessa Hull, Wu Yang, Andrés Viña, Xiaodong Chen, Zhiyun Ouyang, and Hemin Zhang. Published 2016 by Oxford University Press.

to government policies protecting forests and encouraging reforestation (Mather, 2007, Rudel et al., 2005). Therefore, government policies may serve as catalysts that accelerate forest transition (Chen et al., 2009, Daily and Matson, 2008).

Over the past three decades, China has been the fastest-growing economy in the world at the expense of environmental degradation and dramatic natural resources depletion (Liu and Diamond, 2008). But this fast economic development has also meant that people formerly involved in agricultural activities (mainly for their own subsistence) are "modernizing" their livelihoods (Zhang, 2010). Such individuals have become the labor force of non-farm enterprises (He, 2004, Song and Wang, 2001). The human population is also becoming more concentrated, particularly around industrial enterprises and in cities (He, 2004, Song and Wang, 2001). Both of these processes have therefore relieved the pressure on forested areas. According to recent statistics reported by the United Nations Food and Agricultural Organization (FAO), China started to experience net forest expansion at the end of the twentieth century (Food and Agriculture Organization of the United Nations, 2010). In addition, the Chinese government has enacted several national policies. Some have had direct negative consequences on forest cover such as the Great Leap Forward. Others such as the one-child policy have been thought to slow the country's environmental degradation (Liu, 2010). Additionally, by the end of 2012 a total of 2669 protected areas had been established throughout China (Xu et al., 2014).

Protected areas are generally perceived as cornerstones of biodiversity conservation (Armesto et al., 1998, McNeely and Miller, 1983). But human activities from outside as well as within protected areas often persist due to the logistic difficulty of enforcing restrictions on human activity (Dompka, 1996). This shortcoming is particularly conspicuous in the developing world, in which human habitation often continues despite the declaration of protected area boundaries (Han, 2000). This is the case in our model coupled system, Wolong Nature Reserve (chapter 3). Despite its conservation status, Wolong experienced conspicuous forest degradation after its inception in 1963 (Liu et al., 2001). Finally, by the end of the twentieth century the Chinese government had started to implement two of the largest forest conservation and restoration policies on the Earth (Liu et al., 2008, Uchida et al., 2005). The Natural Forest Conservation Program (NFCP) bans logging in natural forests and provides cash for local residents and others to monitor forests to prevent illegal harvesting. The Grain-to-Green Program (GTGP; also referred to as the Sloping Land Conversion Program) encourages farmers to convert from cropland to forest by providing cash, grain, and tree seedlings. As a result of NFCP, commercial harvesting of natural forests halted after 2000 and areal coverage of the program reached around 11 million ha nationwide by 2005 (Liu et al., 2008). By the end of 2009, more than 120 million farmers in 32 million households had participated in GTGP (Liu et al., 2013). The program cumulatively increased vegetative cover by 25 million ha, with 8.8 million ha of cropland being converted to forest and grassland and 14.3 million ha of barren land being afforested (Liu et al., 2013).

Many studies on landscape changes have been done, but very few have simultaneously considered the impacts of policies and natural disasters on landscape dynamics using long-term data before and after these human and natural disturbances occur. In this chapter, we take an in-depth look at landscape dynamics in Wolong over a 40-year period, with a focus on three major land types: cropland, pastureland, and forest. We first present our novel approaches and findings to characterize agricultural land dynamics, a challenge given the difficulties of determining the extent of agricultural lands on heterogeneous landscapes. We then describe our findings on patterns and processes governing forest-cover dynamics in Wolong. We devote a section to a particular focus on the effects of the catastrophic 2008 Wenchuan earthquake (Figure 6.1), a disturbance that reversed the direction of previous land-use change in the reserve. We end with some concluding thoughts about where Wolong might be going in the future with respect to landscape dynamics.

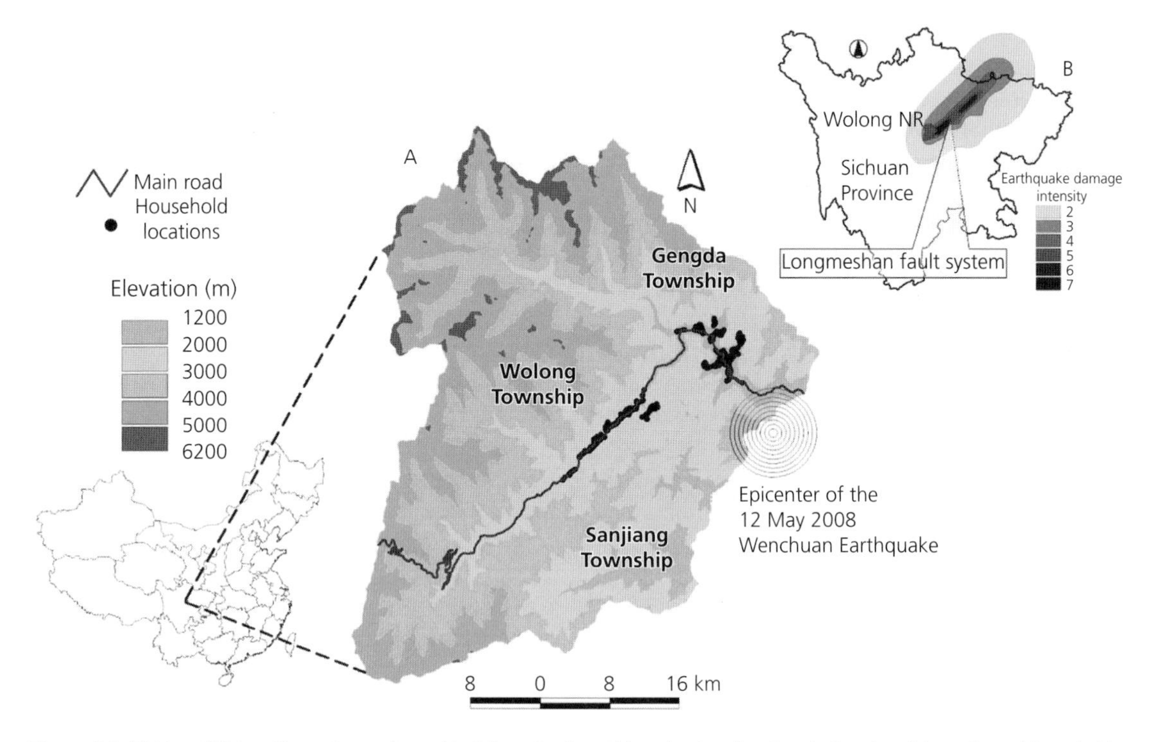

Figure 6.1 (A) Map of Wolong Nature Reserve located in Sichuan Province, China, showing elevation, the location of the main road, households, townships, and the epicenter of the May 12, 2008, Wenchuan earthquake. (B) Distribution of earthquake damage intensity (or instrumental intensity, related to ground motion parameters strongly correlated with potential damage; data obtained from the China Seismological Bureau (2011) along the Longmenshan active fault system.

6.2 Agricultural land dynamics

Wolong is home to nearly 5000 residents. The overwhelming majority of these residents (up to 90%) depend on agricultural production for their livelihoods (Viña et al., 2007). Between 1975 and 2006 the human population increased by ~85% while the number of households tripled (Wolong Administration Bureau Department of Social and Economic Development, 2006). Thus, it could be expected that an increase in human population will drive an increase in the area of agricultural land use. In the case of Wolong, this includes a range of land uses within an agropastoral system in which most farm households maintain a variety of free-ranging and confined livestock along with their fallow rotation crops. Detecting changes in such an agropastoral system is important in order to accurately characterize land transitions. But this task is particularly difficult due to the drastic variability in vegetation types (e.g., crop types and pasture quality), phenologies (e.g., cropping patterns), and human practices (e.g., tillage and burning). Such variation involves minor modifications of land cover too subtle to be easily discriminated with traditional remote sensing techniques (Coppin et al., 2004). In addition, changes in agricultural land may involve many-to-one relations. For example, different vegetation types (e.g., crop types) may revert to a single one (e.g., shrubland), or vice versa. All of these possible changes constitute a challenge for landscape analysis (Seto et al., 2002), particularly in mountain regions where topographic heterogeneity brings additional complexity. Furthermore, as in other smallholder subsistence-market farming systems around the world, agricultural land parcels in Wolong are often small (<1 ha).

To overcome some of these limitations, we applied a procedure for estimating the probability of an area being cropland/pastureland (Viña et al., 2013). We used a fuzzy classification algorithm based on the principle of maximum entropy (Jaynes, 1957). The algorithm was applied to remotely sensed data using the software MaxENT (Phillips et al., 2006). Topographic data consisted of a digital elevation model (DEM) acquired by the Shuttle Radar Topography Mission (SRTM; Berry et al., 2007). Multispectral data included three Landsat Thematic Mapper (TM) images acquired on June 26, 1994, June 13, 2001, and September 18, 2007. The outputs of this procedure correspond to probability maps of the occurrence of cropland/pastureland. Such maps were validated by means of the area under the receiver operating characteristic curve (AUC; Hanley and McNeil, 1982). The AUC values obtained for cropland were 0.90 while those for pastureland were 0.94, denoting a high classification success.

To obtain the areal cover of cropland and pastureland on different dates, it was necessary to find an optimal cumulative threshold value on the probability scale. This procedure was done using the maximum Kappa (Cohen, 1960) approach, following the procedures used in our work on the analysis of panda habitat (Viña et al., 2010). Results show that the optimal threshold for selecting cropland/pastureland from non-cropland/non-pastureland areas was 0.6. With this threshold, it was found that the areas of cropland continuously decreased between 1994 and 2007 (Figure 6.2; Plate 1). Areas of pastureland showed some fluctuations over time (Figure 6.3).

The areas of cropland estimated using the optimal threshold based on the maximum Kappa procedure constitute about three times as much cropland as that reported by the local government

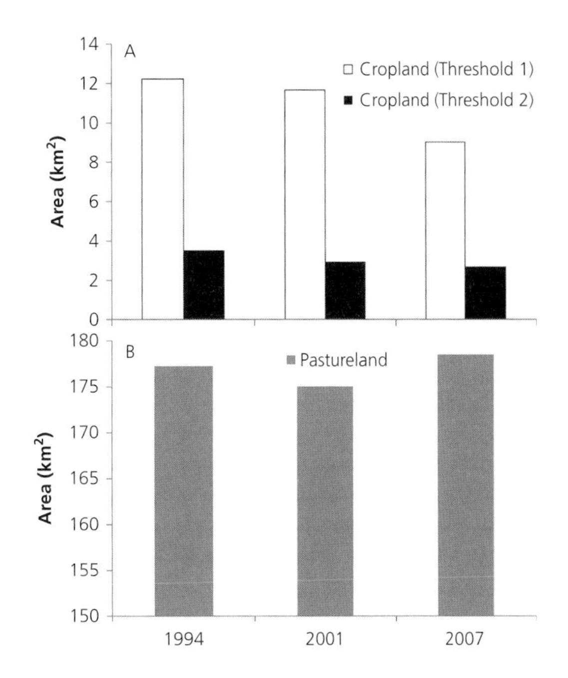

Figure 6.3 Areal changes in (A) cropland and (B) pastureland in Wolong Nature Reserve between 1994 and 2007. Areas of cropland were obtained using two different cumulative thresholds in the cropland probability scale (i.e., threshold 1 = 0.6; threshold 2 = 0.5). See text for details on these thresholds.

(Wolong Nature Reserve, 2008, Yang et al., 2013a). This finding is not unexpected given the difficulty in establishing the exact cropland areas through remote sensing procedures, particularly in mountain regions. In addition, there is a lack of reliability of government reports due mainly to underreporting (Seto et al., 2002). To match the cropland area in 2007

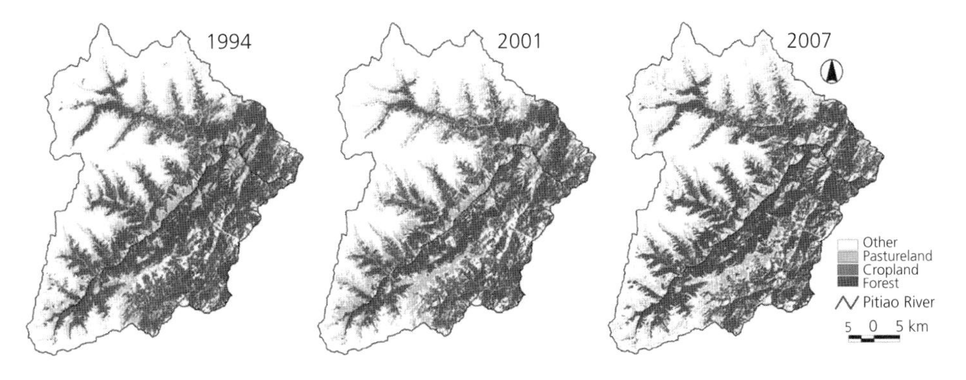

Figure 6.2 Maps of the spatial distribution of three major land-cover types in Wolong Nature Reserve between 1994 and 2007. For color version see the color plates. [PLATE 1]

provided in the government report, we lowered the threshold in the cropland probability to 0.50. With this threshold, it was also shown that cropland areas decreased from 1994 to 2007 (Figure 6.3). Thus, irrespective of the threshold used, cropland area was reduced between 1994 and 2007.

The reduction in cropland and the fluctuations of pastureland in Wolong may be explained by different processes occurring simultaneously. On one hand, economic reform in China has provided many opportunities for rural residents to work in cities (Davin, 1996, Liang and White, 1996, Liang and Ma, 2004, Ma, 2002, Woon, 1999). Thus, while Wolong's economy was still mainly dependent on subsistence agriculture, some local residents had obtained jobs in urban centers outside the reserve (Chen et al., 2012). On the other hand, Wolong experienced a boom in tourism. A dramatic increase in the number of visitors occurred from 20 000 in 1995 to 100 000 in 2000 (Lindberg et al., 2003) and to more than 200 000 in 2006 (He et al., 2008). This increase accompanied the rapid development of tourism infrastructure (e.g., hotels and restaurants), together with road construction, all of which constituted off-farm job opportunities. Finally, since 2000, GTGP has promoted the conversion of cropland into tree plantations (Chen et al., 2010, Viña et al., 2011). The local Grain-to-Bamboo Program (GTBP), mimicking the GTGP by encouraging farmers to convert cropland to bamboo with subsidies, also reduced the amount of cropland. So far, the GTGP and GTBP have converted 3.67 km^2 and 0.82 km^2 of cropland, respectively (Yang et al., 2013a).

6.3 Forest-cover dynamics

In addition to agricultural land dynamics, changes in forest cover have also occurred in Wolong. Vegetation in the reserve is dominated by forests, which occupy around one-third of the reserve. These are composed of evergreen and deciduous broadleaf forests located at lower elevations (up to 2500 m) and subalpine coniferous forests located at higher elevations (up to 3500 m; Schaller et al., 1985). These forest stands have a dense understory dominated by bamboo species (e.g., *Bashania faberi* and *Fargesia robusta*) that are the staple food of the giant pandas (Reid et al., 1991, Schaller et al., 1985, Taylor and Qin, 1987, 1988, 1993a, b).

Forest-cover decline was the dominant pattern of forest-cover change in Wolong over much of the last century. We estimated that forest covered ~50% of Wolong in 1965 but declined to ~33% by 2001 (Viña et al., 2007; Figures 6.4 and 6.5). Timber logging and fuelwood collection were the leading causes of forest loss and degradation that occurred in and around Wolong between 1965 and 1997, but particularly since 1974 (Liu et al., 2001, Viña et al., 2007; Figures 6.4 and 6.5). For instance, forest resources were exploited near households in the 1970s, but as the forests depleted, the local people were forced to cut trees farther away (He et al., 2009). The expansion of local wood extraction activities was driven not only by the need to obtain wood, but also by improved accessibility due to road expansion and improvement. For example, a project supported by the United Nations World Food Programme in the early 1980s to mitigate human impacts on panda habitat resulted in a road expansion in one of Wolong's two townships (Gengda Township). This expansion facilitated the local population to collect wood in an area rarely visited in the 1970s (He et al., 2009, Wolong Administration Bureau, 2004).

In the Wolong and Gengda townships located inside the reserve, commercial logging was practiced until 1975, when all logging beyond domestic needs was legally banned (Wolong Administration Bureau, 2004). In Sanjiang Township, no permanent residents occupy the area inside the reserve, but people living outside the reserve have been conducting legal and illegal forest biomass extraction activities inside the reserve (Wolong Administration Bureau, 2004). Wood is the main material used for traditional house construction. In addition, a substantial amount of fuelwood was consumed by local households for cooking, heating, and cooking pig fodder. The importance of pig fodder is particularly salient, as pigs constitute the main animal protein source for local households (An et al., 2001). Locals prefer deciduous tree species such as red birch, white birch, and large-winged wingnut for fuelwood, and large old-growth spruces and firs for construction. These deciduous and evergreen tree species are dominant canopy components in the overstory of the bamboo forests preferred by wild giant pandas (Schaller et al., 1985). But these species are rarely found in areas near local households.

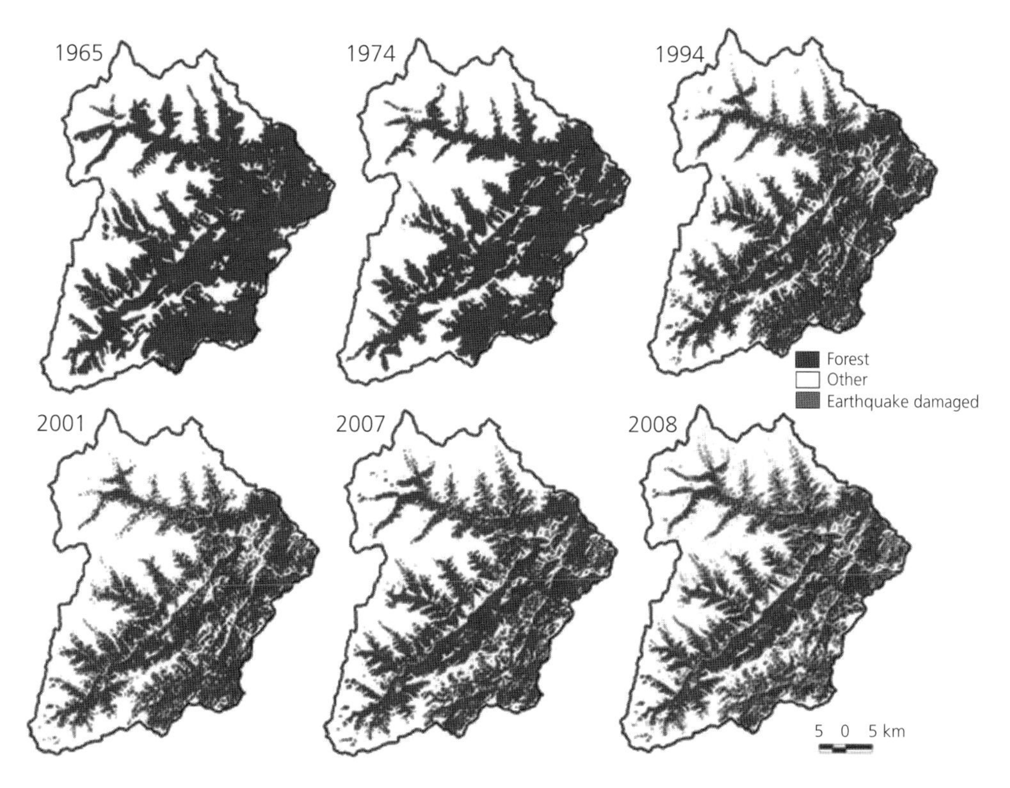

Figure 6.4 Time series of forest cover in Wolong Nature Reserve between 1965 and 2008 (based on Liu et al., 2001, Viña et al., 2007, 2011). See color version in the color plates where forest cover is shown in green, while areas of forest affected by landslides triggered by the May 12, 2008, Wenchuan earthquake are shown in red. Maps reprinted with permission from AAAS (originally published in Liu et al., 2001) and Springer Science and Business Media (originally published in Viña et al., 2011). [PLATE 2]

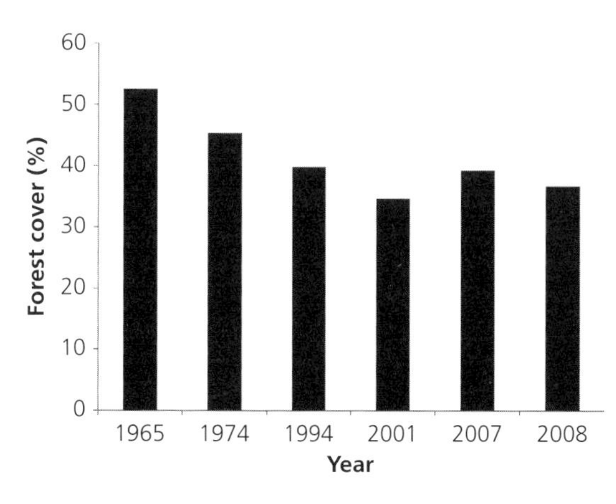

Figure 6.5 Forest-cover dynamics (in percent) in Wolong Nature Reserve. The relative values were obtained from the maps shown in Figure 6.4 (based on Liu et al., 2001, Viña et al., 2007, 2011).

Forest loss and degradation had many negative environmental consequences such as the loss of panda habitat (Liu et al., 2001, Viña et al., 2007). They seem to also be responsible for an overall increase in landslide susceptibility. For instance, since the 1980s landslides have become the most common natural disaster. In the 1990s, the annual number of landslides along the main road (Figure 6.1) in Wolong Nature Reserve ranged between 20 and 50 (Zhou, 1992). In the mid-1990s, an entire hamlet of about two dozen households in Gengda Township was forced to relocate due to nearby landslide risk (H. Xu, Department of Natural Resource Management of Wolong, pers. comm.).

Contrasting the trends of forest cover observed between 1965 and 1997 and since 2001, a forest transition started to become apparent in Wolong. Forest cover stabilized (Viña et al., 2007) between 1997 and

2001, and started to increase after 2001 (Viña et al., 2011; Figures 6.2 and 6.3). This trend reversed more than 40 years of forest loss and degradation observed within reserve borders (Liu et al., 2001, Viña et al., 2007). As stated above, human population in Wolong and Gengda townships almost doubled while the number of households tripled (Wolong Administration Bureau Department of Social and Economic Development, 2006). As households are directly associated with resource consumption (Liu et al., 2003), it is to be expected that the increase of not only human population but particularly of household numbers will have had a negative effect on forest cover. For example, higher populations and numbers of households have greater demand for land and construction materials, as well as fuelwood. Yet, a net increase in forest cover was observed after 2001 (Figures 6.4 and 6.5), together with a reduction of cropland and a small fluctuation of pastureland since the 1990s (Figures 6.2 and 6.3). The forest transition observed in Wolong could be explained by significant monotonic reduction in the use of wood per capita (Chen et al., 2014). This reduction seems to have been accomplished through the implementation of three government policies that constituted a strong catalyst of forest transition in Wolong.

The first policy relates to electricity consumption, since an affordable and stable supply of electricity tends to promote local households to switch from fuelwood to electricity (An et al., 2002). Thus, eight hydropower plants with a total capacity of 34 megawatts were built in the reserve (Yang et al., 2013a). The Gengda Township government improved the electric grid in the 1990s to improve the overall electricity supply system. But most electricity from these hydropower plants was sold to cities. More recently a reserve-wide electricity system improvement was completed and in 2002 a new "ecohydropower" plant was built in Wolong Township (Wolong Administration Bureau, 2004). These improvements increased accessibility to electricity and improved the stability of the electricity supply.

The second policy was the implementation of NFCP. NFCP strictly banned logging and reduced the influence of humans in forested areas, allowing for the natural regeneration of the forest. To implement NFCP, approximately 400 km^2 of forest were assigned to ~1130 households, in varied sizes of groups from one to 16 households per group, who receive a stipend for monitoring activities, while government officials monitor the remaining forest area. Besides creating incentives to participate in forest conservation, the stipends paid to farmers also encourage electricity use (Yang et al., 2013b), reducing their dependency on fuelwood as an energy source (Gössling et al., 2002). Wolong residents have responded positively to NFCP as exemplified by an average annual household fuelwood consumption reduction of approximately 40% between 1999 and 2005 (Chen et al., 2014).

The third policy was the implementation of GTGP. GTGP promoted the conversion of cropland into tree plantations, which are mainly dominated by a fast-growing exotic coniferous species (the Japanese Larch, *Larix kaempferi*). However, while only ~3% of the areas exhibiting forest recovery are located within the GTGP area, little forest recovery can be directly associated with GTGP (Viña et al., 2011). Thus, while GTGP may have exerted a major effect on the trends in cropland (as shown in Section 6.2), it had a minor direct effect on the forest cover. However, GTGP may have had an indirect effect on forests. This policy has been associated with an enhancement of rural-to-urban labor migration (Uchida et al., 2009), which might reduce human pressure on forests and enable natural regeneration of unused cropland.

6.4 Impacts of natural disasters

Landscape dynamics are also affected by natural disasters, such as earthquakes. Wolong is located within the Longmenshan fault zone (see Chapter 12). This zone is an active fault system responsible for the high level of seismic activity that characterizes the region (Di et al., 2010, Qi et al., 2010). On May 12, 2008, a 7.9 M_w earthquake struck (8.0 on the Richter scale), with its epicenter located very close to the eastern boundary of Wolong (Figure 6.1). Followed by more than 30 000 aftershocks, this earthquake was one of the worst natural disasters in China during the last six decades. This classification is in terms of both human casualties and damage to infrastructure. With regard to the former, this disaster left 69 227 people dead, 17 923 people missing, and 374 643 people injured, as of September 2008 (Yin et al., 2009). Inside Wolong, 148 deaths (including 48 local residents and 100 temporary workers and visitors) were reported in the aftermath (Sichuan Department of Forestry, 2008, Wenchuan County People's Government et al., 2008). The earthquake also caused extensive damage to the main road (Figure 6.1) and most buildings. In addition, earthquake-induced landslides affected many forest areas (Figure 6.4; Plate 2). These areas represented an overall reversal in the amount of forest-cover gain observed in the study area during 2001–2007 (Figure 6.5). The damage to forest cover induced by the earthquake was extensive (i.e., more than 10% of the forest cover). But without the implementation of conservation policies, the combined effects of persistent human disturbance and earthquake-induced landslides would have further reduced the forest area by ~15% (Viña et al., 2011). Thus, conservation policies seemed to not only catalyze a forest transition, but also contribute to a reduction of the negative effects of the Wenchuan earthquake on the natural landscapes of Wolong.

In response to the earthquake, the government instituted a series of programs to stimulate economic recovery (Sichuan Department of Forestry, 2008, Wenchuan County People's Government et al., 2008). The massive infrastructure development projects, costing more than US$300 million (Sichuan Department of Forestry, 2008), may have a tremendous impact on landscape dynamics into the future. Also, a series of policies have been developed to stimulate the recovery of the entire earthquake-affected region (Sichuan Department of Forestry, 2008, Wenchuan County People's Government et al., 2008). These policies call for the reconstruction of damaged facilities and promotion of tourism (Li, 2009). The government also put forth plans for relocation of households, particularly from areas that suffered heavy infrastructure damage due to landslides to areas with lower susceptibility to landslides, e.g., with gentle slopes (Li, 2009, Chapter 12). According to the plans, as many as 75% of all households within Wolong may be relocated to areas near the main road (Sichuan Department of Forestry, 2008). Participation in such relocation is voluntary and depends on the level of damage inflicted on the houses. Thus, relocation is not uniform across all households in the reserve, with as yet unknown consequences for the landscape.

6.5 Concluding remarks

The non-sustainable extraction of woody biomass in forest ecosystems has contributed to dramatic deforestation and ecosystem degradation in many places around the world (Davidar et al., 2007, Geist and Lambin, 2002). Deforestation and ecosystem degradation are particularly devastating in mountain regions. Such threats have cascading negative consequences that pervade to the lowland areas, such as erosion, landslides, siltation of waterways, floods, and droughts (Blyth et al., 2002, Bradshaw et al., 2007, Sonesson and Messerli, 2002). Wolong is no different, as it is located in a headwater region that serves urban areas located in the Chengdu Plain, such as the cities of Dujianjiang and Chengdu. Landscape dynamics in Wolong not only affect the biota within its borders (e.g., endangered species such as the giant panda) but also have consequences that go well beyond its borders. Over the second half of the twentieth century, Wolong experienced a continuous loss and degradation of its natural forests. But at the beginning of the twenty-first century, this loss was replaced by an overall gain in forest cover (i.e., forest transition), mainly as a result of the implementation of conservation policies. Simultaneously with this forest transition there has been a reduction in the areas of cropland and a small fluctuation of pastureland.

However, the gains in forest cover obtained after the implementation of government policies were diminished by widespread landslides triggered by a strong earthquake. Over the long run, the fate of the forests in the reserve will depend on the responses of local people and the government to the earthquake. If all households get involved in off-farm economic activities such as tourism, Wolong could witness a rapid acceleration in the abandonment of farmland and subsequent regeneration of forest cover. This prediction could especially come to fruition if NFCP and GTGP continue to be implemented. But while this may lead to accelerated achievement of conservation goals, it may also be disadvantageous for the people who have already suffered the most in the wake of the Wenchuan earthquake. For instance, the benefits from tourism in the reserve were unequally distributed, with the majority of the benefits accruing to external investors, rather than local residents (He et al., 2008). In addition, the benefits were also non-uniform among local households due to differences in the level of participation, many of which were geographically driven (Liu et al., 2012). Hopefully, the post-earthquake government efforts will avoid these pitfalls. It would be ideal for all households in the reserve to have equal opportunities to access the benefits of economic development and to secure livelihoods that do not incur further ecological degradation of the unique natural landscapes in Wolong.

Our analysis of landscape dynamics in Wolong has shown that while economic development may directly contribute to land use and forest transitions, government intervention (in the form of conservation policies) also has a strong effect on both of these processes. Yet, no single policy prescription can be considered a magic bullet. Instead, the simultaneous implementation of conservation policies (e.g., establishment of the protected area, NFCP, GTGP, GTBP, and hydropower generation) is what facilitates reaching environmental goals while also sustaining and even improving human livelihoods.

6.6 Summary

Much of the earth's land surface is currently experiencing environmental degradation due to myriad human impacts, resulting in profound land-use and land-cover changes. This chapter described the spatiotemporal dynamics of major land-use and land-cover types in Wolong Nature Reserve over more than 40 years, and their main drivers. Forest cover declined from 50% in 1965 to 33% by 2001, due mainly to timber harvesting and fuelwood collection. However, despite the continuous growth of human population and household proliferation, this trend was reversed during the first decade of the twenty-first century. Forest cover increased to 38% of the reserve area by 2007, mainly in response to the implementation of two national conservation policies—the Natural Forest Conservation Program (NFCP) and the Grain to Green Program (GTGP)—and a local policy of hydropower generation. In addition, areas under cropland simultaneously decreased by 25–29% due to off-farm labor opportunities (e.g., tourism development and rural-to-urban labor migration) and the implementation of GTGP and the Grain to Bamboo Program (GTBP). Areas under pastureland exhibited small fluctuations. These results suggest that besides experiencing a forest transition, Wolong may have also been experiencing a land-use transition. However, these dynamics were substantially modified in 2008 by the Wenchuan earthquake. The earthquake induced numerous landslides that affected not only infrastructure but also reduced forest cover to 35% of the reserve area. Over the next few decades, the fate of Wolong's landscapes will depend on the complex interactions among local people, government policies, and telecoupling processes (e.g., rural-to-urban labor migration and tourism).

References

Aide, T.M., Zimmerman, J.K., Herrera, L., et al. (1995) Forest recovery in abandoned tropical pastures in Puerto Rico. *Forest Ecology and Management*, **77**, 77–86.

Alberti, M., Marzluff, J.M., Shulenberger, E., et al. (2003) Integrating humans into ecology: opportunities and challenges for studying urban ecosystems *BioScience*, **53**, 1169–79.

An, L., Liu, J.G., Ouyang, Z.Y., et al. (2001) Simulating demographic and socioeconomic processes on household level and implications for giant panda habitats. *Ecological Modelling*, **140**, 31–49.

An, L., Lupi, F., Liu, J., et al. (2002) Modeling the choice to switch from fuelwood to electricity: implications for giant panda habitat conservation. *Ecological Economics*, **42**, 445–57.

Armesto, J.J., Rozzi, R., Smith-Ramirez, C., and Arroyo, M.T.K. (1998) Conservation targets in South American temperate forests. *Science*, **282**, 1271–72.

Bentley, J.W. (1989) Bread forests and new fields: the ecology of reforestation and forest clearing among small-woodland owners in Portugal. *Journal of Forest History*, **33**, 188–95.

Berry, P.A.M., Garlick, J.D., and Smith, R.G. (2007) Near-global validation of the SRTM DEM using satellite radar altimetry. *Remote Sensing of Environment*, **106**, 17–27.

Blyth, S., Groombridge, B., Lysenko, I., et al. (2002) *Mountain Watch: Environmental Change & Sustainable Development in Mountains*. UNEP World Conservation Monitoring Centre and UNEP Mountain Programme, Cambridge, UK.

Bowen, M.E., McAlpine, C.A., House, A.P.N., and Smith, G.C. (2007) Regrowth forests on abandoned agricultural land: a review of their habitat values for recovering forest fauna. *Biological Conservation*, **140**, 273–96.

Bradshaw, C.J.A., Sodhi, N.S., Peh, K.S.H., and Brook, B.W. (2007) Global evidence that deforestation amplifies flood risk and severity in the developing world. *Global Change Biology*, **13**, 2379–95.

Carr, D.L. (2004) Proximate population factors and deforestation in tropical agricultural frontiers. *Population and Environment*, **25**, 585–612.

Carr, D.L. (2005) Forest clearing among farm households in the Maya Biosphere Reserve. *Professional Geographer*, **57**, 157–68.

Chapron, G., Kaczensky, P., Linnell, J.D.C., et al. (2014) Recovery of large carnivores in Europe's modern human-dominated landscapes. *Science*, **346**, 1517–19.

Chen, X., Frank, K.A., Dietz, T., and Liu, J. (2012) Weak ties, labor migration, and environmental impacts: toward a sociology of sustainability. *Organization & Environment*, **25**, 3–24.

Chen, X., Lupi, F., Viña, A., et al. (2010) Using cost-effective targeting to enhance the efficiency of conservation investments in payments for ecosystem services. *Conservation Biology*, **24**, 1469–78.

Chen, X., Viña, A., Shortridge, A., et al. (2014) Assessing the effectiveness of payments for ecosystem services: an agent-based modeling approach. *Ecology and Society*, **19**, 7.

Chen, X.D., Lupi, F., He, G.M., and Liu, J.G. (2009) Linking social norms to efficient conservation investment in payments for ecosystem services. *Proceedings of the National Academy of Sciences of the United States of America*, **106**, 11812–17.

China Seismological Bureau (2011) *China Seismic Information*. http://www.csi.ac.cn (in Chinese).

Cohen, J. (1960) A coefficient of agreement for nominal scales. *Education and Psychological Measurement*, **20**, 37–46.

Collins, J.P., Kinzig, A., Grimm, N.B., et al. (2000) A new urban ecology—modeling human communities as integral parts of ecosystems poses special problems for the development and testing of ecological theory. *American Scientist*, **88**, 416–25.

Coppin, P., Jonckheere, I., Nackaerts, K., et al. (2004) Digital change detection methods in ecosystem monitoring: a review. *International Journal of Remote Sensing*, **25**, 1565–96.

Daily, G.C. and Matson, P.A. (2008) Ecosystem services: from theory to implementation. *Proceedings of the National Academy of Sciences of the United States of America*, **105**, 9455–56.

Davidar, P., Arjunan, M., Mammen, P.C., et al. (2007) Forest degradation in the Western Ghats biodiversity hotspot: resource collection, livelihood concerns and sustainability. *Current Science*, **93**, 1573–78.

Davin, D. (1996) Migration and urbanisation in China. *Pacific Review*, **9**, 131–32.

Di, B.F., Zeng, H.J., Zhang, M.H., et al. (2010) Quantifying the spatial distribution of soil mass wasting processes after the 2008 earthquake in Wenchuan, China—a case study of the Longmenshan area. *Remote Sensing of Environment*, **114**, 761–71.

Dompka, V., ed. (1996) *Human Population, Biodiversity and Protected Areas: Science and Policy Issues*. American Association for the Advancement of Science, Washington, DC.

Foley, J.A., DeFries, R., Asner, G.P., et al. (2005) Global consequences of land use. *Science*, **309**, 570–74.

Food and Agriculture Organization of the United Nations (2010) *Global Forest Resources Assessment FRA 2010. Country Report. China*. FAO, Rome, Italy.

Foster, A.D. and Rosenzweig, M.R. (2003) Economic growth and the rise of forests. *Quarterly Journal of Economics*, **118**, 601–37.

Foster, D.R., Motzkin, G., and Slater, B. (1998) Land-use history as long-term broad-scale disturbance: regional forest dynamics in central New England. *Ecosystems*, **1**, 96–119.

Geist, H.J. and Lambin, E.F. (2002) Proximate causes and underlying driving forces of tropical deforestation. *BioScience*, **52**, 143–50.

Gössling, S., Hansson, C.B., Hörstmeier, O., and Saggel, S. (2002) Ecological footprint analysis as a tool to assess tourism sustainability. *Ecological Economics*, **43**, 199–211.

Han, N., ed. (2000) *Analyses and Suggestions for Management Policies of China's Protected Areas*. Scientific and Technical Documents Publishing House, Beijing, China (in Chinese).

Hanley, J.A. and McNeil, B.J. (1982) The meaning and use of the area under a receiver operating characteristic (ROC) curve. *Radiology*, **143**, 29–36.

Hardie, I.W. and Parks, P.J. (1996) Program enrollment and acreage response to reforestation cost-sharing programs. *Land Economics*, **72**, 248–60.

Hart, J.F. (1968) Loss and abandonment of cleared farm land in eastern united-states. *Annals of the Association of American Geographers*, **58**, 417–40.

He, G., Chen, X., Bearer, S., et al. (2009) Spatial and temporal patterns of fuelwood collection in Wolong Nature Reserve: implications for panda conservation. *Landscape and Urban Planning*, **92**, 1–9.

He, G., Chen, X., Liu, W., et al. (2008) Distribution of economic benefits from ecotourism: a case study of Wolong Nature Reserve for giant pandas in China. *Environmental Management*, **42**, 1017–25.

He, L. (2004) Options for accelerating surplus rural labor migration in China. *Problems of Agricultural Economy*, **4**. (in Chinese).

Jaynes, E.T. (1957) Information theory and statistical mechanics. *The Physical Review*, **106**, 620–30.

Kaimowitz, D. (1997) Factors determining low deforestation: the Bolivian Amazon. *Ambio*, **26**, 537–40.

Klooster, D. (2003) Forest transitions in Mexico: institutions and forests in a globalized countryside. *Professional Geographer*, **55**, 227–37.

Lambin, E.F. and Meyfroidt, P. (2010) Land use transitions: socio-ecological feedback versus socio-economic change. *Land Use Policy*, **27**, 108–18.

Le Saout, S., Hoffmann, M., Shi, Y., et al. (2013) Protected areas and effective biodiversity Conservation. *Science*, **342**, 803–05.

Li, D. (2009) *Wolong Reconstruction Plan*. Beijing University, Beijing, China (in Chinese).

Liang, Z. and Ma, Z.D. (2004) China's floating population: new evidence from the 2000 census. *Population and Development Review*, **30**, 467–88.

Liang, Z. and White, M.J. (1996) Internal migration in China: 1950–88. *Demography*, **33**, 375–84.

Lindberg, K., Tisdell, C., and Xue, D. (2003) Ecotourism in China's nature reserves. In A.A. Lew, ed., *Tourism in China*, pp. 103–25. Haworth Hospitality Press, New York, NY.

Liu, J. (2010) China's road to sustainability. *Science*, **328**, 50.

Liu, J. (2014) Forest sustainability in China and implications for a telecoupled world. *Asia and the Pacific Policy Studies*, **1**, 230–50.

Liu, J., Daily, G.C., Ehrlich, P.R., and Luck, G.W. (2003) Effects of household dynamics on resource consumption and biodiversity. *Nature*, **421**, 530–33.

Liu, J. and Diamond, J. (2008) Revolutionizing China's environmental protection. *Science*, **319**, 37–38.

Liu, J., Li, S., Ouyang, Z., et al. (2008) Ecological and socioeconomic effects of China's policies for ecosystem services. *Proceedings of the National Academy of Sciences of the United States of America*, **105**, 9477–82.

Liu, J., Linderman, M., Ouyang, Z., et al. (2001) Ecological degradation in protected areas: the case of Wolong Nature Reserve for giant pandas. *Science*, **292**, 98.

Liu, J., Mooney, H., Hull, V., et al. (2015) Systems integration for global sustainability. *Science*, **347**, 1–9.

Liu, J., Ouyang, Z., Yang, W., et al. (2013) Evaluation of ecosystem service policies from biophysical and social perspectives: the case of China. In S.A. Levin, ed., *Encyclopedia of Biodiversity* (**vol. 3**), pp. 372–84. Academic Press, Waltham, MA.

Liu, W., Vogt, C.A., Luo, J., et al. (2012) Drivers and socioeconomic impacts of tourism participation in protected areas. *PLoS ONE*, **7**, e35420.

Ma, Z.D. (2002) Social-capital mobilization and income returns to entrepreneurship: the case of return migration in rural China. *Environment and Planning A*, **34**, 1763–84.

Mather, A.S. (1992) The forest transition. *Area*, **24**, 367–79.

Mather, A.S. (2004) Forest transition theory and the reforesting of Scotland. *Scottish Geographical Journal*, **120**, 83–98.

Mather, A.S. (2007) Recent Asian forest transitions in relation to forest-transition theory. *International Forestry Review*, **9**, 491–502.

Mather, A.S. and Fairbairn, J. (2000) From floods to reforestation: the forest transition in Switzerland. *Environment and History*, **6**, 399–421.

Mather, A.S., Fairbairn, J. and Needle, C.L. (1999) The course and drivers of the forest transition: the case of France. *Journal of Rural Studies*, **15**, 65–90.

Mather, A.S., Needle, C.L., and Coull, J.R. (1998) From resource crisis to sustainability: the forest transition in Denmark. *International Journal of Sustainable Development and World Ecology*, **5**, 182–93.

McNeely, J.A. and Miller, K.R. (1983) *National Parks and Protected Areas*. UN Economic and Social Commission for Asia and the Pacific, Bangkok, Thailand.

Myers, N. (1990) The world's forests and human populations: the environmental interconnections. *Population and Development Review*, **16**, 237–51.

Nagendra, H., Karmacharya, M., and Karna, B. (2005) Evaluating forest management in Nepal: views across space and time. *Ecology and Society*, **10**(1), 24.

Pahari, K. and Murai, S. (1999) Modelling for prediction of global deforestation based on the growth of human population. *ISPRS Journal of Photogrammetry and Remote Sensing*, **54**, 317–24.

Pan, W.K.Y. and Bilsborrow, R.E. (2005) The use of a multilevel statistical model to analyze factors influencing land use: a study of the Ecuadorian Amazon. *Global and Planetary Change*, **47**, 232–52.

Perz, S.G. (2007) Grand theory and context-specificity in the study of forest dynamics: forest transition theory and other directions. *Professional Geographer*, **59**, 105–14.

Perz, S.G. and Skole, D.L. (2003) Secondary forest expansion in the Brazilian Amazon and the refinement of forest transition theory. *Society & Natural Resources*, **16**, 277–94.

Phillips, S.J., Anderson, R.P., and Schapire, R.E. (2006) Maximum entropy modeling of species geographic distributions. *Ecological Modelling*, **190**, 231–59.

Pimm, S.L., Jenkins, C.N., Abell, R., et al. (2014) The biodiversity of species and their rates of extinction, distribution, and protection. *Science*, **344**, 1246752–1246751–10.

Qi, S.W., Xu, Q.A., Lan, H.X., et al. (2010) Spatial distribution analysis of landslides triggered by 2008.5.12 Wenchuan Earthquake, China. *Engineering Geology*, **116**, 95–108.

Reid, D.G., Taylor, A.H., Hu, J.C., and Qin, Z.S. (1991) Environmental influences on bamboo *Bashania-Fangiana* growth and implications for giant panda conservation. *Journal of Applied Ecology*, **28**, 855–68.

Rudel, T.K. (1998) Is there a forest transition? Deforestation, reforestation, and development. *Rural Sociology*, **63**, 533–52.

Rudel, T.K. (2005) *Tropical Forests*. Columbia University Press, New York, NY.

Rudel, T.K., Coomes, O.T., Moran, E., et al. (2005) Forest transitions: towards a global understanding of land use change. *Global Environmental Change: Human and Policy Dimensions*, **15**, 23–31.

Schaller, G.B., Hu, J., Pan, W., and Zhu, J. (1985) *The Giant Pandas of Wolong*. University of Chicago Press, Chicago, IL.

Seppelt, R., Manceur, A., Liu, J., et al. (2014) Synchronized peak rate years of global resources use. *Ecology and Society*, **19**, 50.

Seto, K.C., Woodcock, C.E., Song, C., et al. (2002) Monitoring land-use change in the Pearl River Delta using Landsat TM. *International Journal of Remote Sensing*, **23**, 1985–2004.

Sichuan Department of Forestry (2008) *Overall Planning for Post-Wenchuan Earthquake Restoration and Reconstruction in Wolong National Nature Reserve*. Sichuan Department of Forestry, Chengdu, China (in Chinese).

Sonesson, M. and Messerli, B. (2002) The abisko agenda: research for mountain area development. *Ambio*, Special Report Number **11**, 3–103.

Song, J. and Wang, E. (2001) Model and development trend of rural surplus labor migration in China. *Population Science of China*, **6**, 5. (in Chinese).

Taylor, A.H. and Qin, Z. (1987) Culm dynamics and dry-matter production of bamboos in the Wolong and Tangjiahe Giant Panda Reserves, Sichuan, China. *Journal of Applied Ecology*, **24**, 419–33.

Taylor, A.H. and Qin, Z. (1988) Regeneration from seed of *Sinarundinaria-Fangiana*, a bamboo, in the Wolong Giant Randa Reserve, Sichuan, China. *American Journal of Botany*, **75**, 1065–73.

Taylor, A.H. and Qin, Z. (1993a) Bamboo regeneration after flowering in the Wolong Giant Panda Reserve, China. *Biological Conservation*, **63**, 231–34.

Taylor, A.H. and Qin, Z. (1993b) Aging bamboo culms to assess bamboo population-dynamics in panda habitat. *Environmental Conservation*, **20**, 76–79.

Totman, C. (1986) Plantation forestry in early-modern Japan: economic aspects of its emergence. *Agricultural History*, **60**, 23–51.

Uchida, E., Rozelle, S., and Xu, J.T. (2009) Conservation payments, liquidity constraints, and off-farm labor: impact of the Grain-for-Green Program on rural households in China. *American Journal of Agricultural Economics*, **91**, 70–86.

Uchida, E., Xu, J.T., and Rozelle, S. (2005) Grain for green: cost-effectiveness and sustainability of China's conservation set-aside program. *Land Economics*, **81**, 247–64.

Viña, A., Bearer, S., Chen, X., et al. (2007) Temporal changes in giant panda habitat connectivity across boundaries of Wolong Nature Reserve, China. *Ecological Applications*, **17**, 1019–30.

Viña, A., Chen, X., Mcconnell, W.J., et al. (2011) Effects of natural disasters on conservation policies: the case of the 2008 Wenchuan Earthquake, China. *Ambio*, **40**, 274–84.

Viña, A., Chen, X., Yang, W., et al. (2013) Improving the efficiency of conservation policies with the use of surrogates derived from remotely sensed and ancillary data. *Ecological Indicators*, **26**, 103–11.

Viña, A., Tuanmu, M.N., Xu, W., et al. (2010) Range-wide analysis of wildlife habitat: implications for conservation. *Biological Conservation*, **143**, 1960–69.

Walters, B.B. (1997) Human ecological questions for tropical restoration: experiences from planting native upland trees and mangroves in the Philippines. *Forest Ecology and Management*, **99**, 275–90.

Wenchuan County People's Government, Guangdong Province People's Government and Guangdong Province Institute of Planning and Design (2008) *Overall Planning for Post-earthquake Restoration and Reconstruction in Wenchuan County* (in Chinese).

Wolong Administration Bureau (2004) *The Chronicle of Wolong Nature Reserve*. Sichuan Science and Technology Press, Chengdu, China (in Chinese).

Wolong Administration Bureau Department of Social and Economic Development (2006) *Annual Report on the Rural Economy of Wolong Nature Reserve*. Wolong Administration Bureau, Wolong, China (in Chinese).

Wolong Nature Reserve (2008) *Wolong Nature Reserve Annual Agricultural Report, 2008* (in Chinese).

Woon, Y.F. (1999) Labor migration in the 1990s—Homeward orientation of migrants in the Pearl River Delta region and its implications for interior China. *Modern China*, **25**, 475–512.

Xu, W., Viña, A., Qi, Z., et al. (2014) Evaluating conservation effectiveness of nature reserves established for surrogate species: case of a giant panda nature reserve in Qinling Mountains, China. *Chinese Geographical Science*, **24**, 60–70.

Yang, W., Dietz, T., Liu, W., et al. (2013b) Going beyond the Millennium Ecosystem Assessment: an index system of human dependence on ecosystem services. *PLoS ONE*, **8**, e64581.

Yang, W., Liu, W., Viña, A., et al. (2013a) Performance and prospects of payments for ecosystem services programs: evidence from China. *Journal of Environmental Management*, **127**, 86–95.

Yin, Y.P., Wang, F.W., and Sun, P. (2009) Landslide hazards triggered by the 2008 Wenchuan earthquake, Sichuan, China. *Landslides*, **6**, 139–52.

Zhang, X.D. (2010) A path to modernization: a review of documentaries on migration and migrant labor in China. *International Labor and Working-Class History*, **77**(1), 174–89.

Zhou, S. (1992) An analysis on the landslide calamity and its countermeasures in Wolong Nature Reserve. *Ecological Economy*, **3**, 49–51. (in Chinese).

CHAPTER 7

Panda Habitat Transition

Andrés Viña, Mao-Ning Tuanmu, and Jianguo Liu

7.1 Introduction

Human activities degrade ecosystems and threaten the long-term survival of many wildlife species around the world (Araujo et al., 2006, Botkin et al., 2007, Thuiller et al., 2008). In response, many conservation efforts have emerged to stop and reverse the degradation of natural ecosystems. The establishment of protected areas is considered one of the most effective efforts (Andam et al., 2008, Hannah et al., 2007, Jenkins and Joppa, 2009). Currently, there are more than 170 000 protected areas (Le Saout et al., 2013). Many are experiencing conservation successes. Increases in forest cover and other natural vegetation types (e.g., grasslands; Brereton et al., 2008, Norton et al., 2012) have translated into an increase in suitable habitat for wildlife (Bruner et al., 2001, Fall and Jackson, 1998, Messmer et al., 1997). As a result of such successful conservation actions, many areas around the world have experienced increases in wildlife habitat and populations (Enserink and Vogel, 2006, Gehring et al., 2010a, b, Pyare et al., 2004). In some cases, populations are larger than a century ago. These areas are in countries such as Mexico and the United States in North America (Dobkin et al., 1998, Martin-Rivera et al., 2001, Taylor et al., 2005), the United Kingdom and Spain in Europe (Aebischer and Ewald, 2010, Kuijper et al., 2009, Lozano et al., 2007, Smart et al., 2013), Ghana and the Republic of Congo in Africa (Adum et al., 2013, Robbins et al., 2011), Colombia and Costa Rica in Latin America (Sanchez-Cuervo et al., 2012, Timm et al., 2009), and China and Cambodia in Asia (O'Kelly et al., 2012, Wang and Li, 2008). However, the limitation or complete prohibition of natural resource use inside protected areas has negatively affected the livelihoods of people living in these areas (Adams et al., 2004, McShane et al., 2011). Consequently, people-oriented conservation activities are rapidly becoming widespread, including programs of payments for ecosystem services (PES). These programs provide alternative livelihood options that reduce the pressure on biodiversity and lead to sustainable use of natural resources. Thus, the programs have the dual goal of simultaneously protecting biodiversity while sustaining and even improving human livelihoods (Berkes, 2004, Hughes and Flintan, 2001).

China is the most populous nation and one of the most biologically diverse countries in the world (Liu, 2010, Liu and Diamond, 2005). China exemplifies the challenging balancing act of biodiversity conservation while supporting human livelihoods. In response to biodiversity loss, the number and spatial coverage of protected areas in China have increased exponentially, particularly since the 1980s (Liu and Raven, 2010). The first nature reserve was established in 1956. By the end of 2012, a total of 2669 nature reserves had been established, covering 15% of China's land surface (Xu et al., 2014). Conventional top-down management is prevalent in these protected areas (Liu and Diamond, 2008), in which the government dictates management decisions without much input from local communities. As a result, the livelihood of tens of millions of poor rural people living in and around protected areas was negatively affected (An et al., 2001, Xu and Melick, 2007). Thus, due to an inadequate consideration of local people's dependence on natural resources, conservation failures are common in China's protected areas, even in flagship reserves (Liu et al., 2001).

Pandas and People. Edited by Jianguo Liu, Vanessa Hull, Wu Yang, Andrés Viña, Xiaodong Chen, Zhiyun Ouyang, and Hemin Zhang. © Oxford University Press 2016. Published 2016 by Oxford University Press.

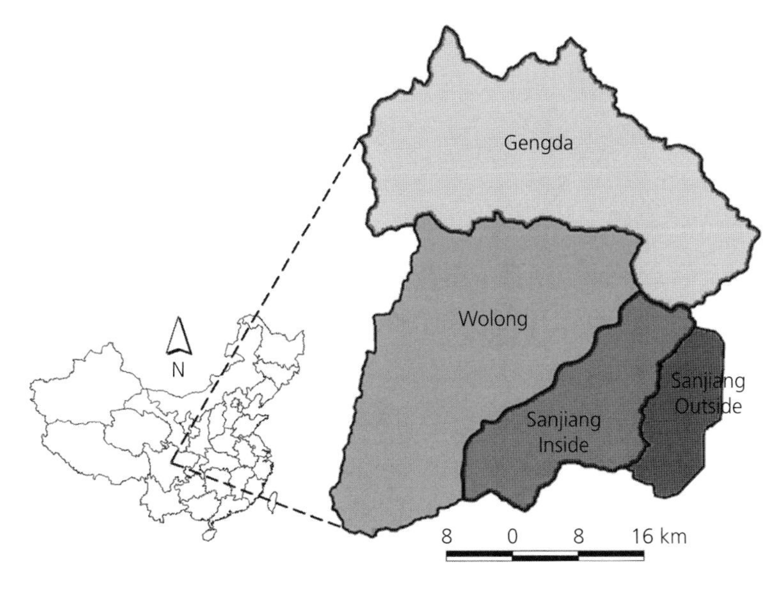

Figure 7.1 Map of Wolong Nature Reserve in Sichuan Province, China, showing the location of Wolong and Gengda townships completely within the reserve and of Sanjiang Township with portions inside and outside the reserve.

To enhance the long-term effectiveness of conservation actions without negatively affecting local livelihoods, it is essential to understand the complexity of human–nature interactions and their effects on the spatiotemporal dynamics of natural ecosystems. Such an understanding can be attained through the use of a coupled human and natural system (CHANS) framework (Alberti et al., 2011, An et al., 2014, Carter et al., 2014, Liu et al., 2007a, b, 2013a). Our coupled systems study in Wolong Nature Reserve in Sichuan Province, China (Figure 7.1) is a good example. Our long-term study in this system assesses the impacts of conservation actions on both humans and the environment, as represented by the habitat of an iconic endangered species—the giant panda.

In this chapter, we discuss the relationship between panda habitat dynamics, conservation efforts, and other human activities in Wolong Nature Reserve. Specifically, we present (1) methodological advances in panda habitat modeling, (2) patterns of panda habitat dynamics, (3) drivers of panda habitat dynamics, and (4) socioeconomic and environmental effects of panda habitat transition. We conclude with a discussion of the lessons learned and the implications of our research for management of this and many other coupled systems worldwide.

7.2 Methodological advances in panda habitat modeling

Habitat for the giant panda is defined according to four main factors: presence of understory bamboo, forest cover, elevation, and slope (Johnson et al., 1988, Liu et al., 1999, Reid et al., 1989, Schaller et al., 1985). As an extreme dietary specialist, 99% of the panda diet is composed of bamboo (Schaller et al., 1985). Thus, the presence of understory bamboo is a crucial component of giant panda habitat. Giant pandas require forests for shelter. Thus, the quality, quantity, spatial distribution, and degree of fragmentation of panda habitat are also affected if forests are degraded (Liu et al., 1999, 2001, Liu and Viña, 2014). The panda's elevational range is constrained between 1500 m and 3250 m, with an optimal range between 2500 m and 3000 m (Liu et al., 1999). They prefer flat areas and slopes of less than 45° for ease of movement, with optimal slopes of less than 15° (Liu et al., 1999). We initially did not have adequate information about bamboo distribution. Thus, we used a three-factor scheme that combines information on forest cover (derived from digital land-cover classification from satellite imagery) with elevation and slope (derived from digital elevation models; Figure 7.2). This proxy

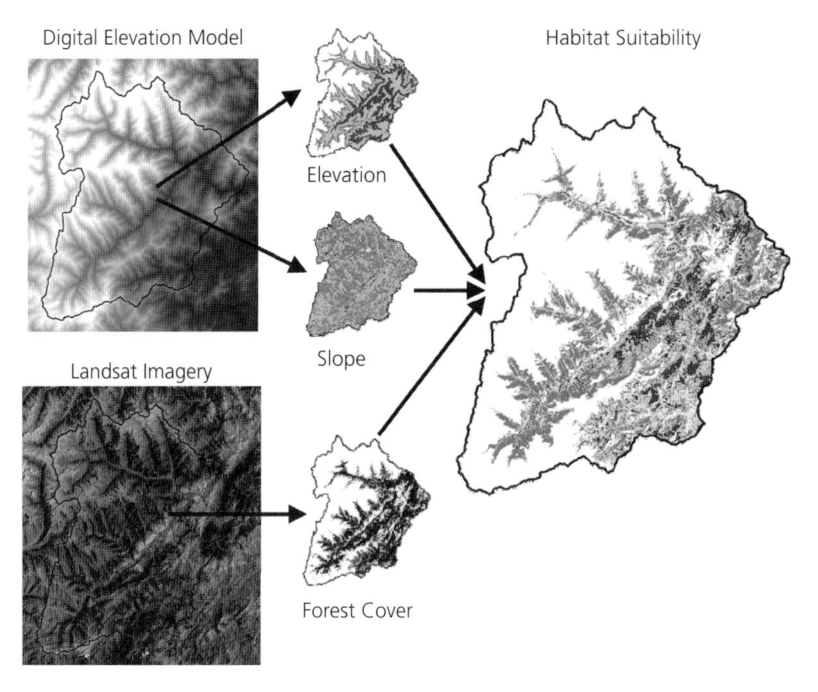

Figure 7.2 Discrete panda habitat suitability model based on forest cover obtained from the classification of multispectral satellite imagery (e.g., Landsat TM) combined with information on topography (i.e., elevation and slope) derived from a digital elevation model (based on Liu et al., 1999, 2001, Viña et al., 2007).

was used for modeling both the distribution and the temporal dynamics of panda habitat (Liu et al., 2001, Viña et al., 2007, Xu et al., 2006).

The three-factor scheme is suitable for assessing habitat loss, given that most natural forests in the region contain understory bamboo. Nonetheless, this method is not completely suitable for assessing habitat gain. Even if there is a gain in forest cover, an area is unsuitable for the panda if it lacks understory bamboo (Reid and Hu, 1991, Schaller et al., 1985, Viña et al., 2007). In other words, while an existing suitable habitat area may become unsuitable after forest loss, an unsuitable area may not become suitable after the forest recovers if bamboo has not recovered. As such, information on understory bamboo is required for a complete characterization of panda habitat (Linderman et al., 2004, 2005).

The detection of understory bamboo using information on the spectral reflectance of the land surface derived from multispectral satellite sensor systems is challenging (Linderman et al., 2004, 2005). The optical response of the vegetation captured by a satellite sensor is a complex non-linear combination of overstory and understory canopy components (Borel and Gerstl, 1994). In fact, characterizing bamboo distribution was one of the central challenges of panda habitat assessment. We developed a novel procedure to overcome this challenge (Tuanmu et al., 2011, Viña et al., 2008). We then successfully applied the method to assessing giant panda habitat in Wolong (Tuanmu et al., 2011, Viña et al., 2008) and across the entire geographic range of the species (Viña et al., 2010). This procedure is based on phenological signatures and phenology metrics in combination with panda feces locations obtained during field surveys. We derived phenological characteristics from a vegetation index (i.e., the Wide Dynamic Range Vegetation Index) acquired with a high temporal resolution by the Moderate Resolution Imaging Spectroradiometer (MODIS). This procedure is particularly useful for monitoring changes in panda habitat. The method captures information on the distribution of the most important determinants of panda habitat, i.e., forest cover and understory bamboo (Viña et al., 2008; Figure 7.3).

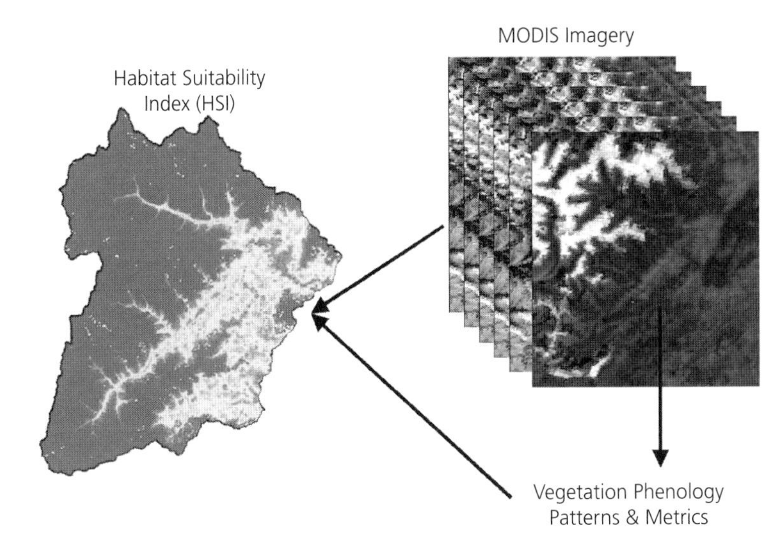

Figure 7.3 Continuous panda habitat suitability model based on phenological signatures and metrics derived from a time series of vegetation index imagery acquired by the Moderate Resolution Imaging Spectroradiometer (MODIS) combined with information on panda habitat occurrence locations (based on Tuanmu et al., 2010, 2011, Viña et al., 2008, 2010).

The model results in a spatially explicit habitat suitability index (HSI) ranging from 0 to 1, with higher values indicating more suitable habitat. We found that the model exhibited high accuracy at estimating panda habitat suitability. The values for the area under the receiver operating characteristic curve (AUC) were 0.853 and 0.855 for 2001 and 2007, respectively (Tuanmu et al., 2011).

7.3 Patterns of panda habitat dynamics

To characterize the changes in panda habitat in Wolong over a 40-year period (beginning when the reserve was established in 1963), it was necessary to apply an integrative approach. To do so, we implemented the habitat assessment by combining the three-factor scheme with the vegetation phenology. This integration was necessary given the lack of satellite imagery with sufficiently high spatial and temporal resolution data prior to the launch of the MODIS sensor system in 2000. We found that panda habitat degraded in the reserve between the 1960s and the late 1990s, despite its protected status (Liu et al., 2001, Viña et al., 2007; Figure 7.4). Yet, between 2001 and 2007, a significant increase in forest cover was observed in the reserve (Viña

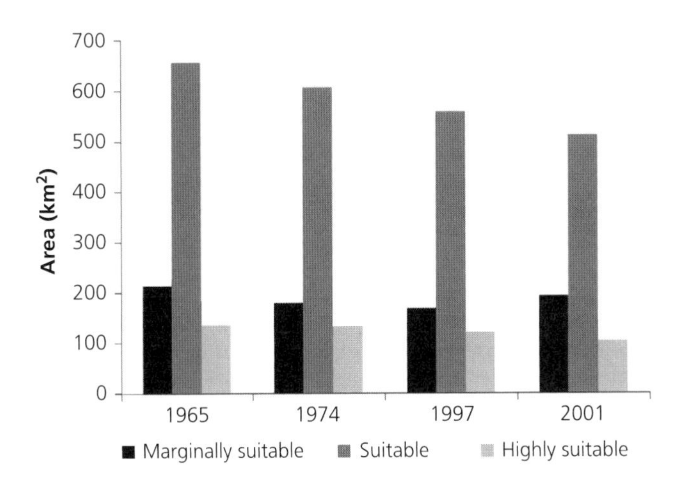

Figure 7.4 Spatiotemporal dynamics of panda habitat in Wolong Nature Reserve based on the discrete three-factor scheme that combines information on forest cover, elevation, and slope (see Figure 7.2). Based on Liu et al. (2001), Viña et al. (2007).

et al., 2007, 2011). However, this forest recovery could not be directly associated with a recovery of panda habitat because, as stated above, an increase in forest cover does not necessarily translate into an increase in panda habitat.

To assess the changes in panda habitat between 2001 and 2007, we used the novel procedure based on vegetation phenology to develop maps of HSI in 2001 and 2007 (Figure 7.5; Plate 3). The temporal changes of HSI during 2001–2007 were then assessed using the delta of HSI (i.e., $\Delta HSI = HSI_{2007} - HSI_{2001}$) for each pixel across three townships. Two townships are completely within the reserve (Wolong and Gengda) while Sanjiang Township has areas inside and outside the reserve (Figure 7.1). Our results show that between 2001 and 2007 there was a net gain of suitable habitat throughout the study area (Figure 7.6; Plate 4). However, when analyzed

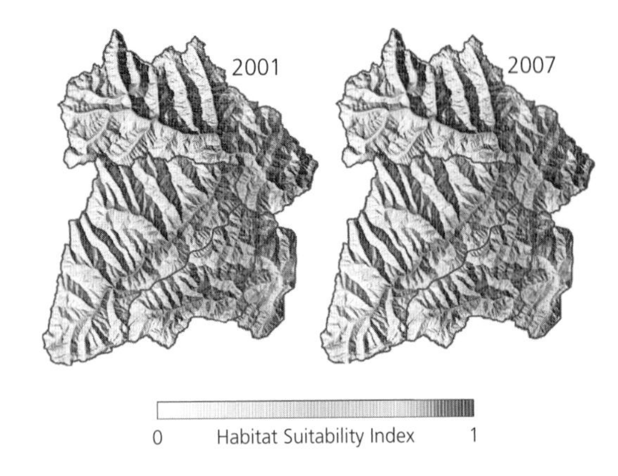

Figure 7.5 Topographic shaded relief maps showing the spatial distribution of panda habitat suitability index in the Wolong Nature Reserve in 2001 and 2007, based on phenological patterns derived from MODIS imagery (see Figure 7.3). Also shown are the locations of Wolong, Gengda, and Sanjiang townships inside and outside the reserve (see Figure 7.1). For color version see color plates. [PLATE 3]

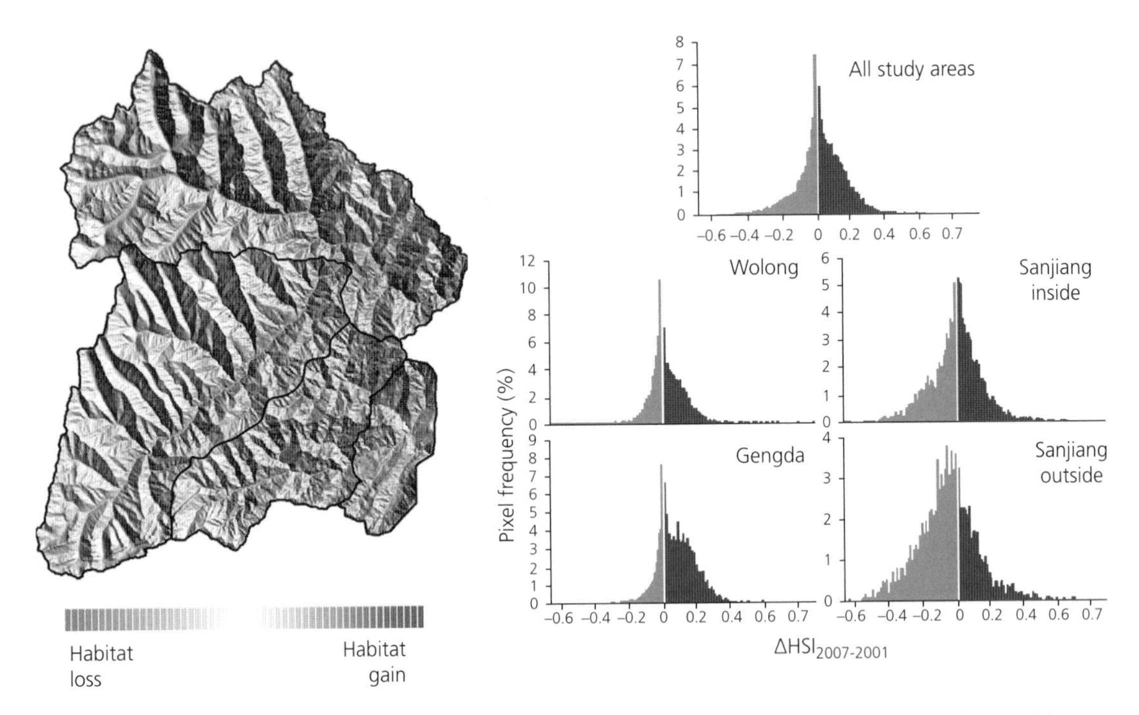

Figure 7.6 The left panel shows a topographic shaded relief map exhibiting the spatiotemporal dynamics of giant panda habitat suitability between 2001 and 2007 in the Wolong Nature Reserve. Also shown are the locations of Wolong, Gengda, and Sanjiang townships inside and outside the reserve (see Figure 7.1). Histograms in the right panel show the frequency distribution of the delta in habitat suitability index (HSI) pixel values across the entire study area and in each township. For color version see color plates. As with the colors shown in the map in the left panel, the red color represents negative deltas (i.e., loss of habitat suitability) while the blue color represents positive deltas (i.e., gains of habitat suitability). For clarity, the frequency of pixels exhibiting no change (i.e., HSI delta = 0) was omitted from the histograms. [PLATE 4]

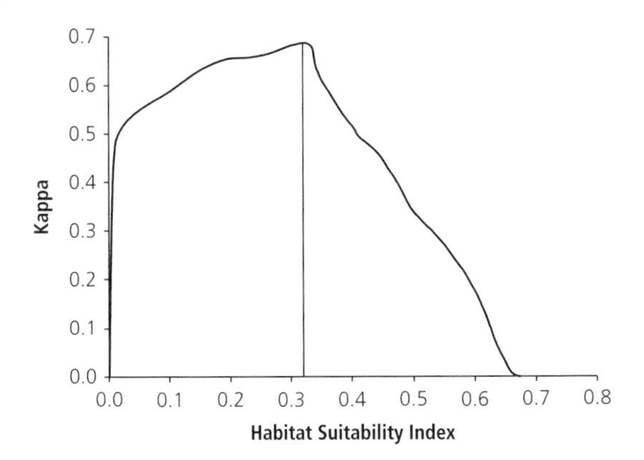

Figure 7.7 Procedure based on the maximum Kappa for selecting the optimal HSI threshold value for separating panda habitat from non-habitat pixels.

Table 7.1 Areal change (total and annual rates) of suitable panda habitat between 2001 and 2007 in the three townships comprising the study area. Positive values correspond to habitat gains, while negative values correspond to habitat losses.

Region	Habitat change (%)	Rate (% per year)
Wolong Township	6.7	1.1
Gengda Township	3.9	0.7
Sanjiang Township (inside the reserve)	−0.6	−0.1
Sanjiang Township (outside the reserve)	−12.4	−2.1
Entire study area	2.2	0.4

on a per-township basis, the gains were conspicuous in Wolong and Gengda townships while Sanjiang Township exhibited a net loss, particularly in the area outside the reserve (Figure 7.6).

The areal change in panda habitat was also assessed by applying a threshold to convert the continuous HSI scale into a binary outcome (i.e., habitat vs non-habitat). Using the maximum Kappa value, which is a chance-corrected measure of agreement (Cohen, 1960), we selected the optimal threshold for separating habitat from non-habitat pixels (Figure 7.7). After applying this threshold, it was observed that there was an overall gain in the area of panda habitat across the entire study area (Figure 7.8). This overall gain was driven mainly by the gains in Wolong and Gengda townships (Table 7.1).

7.4 Drivers of panda habitat dynamics

A number of drivers shaped patterns of panda habitat change. Wolong's local residents depend on local natural resources as they carry out diverse activities. Examples include farming, raising livestock, timber harvesting, fuelwood collection, herbal medicine collection, road construction, and tourism (An et al., 2001, 2002, He et al., 2008, Liu et al., 2012, Viña et al., 2007). These human activities are associated with the loss of forest cover (Chapter 6) and its resultant degradation of panda habitat observed in the reserve between the mid-1960s and the late 1990s (Liu et al., 2001, Viña et al., 2007).

During the 2000s, the Chinese government implemented one of the largest PES programs in the world, the Natural Forest Conservation Program (NFCP). NFCP was enacted mainly in response to the severe droughts of 1997 and devastating floods of 1998, both of which affected the Yangtze and Yellow River basins. The policy seeks to protect and restore natural forests through timber harvesting bans and afforestation (Liu et al., 2008, 2013b). The program provides payments typically to forest enterprises and local governments to compensate their economic losses due to a shift from timber harvesting to afforestation and forest conservation. With NFCP payments, ~0.7 million former logging and timber-processing workers nationwide have retired. These individuals have either obtained jobs in other economic sectors or have been hired in tree plantation and forest monitoring activities (Yin and Yin, 2010). Since NFCP implementation, a considerable amount of natural forest has been protected and many new forested areas have formed (Liu et al., 2008, 2013b, Yin and Yin, 2010). In addition to forest monitoring by government officials, NFCP in Wolong also included monitoring efforts by local households (Yang et al., 2013a, b; Chapter 11). All households within a group suffer payment reductions if anthropogenic damage is found in their comonitored forest parcel.

NFCP has been credited with a significant increase in forest cover in Wolong during 2001–2007 (Chapter 6; see also Viña et al., 2007, 2011). This success can be partly attributed to its decentralized implementation, which included the participation of local households in monitoring activities (Yang et al., 2013a, b). This increase in forest cover also translated into an increase in the panda habitat. Spatial variability occurs in the implementation of NFCP in Wolong (i.e., government vs household monitoring, and differences in payment level). This variation makes Wolong an excellent place to examine how different NFCP implementations affect conservation outcomes (Tuanmu, 2012). Before NFCP, there was a conspicuous areal loss of forest cover, thus a degradation of suitable habitat within and around reserve boundaries (Liu et al., 2001, Viña et al., 2007). However, this trend was reversed after the implementation of NFCP. The study area exhibited an overall gain not only in forest cover (Viña et al., 2011) but also in suitable panda habitat (Figures 7.7 and 7.8, respectively). Also, there was a conspicuous effect of different compensation levels on panda habitat recovery outcomes. For instance, while Wolong and Gengda townships exhibited net gains, Sanjiang Township exhibited net losses, particularly the area outside the reserve (Table 7.1). Households in Wolong and Gengda townships receive twice the NFCP payment for monitoring activities than households in Sanjiang Township (Tuanmu, 2012). Thus the gains in suitable habitat observed in Wolong and Gengda townships as opposed to the losses in Sanjiang Township suggest that a higher economic compensation constitutes a higher incentive for conservation activities. Furthermore, while Sanjiang Township exhibited a net loss of panda habitat, this loss was conspicuously higher in the area outside the reserve than in the area inside. This difference suggests that the reserve boundary (thus the top-down management strategy), while not enhancing panda habitat recovery, seemed to have acted as an effective deterrent of further panda habitat losses. Therefore, the combination of an incentive-based decentralized NFCP implementation together with top-down reserve management approaches seem to produce better conservation outcomes than either one applied in isolation (Tuanmu, 2012).

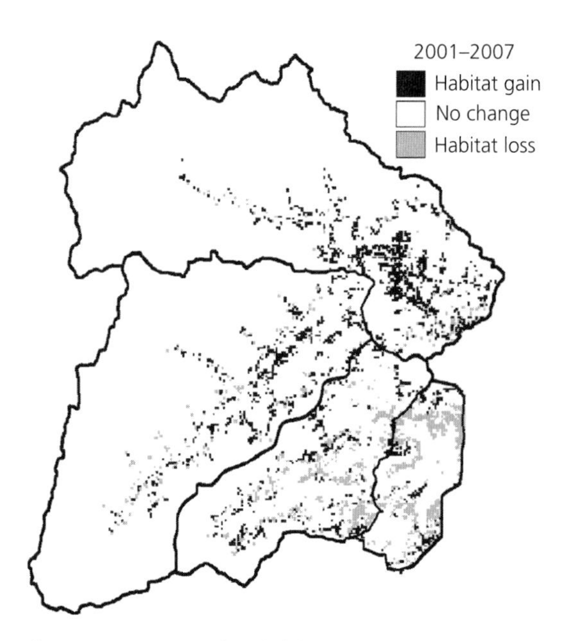

Figure 7.8 Binary map of panda habitat/non-habitat pixels based on the maximum Kappa procedure (see Figure 7.7). Also shown are the locations of Wolong, Gengda, and Sanjiang townships inside and outside the reserve (see Figure 7.1).

There is synchronicity in the habitat trend reversal and NFCP implementation, suggesting a positive effect of NFCP on panda habitat. However, it may also be argued that habitat dynamics before vs after NFCP implementation might have been potentially affected by other factors acting at the same time. However, although other conservation policies may also constitute drivers of habitat change, their contribution to the panda habitat recovery in Wolong was small compared to that of NFCP. For example, another national conservation policy implemented in the reserve in 2000 is the Grain to Green Program (GTGP). This program provides local farmers with cash or grain subsidies to convert cropland on steep slopes into forest or grassland (Liu et al., 2008, 2013b, Viña et al., 2011). GTGP may produce long-term benefits through indirect effects, e.g., promoting out-migration of agricultural surplus labor (Uchida et al., 2009). However, GTGP's direct contribution to panda habitat recovery was negligible. The tree seedlings and saplings planted through this program do not constitute suitable panda habitat (Bearer et al., 2008), at least within the time frame assessed. In addition, areas of GTGP in the reserve correspond to less than 1% of the total reserve area. Even if all the areas are suitable for the pandas, the relative contribution to panda habitat is small. In addition, potential panda habitat gains obtained through GTGP may not be discernible given the accuracy of the panda habitat model (AUC value = 0.85). Other drivers of change such as demographic dynamics and associated changes in resource consumption also do not explain the reverse in the panda habitat trends observed after NFCP implementation. For instance, between 2001 and 2007, human population increased 6% and the number of households increased 23% in the three townships (Wenchuan Statistics Bureau, 2008). Both of these demographic increases are associated with an increase in resource consumption (Liu et al., 2003). Such an increase would translate into further degradation of forest cover and panda habitat (An et al., 2001), an outcome that was not realized in the reserve.

7.5 Socioeconomic and environmental effects of panda habitat change

The degradation and subsequent recovery of giant panda habitat had and will have different effects on the environmental conditions in the reserve. On one hand, knowledge of the continual habitat degradation brought stronger enforcement of conservation actions. Examples included stronger controls at the two check points along the main road crossing the reserve, as well as additional monitoring activities by the reserve staff in areas outside NFCP parcels assigned to local households. The strengthening of government sanctions, together with NFCP and other conservation programs (e.g., GTGP), contributed to the reduction of household fuelwood consumption. At the same time, these measures also facilitated the adoption of alternative energy sources such as electricity. For instance, the amount of fuelwood consumed per household after NFCP (2001–2005) declined by an average of 40% while electricity consumption doubled compared to before NFCP implementation (1998–1999) (Chen et al., 2014, Yang et al., 2013a).

While stronger conservation actions contributed to the recovery of giant panda habitat, their implementation had both positive and negative consequences on the local people. On the one hand, stronger restrictions were put in place on the use of forest resources (e.g., timber harvesting, collection of fuelwood, bamboo shoots, medicinal herbs, and mushrooms, and wildlife hunting). Such restrictions not only negatively affect residents economically, but also pose a threat to the local culture of ethnic minorities because many of these activities have been part of their cultural identity for generations (Yang et al., 2013a). Such restrictions, therefore, bring conflicts between local people and the reserve administration, which hinder the successful implementation of conservation actions. On the positive side, the recovery of panda habitat together with the success of breeding pandas in captivity has placed Wolong in the national and international spotlight, which has translated into a higher influx of tourists. The number of tourists visiting the reserve increased tenfold between 1996 (~20 000) and 2006 (~200 000). A sevenfold increase in household participation in tourism activities also occurred, from 4% in 1998 to 28% in 2007 (Liu et al., 2012). This trend has contributed to an increase in mean household income during the same period (Yang et al., 2013a).

Feedbacks also occur among the human and natural subsystems. Panda habitat comprises several types of forest ecosystems (Reid and Hu, 1991, Taylor and Qin, 1993) that are also habitat of many other wildlife species. The recovery of panda habitat also represents the recovery of the habitat of other wildlife species, some of which are crop/livestock raiders (e.g., wild boar (*Sus scrofa*) and sambar (*Rusa unicolor*)). Therefore, the recovery of panda habitat has also brought an increase in the number of instances of crop and livestock raiding. Such occurrences have negative economic effects on local people because the issue has not received enough attention by government officials and farmers have not received sufficient economic compensation for the damages (Yang et al., 2013a). Such oversights may have negative consequences for the continued success of conservation actions because participation by local people may diminish over time if they do not receive enough compensation for their losses. Or farmers may abandon more cropland to reduce crop damage, thus reducing their capability to earn income.

7.6 Lessons learned and their implications for biodiversity conservation

Between the establishment of the reserve in 1963 and the end of the twentieth century, the panda habitat of Wolong exhibited a negative trend. This trend was reversed during the first decade of the twenty-first century, when the panda habitat started to recover. Such habitat transition is concurrent with an observed forest transition (Viña et al., 2011; Chapter 6) and started to occur after the implementation of NFCP. This synchronicity suggests that NFCP has an overall positive effect not only on forests but also on panda habitat. Three potential reasons may explain the effectiveness of NFCP. First, direct payments to local residents compensate the costs of forgoing resource-depleting activities (e.g., timber harvesting). Thus, direct payments may create a strong incentive for conservation activities (Engel et al., 2008, Ferraro and Kiss, 2002). Second, besides constituting a strong conservation incentive, the payments may also encourage additional responses. Examples include making the switch from fuelwood to electricity more affordable and reducing the dependence on fuelwood as the primary energy source (An et al., 2002; Chapter 10). And third, by assigning monitoring duties for forest parcels to household groups, the reserve administration induced a shared responsibility in monitoring activities. This design feature enhanced rule compliance through social norms (Chen et al., 2009, Dietz et al., 2003). As a result, local people dutifully perform their monitoring activities not only to obtain the economic compensation but also to avoid payment reductions that could harm their social relations with other group members (Yang et al., 2013b; Chapter 11). They also recoil from harvesting timber or collecting fuelwood in the parcels monitored by other groups to avoid harming their social relations with the people in those groups (Yang et al., 2013b).

Our results suggest that complementary conservation instruments explicitly incorporating local residents (e.g., through the combination of PES, decentralized management, and top-down regulations) may offer better conservation outcomes than the implementation of a single instrument. In addition to restoring forest cover (Viña et al., 2011), NFCP implementation was also effective in restoring panda habitat. The successful engagement of local residents in forest monitoring activities through adequate economic compensation enhanced such effectiveness. The involvement of local residents in NFCP monitoring efforts is, therefore, generating greater overall conservation benefits than in most other areas of China, where local residents are not involved in monitoring (Yin and Yin, 2010). Thus, this type of household involvement in conservation activities should be encouraged in other parts of China and the world.

7.7 Summary

Conservation policies such as payments for ecosystem services (PES) have emerged to address local people's needs. Some studies have assessed the effectiveness of PES programs at reducing the degradation of natural ecosystems. Few studies, however, have assessed their effectiveness in conserving wildlife habitat. In this chapter, we evaluated panda habitat dynamics using multiple satellite sensor systems (including the Landsat series and MODIS) together with field data and novel remote sensing procedures. The procedures included generating phenological signatures and phenology metrics derived from a vegetation index (i.e., the Wide Dynamic Range Vegetation Index) using imagery acquired with a high temporal resolution by MODIS. They allowed us to analyze the distribution of understory bamboo—the crucial component characterizing panda habitat not included in previous habitat models. Our method showed an overall gain in panda habitat of 2.2% between 2001 and 2007. The most significant driver that may explain this change is the Natural Forest Conservation Program (NFCP), implemented in 2001, that monitors and prevents illegal timber harvesting. In some areas of Wolong, NFCP involves local households and provides them with economic compensation. Compensation was not equal across space, and habitats were more improved in townships where people received higher compensation. Our findings demonstrate the value of including understory vegetation estimates in habitat modeling for evaluation of conservation programs. In addition, our results suggest that conservation actions that successfully engage local residents and provide adequate economic compensation generate higher environmental gains than simple bans on resource use.

References

Adams, W.M., Aveling, R., Brockington, D., et al. (2004) Biodiversity conservation and the eradication of poverty. *Science*, **306**, 1146–49.

Adum, G.B., Eichhorn, M.P., Oduro, W., et al. (2013) Two-stage recovery of amphibian assemblages following selective logging of tropical forests. *Conservation Biology*, 27, 354–63.

Aebischer, N.J. and Ewald, J.A. (2010) Grey partridge *Perdix perdix* in the UK: recovery status, set-aside and shooting. *Ibis*, **152**, 530–42.

Alberti, M., Asbjornsen, H., Baker, L.A., et al. (2011) Research on coupled human and natural systems (CHANS): approach, challenges, and strategies. *Bulletin of the Ecological Society of America*, **92**, 218–28.

An, L., Liu, J., Ouyang, Z., et al. (2001) Simulating demographic and socioeconomic processes on household level and implications for giant panda habitats. *Ecological Modelling*, 140, 31–49.

An, L., Lupi, F., Liu, J., et al. (2002) Modeling the choice to switch from fuelwood to electricity: implications for giant panda habitat conservation. *Ecological Economics*, **42**, 445–57.

An, L., Zvoleff, A., Liu, J., and Axinn, W. (2014) Agent-based modeling in coupled human and natural systems (CHANS): lessons from a comparative analysis. *Annals of the Association of American Geographers*, **104**, 723–45.

Andam, K.S., Ferraro, P.J., Pfaff, A., et al. (2008) Measuring the effectiveness of protected area networks in reducing deforestation. *Proceedings of the National Academy of Sciences of the United States of America*, **105**, 16089–94.

Araujo, M.B., Thuiller, W., and Pearson, R.G. (2006) Climate warming and the decline of amphibians and reptiles in Europe. *Journal of Biogeography*, **33**, 1712–28.

Bearer, S., Linderman, M., Huang, J., et al. (2008) Effects of fuelwood collection and timber harvesting on giant panda habitat use. *Biological Conservation*, **141**, 385–93.

Berkes, F. (2004) Rethinking community-based conservation. *Conservation Biology*, **18**, 621–30.

Borel, C.C. and Gerstl, S.A.W. (1994) Nonlinear spectral mixing models for vegetative and soil surfaces. *Remote Sensing of Environment*, **47**, 403–16.

Botkin, D.B., Saxe, H., Araujo, M.B., et al. (2007) Forecasting the effects of global warming on biodiversity. *Bioscience*, **57**, 227–36.

Brereton, T.M., Warren, M.S., Roy, D.B., and Stewart, K. (2008) The changing status of the Chalkhill Blue butterfly *Polyommatus coridon* in the UK: the impacts of conservation policies and environmental factors. *Journal of Insect Conservation*, **12**, 629–38.

Bruner, A.G., Gullison, R.E., Rice, R.E., and Da Fonseca, G.A. (2001) Effectiveness of parks in protecting tropical biodiversity. *Science*, **291**, 125–28.

Carter, N.H., Viña, A., Hull, V., et al. (2014) Coupled human and natural systems approach to wildlife research and conservation. *Ecology and Society*, 19, 43.

Chen, X., Lupi, F., He, G., and Liu, J. (2009) Linking social norms to efficient conservation investment in payments for ecosystem services. *Proceedings of the National Academy of Sciences of the United States of America*, **106**, 11812–17.

Chen, X., Viña, A., Shortridge, A., et al. (2014) Assessing the effectiveness of payments for ecosystem services: an agent-based modeling approach. *Ecology and Society*, 19, 7.

Cohen, J. (1960) A coefficient of agreement for nominal scales. *Education and Psychological Measurement*, **20**, 37–46.

Dietz, T., Ostrom, E., and Stern, P.C. (2003) The struggle to govern the commons. *Science*, **302**, 1907–12.

Dobkin, D.S., Rich, A.C., and Pyle, W.H. (1998) Habitat and avifaunal recovery from livestock grazing in a riparian meadow system of the northwestern Great Basin. *Conservation Biology*, **12**, 209–21.

Engel, S., Pagiola, S., and Wunder, S. (2008) Designing payments for environmental services in theory and practice: an overview of the issues. *Ecological Economics*, **65**, 663–74.

Enserink, M. and Vogel, G. (2006) Wildlife conservation—the carnivore comeback. *Science*, **314**, 746–49.

Fall, M.W. and Jackson, W.B. (1998) A new era of vertebrate pest control? An introduction. *International Biodeterioration & Biodegradation*, 42, 85–91.

Ferraro, P.J. and Kiss, A. (2002) Direct payments to conserve biodiversity. *Science*, 298, 1718–19.

Gehring, T.M., Vercauteren, K.C., and Landry, J.M. (2010a) Livestock protection dogs in the 21st Century: is an ancient tool relevant to modern conservation challenges? *Bioscience*, 60, 299–308.

Gehring, T.M., Vercauteren, K.C., Provost, M.L., and Cellar, A.C. (2010b) Utility of livestock-protection dogs for deterring wildlife from cattle farms. *Wildlife Research*, 37, 715–21.

Hannah, L., Midgley, G., Andelman, S., et al. (2007) Protected area needs in a changing climate. *Frontiers in Ecology and the Environment*, 5, 131–38.

He, G., Chen, X., Liu, W., et al. (2008) Distribution of economic benefits from ecotourism: a case study of Wolong Nature Reserve for Giant Pandas in China. *Environmental Management*, 42, 1017–25.

Hughes, R. and Flintan, F. (2001) *Integrating Conservation and Development Experience: A Review and Bibliography of the ICDP Literature*. International Institute for Environment and Development, London, UK.

Jenkins, C.N. and Joppa, L. (2009) Expansion of the global terrestrial protected area system. *Biological Conservation*, 142, 2166–74.

Johnson, K.G., Schaller, G.B., and Hu, J. (1988) Responses of giant pandas to a bamboo die-off. *National Geographic Research*, 4, 161–77.

Kuijper, D.P.J., Oosterveld, E., and Wymenga, E. (2009) Decline and potential recovery of the European grey partridge (*Perdix perdix*) population: a review. *European Journal of Wildlife Research*, 55, 455–63.

Le Saout, S., Hoffmann, M., Shi, Y., et al. (2013) Protected areas and effective biodiversity conservation. *Science*, 342, 803–05.

Linderman, M.A., Bearer, S., An, L., et al. (2005) The effects of understory bamboo on broad-scale estimates of giant panda habitat. *Biological Conservation*, 121, 383–90.

Linderman, M.A., Liu, J., Qi, J., et al. (2004) Using artificial neural networks to map the spatial distribution of understorey bamboo from remote sensing data. *International Journal of Remote Sensing*, 25, 1685–700.

Liu, J. (2010) China's road to sustainability. *Science*, 328, 974–74.

Liu, J., Daily, G.C., Ehrlich, P.R., and Luck, G.W. (2003) Effects of household dynamics on resource consumption and biodiversity. *Nature*, 421, 530–33.

Liu, J. and Diamond, J. (2005) China's environment in a globalizing world. *Nature*, 435, 1179–86.

Liu, J. and Diamond, J. (2008) Revolutionizing China's environmental protection. *Science*, 319, 37–38.

Liu, J., Dietz, T., Carpenter, S.R., et al. (2007a) Complexity of coupled human and natural systems. *Science*, 317, 1513–16.

Liu, J., Dietz, T., Carpenter, S.R., et al. (2007b) Coupled human and natural systems. *Ambio*, 36, 639–49.

Liu, J., Hull, V., Batistella, M., et al. (2013a) Framing sustainability in a telecoupled world. *Ecology and Society*, 18, 26.

Liu, J., Li, S., Ouyang, Z., et al. (2008) Ecological and socioeconomic effects of China's policies for ecosystem services. *Proceedings of the National Academy of Sciences of the United States of America*, 105, 9477–82.

Liu, J., Linderman, M., Ouyang, Z., et al. (2001) Ecological degradation in protected areas: the case of Wolong Nature Reserve for giant pandas. *Science*, 292, 98–101.

Liu, J., Ouyang, Z., Taylor, W.W., et al. (1999) A framework for evaluating the effects of human factors on wildlife habitat: the case of giant pandas. *Conservation Biology*, 13, 1360–70.

Liu, J., Ouyang, Z., Yang, W., et al. (2013b) Evaluation of ecosystem service policies from biophysical and social perspectives: the case of China. In S.A. Levin, ed., *Encyclopedia of Biodiversity* (vol. 3), pp. 372–84. Academic Press, Waltham, MA.

Liu, J. and Raven, P.H. (2010) China's environmental challenges and implications for the world. *Critical Reviews in Environmental Science and Technology*, 40, 823–51.

Liu, J. and Viña, A. (2014) Pandas, plants and people. *Annals of the Missouri Botanical Garden*, 100, 108–25.

Liu, W., Vogt, C.A., Luo, J., et al. (2012) Drivers and socioeconomic impacts of tourism participation in protected areas. *PLoS ONE*, 7, e35420.

Lozano, J., Virgos, E., Cabezas-Diaz, S., and Mangas, J.G. (2007) Increase of large game species in Mediterranean areas: is the European wildcat (*Felis silvestris*) facing a new threat? *Biological Conservation*, 138, 321–29.

Martin-Rivera, M., Ibarra-Flores, F., Guthery, F.S., et al. (2001) Habitat improvement for wildlife in north-central Sonora, Mexico. In E.D. McArthur and D. J. Fairbanks, eds, *Shrubland Ecosystem Genetics and Biodiversity*, pp. 356–60. US Department of Agriculture, Forest Service, Rocky Mountain Research Station, Provo, UT.

Mcshane, T.O., Hirsch, P.D., Trung, T.C., et al. (2011) Hard choices: making trade-offs between biodiversity conservation and human well-being. *Biological Conservation*, 144, 966–72.

Messmer, T.A., George, S.M., and Cornicelli, L. (1997) Legal considerations regarding lethal and nonlethal approaches to managing urban deer. *Wildlife Society Bulletin*, 25, 424–29.

Norton, L.R., Maskell, L.C., Smart, S.S., et al. (2012) Measuring stock and change in the GB countryside for policy—key findings and developments from the Countryside Survey 2007 field survey. *Journal of Environmental Management*, **113**, 117–27.

O'Kelly, H.J., Evans, T.D., Stokes, E.J., et al. (2012) Identifying conservation successes, failures and future opportunities: assessing recovery potential of wild ungulates and tigers in eastern Cambodia. *PLoS ONE*, 7, e40482.

Pyare, S., Cain, S., Moody, D., et al. (2004) Carnivore recolonisation: reality, possibility and a non-equilibrium century for grizzly bears in the southern Yellowstone ecosystem. *Animal Conservation*, **7**, 71–77.

Reid, D.G. and Hu, J. (1991) Giant panda selection between *Bashania fangiana* bamboo habitats in Wolong Reserve, Sichuan, China. *Journal of Applied Ecology*, **28**, 228–43.

Reid, D.G., Hu, J., Sai, D., et al. (1989) Giant panda *Ailuropoda melanoleuca* behavior and carrying-capacity following a bamboo die-off. *Biological Conservation*, **49**, 85–104.

Robbins, M.M., Gray, M., Fawcett, K.A., et al. (2011) Extreme conservation leads to recovery of the Virunga Mountain gorillas. *PLoS ONE*, 6, e19788.

Sanchez-Cuervo, A.M., Aide, T.M., Clark, M.L., and Etter, A. (2012) Land cover change in Colombia: surprising forest recovery trends between 2001 and 2010. *PLoS ONE*, 7, e43943.

Schaller, G.B., Hu, J., Pan, W., and Zhu, J. (1985) *The Giant Pandas of Wolong*. University of Chicago Press, Chicago, IL.

Smart, J., Bolton, M., Hunter, F., et al. (2013) Managing uplands for biodiversity: do agri-environment schemes deliver benefits for breeding lapwing *Vanellus vanellus*? *Journal of Applied Ecology*, **50**, 794–804.

Taylor, A.H. and Qin, Z. (1993) Bamboo regeneration after flowering in the Wolong Giant Panda Reserve, China. *Biological Conservation*, **63**, 231–34.

Taylor, M.F.J., Suckling, K.F., and Rachlinski, J.J. (2005) The effectiveness of the Endangered Species Act: a quantitative analysis. *Bioscience*, **55**, 360–67.

Thuiller, W., Albert, C., Araujo, M.B., et al. (2008) Predicting global change impacts on plant species' distributions: future challenges. *Perspectives in Plant Ecology, Evolution and Systematics*, **9**, 137–52.

Timm, R.M., Lieberman, D., Lieberman, M., and McClearn, D. (2009) Mammals of Cabo Blanco: history, diversity, and conservation after 45 years of regrowth of a Costa Rican dry forest. *Forest Ecology and Management*, **258**, 997–1013.

Tuanmu, M.N. (2012) *Spatiotemporal Dynamics of Giant Panda Habitat: Implications for Panda Conservation under a Changing Environment*. Doctoral Dissertation, Michigan State University, East Lansing, MI.

Tuanmu, M.N., Viña, A., Bearer, S., et al. (2010) Mapping understory vegetation using phenological characteristics derived from remotely sensed data. *Remote Sensing of Environment*, **114**, 1833–44.

Tuanmu, M.N., Viña, A., Roloff, G.J., et al. (2011) Temporal transferability of wildlife habitat models: implications for habitat monitoring. *Journal of Biogeography*, **38**, 1510–23.

Uchida, E., Rozelle, S., and Xu, J. (2009) Conservation payments, liquidity constraints, and off-farm labor: impact of the Grain-for-Green Program on rural households in China. *American Journal of Agricultural Economics*, **91**, 70–86.

Viña, A., Bearer, S., Chen, X., et al. (2007) Temporal changes in giant panda habitat connectivity across boundaries of Wolong Nature Reserve, China. *Ecological Applications*, **17**, 1019–30.

Viña, A., Bearer, S., Zhang, H., et al. (2008) Evaluating MODIS data for mapping wildlife habitat distribution. *Remote Sensing of Environment*, **112**, 2160–69.

Viña, A., Chen, X., McConnell, W., et al. (2011) Effects of natural disasters on conservation policies: the case of the 2008 Wenchuan Earthquake, China. *Ambio*, 40, 274–84.

Viña, A., Tuanmu, M.N., Xu, W., et al. (2010) Range-wide analysis of wildlife habitat: implications for conservation. *Biological Conservation*, **143**, 1960–63.

Wang, G.M. and Li, X.H. (2008) Population dynamics and recovery of endangered crested ibis (*Nipponia nippon*) in Central China. *Waterbirds*, **31**, 489–94.

Wenchuan Statistics Bureau (2008) *Wenchuan County Statistical Yearbook 2007*. Wenchuan Statistic Bureau, Wenchuan, Sichuan Province, China (in Chinese).

Xu, J. and Melick, D.R. (2007) Rethinking the effectiveness of public protected areas in southwestern China. *Conservation Biology*, **21**, 318–28.

Xu, W., Viña, A., Qi, Z., et al. (2014) Evaluating conservation effectiveness of nature reserves established for surrogate species: case of a giant panda nature reserve in the Qinling Mountains. *Chinese Geographical Science*, **24**, 60–70.

Xu, W., Ouyang, Z., Viña, A., et al. (2006) Designing a conservation plan for protecting the habitat for giant pandas in the Qionglai mountain range, China. *Diversity and Distributions*, **12**, 610–19.

Yang, W., Liu, W., Viña, A., et al. (2013a) Performance and prospects of payments for ecosystem services programs: evidence from China. *Journal of Environmental Management*, **127**, 86–95.

Yang, W., Liu, W., Viña, A., et al. (2013b) Nonlinear effects of group size on collective action and resource outcomes. *Proceedings of the National Academy of Sciences of the United States of America*, **110**, 10916–21.

Yin, R. and Yin, G. (2010) China's primary programs of terrestrial ecosystem restoration: initiation, implementation, and challenges. *Environmental Management*, **45**, 429–41.

CHAPTER 8

Demographic Decisions and Cascading Consequences

Li An, Wu Yang, Zai Liang, Ashton Shortridge, and Jianguo Liu

8.1 Introduction

Global human population has been increasing rapidly in the last several decades from 2.5 billion in 1950 (U.S. Census Bureau, 2009) to more than 7.2 billion in 2014 (Population Reference Bureau, 2014). The vast majority of such population growth has taken place in developing countries. The global decline in mortality rate along with the high (although declining) fertility rates in developing countries may explain this rapid population growth (Bilsborrow et al., 2001, The World Bank, 2015). Such high population size and growth have directly or indirectly caused many socioeconomic and environmental problems across local to global scales (Cohen, 2003, de Sherbinin et al., 2007, Vitousek, 1994). These problems include biodiversity loss, ecosystem degradation, habitat fragmentation, hunger, and social unrest. They are prevalent in many parts of the world, even in "protected areas" (Curran et al., 2004, Liu et al., 2001). The population issue has gained attention at least as far back as the famous book *An Essay on the Principle of Population* (Malthus, 1798). Over the last several decades, many calls to curb the population explosion have been heard (e.g., Ehrlich, 1968, Ehrlich and Ehrlich, 2006, Meadows et al., 1972, O'Neill et al., 2010).

Many scholars have proposed models or theories regarding demographic dynamics (e.g., changes to birth and death rates; see Bilsborrow, 2002, Boserup, 2005) and their subsequent environmental consequences. For example, the vicious circle model (VCM; Brown, 2003, Marcoux, 1999) posits positive feedback loops among resource depletion, growing poverty, and high fertility. The widely cited IPAT model states that the environmental impacts (I) are the product of population (P), affluence (A), and technology (T) (Ehrlich and Holdren, 1972). Some scholars hold more optimistic views toward such acute human–environment crises in light of many mediating factors. Examples include agricultural intensification, technological advancement, institutional or cultural adaptation, and market substitution (Boserup, 1965, Simon, 1990).

There are many models and theories concerning population–environment relationships (e.g., Bilsborrow, 2002, Bilsborrow et al., 2001, Boserup, 1965, de Sherbinin et al., 2008, O'Neill et al., 2010). Many of these approaches rely on aggregated population measures (size, growth rate, fertility, etc.). Increasingly, other dimensions of demographics have been shown to deserve more attention. Examples include household numbers, demographic compositional measures (e.g., age and gender structure), and individual-level demographic decisions. For instance, Liu et al. (2003) found that in many parts of the world, even if population size declined, the number of households increased substantially. The resultant smaller households tended to have lower resource-use efficiency and posed serious threats to the environment. This thread of microlevel (primarily household) research in coupled human and natural systems (CHANS) is found in a large number of empirical studies (for a review on this topic, see de Sherbinin et al., 2007). Furthermore, it is considered essential to incorporate gender and age

Pandas and People. Edited by Jianguo Liu, Vanessa Hull, Wu Yang, Andrés Viña, Xiaodong Chen, Zhiyun Ouyang, and Hemin Zhang. Published 2016 by Oxford University Press.

differences between individual household members and understand intrahousehold processes when conducting coupled systems research (de Sherbinin et al., 2008).

Addressing the above demographic dimensions poses substantial challenges to traditional approaches for coupled systems research. Such approaches tend to focus on either human systems or natural systems (An, 2012) while holding the other as "exogenous" context or as background. Inadequate attention to the complex reciprocal relationships between human and natural systems is rooted in the division between natural and social sciences. Such a shortcoming has hindered understanding of complexity in these systems.

To address this divide, the framework of CHANS has been developed (Liu et al., 2007a, b). It integrates knowledge and data across disciplines and across spatial, temporal, and organizational scales (see An, 2012, An et al., 2014, Liu et al., 2015). The framework emphasizes a range of complex features. Examples include reciprocal effects and feedback loops, heterogeneity, non-linearity and thresholds, legacy effects and time lags, surprises, and resilience (Liu et al., 2007a, b). These features were identified in the six sites examined by Liu et al. (2007a). Others have found these same features in other sites such as the Amazon (Malanson et al., 2006), Chitwan National Park in Nepal (Carter et al., 2014; Chapter 16), southern Yucatán in Mexico (Manson, 2005), and Northern Ecuador (Walsh et al., 2008).

Characterizing these complex features requires systems modeling approaches that can represent their characteristic structures and processes. Agent-based modeling (ABM) is a useful tool for coupled systems research largely because ABMs represent agents (decision-making entities such as individuals and households) and their environment in a computer model where agents interact with one another and with the environment. Models are typically dynamic: they run over multiple time steps, enabling agents to develop and interact with other agents as well as the environment. This agent-environment platform endows ABMs with a strong capacity to integrate data across spatial, temporal, and organizational scales, and to capture the decision-making processes of individual agents (An, 2012). Thus an ABM is capable of predicting or explaining "emergent higher-level phenomena by tracking the actions of multiple low-level 'agents' that constitute, or at least impact, the system behaviors" (An et al., 2005). ABMs therefore offer great promise in CHANS research, as they can be parameterized with known information about individual agent behavior and interactions, and then general, emergent effects can be identified by studying outcomes. ABMs also show potential for scenario testing, as the effects of input rules and parameters can be evaluated. On the other hand, ABMs do not rule out the usefulness of traditional top-down approaches such as regression analysis, which actually may complement ABMs (e.g., through providing some decision rules for ABMs). In many instances, such traditional approaches even manifest greater effectiveness in capturing the corresponding human-nature relationships (An et al., 2005).

Many studies have focused on effects of demographic decisions on population size, but little attention has been paid to their cascading effects such as those on household number and the environment. In this chapter, we intend to fill this gap using interdisciplinary data from our model coupled system of Wolong Nature Reserve (Chapter 3). We set out to (1) illustrate demographic changes, (2) provide an overview of our ABM to understand detailed human–nature interactions, and (3) present simulation results using the ABM. Our simulations explore cascading effects of demographic decisions on complex features of coupled systems, including population size, household number, and panda habitat. In this chapter, we used the three-factor scheme approach (i.e., forest cover, elevation, and slope layers; see details in Chapter 7) to assess panda habitat.

8.2 Demographic dynamics

Demographic decisions, characterized by changes in fertility, marriage age, birth interval, migration, and a range of other factors (An et al., 2006, 2011, An and Liu, 2010) can translate into many demographic and environmental consequences. Examples include changes in population size, composition, number of households, household size, and panda habitat. In Wolong (as well as in many other parts of the world), fertility has declined in the past few decades. But given the large population base and relatively stable mortality rate, stable or even decreased fertility may still lead to considerable population increase (Figure 3.3).

The human population in Wolong increased from 3659 in 1982 to 4318 in 1997 and 4933 people in 2012 (Figure 3.3). Migration into Wolong is still strictly controlled due to Wolong's status as a nature reserve and via the *hukou* or household registration system implemented throughout China. In comparison, the number of households increased much faster than the population size over the same period. The number of households jumped from 631 in 1982 to 920 in 1996 to 1436 in 2012, representing an increase of 3.05% per year during 1982–1996 and 3.3% per year from 1996 to 2012 (Figure 3.3). Household size (number of people in a household) decreased from 5.8 in 1982 to 4.59 in 1997 to 3.44 in 2012 (Figure 8.1). There are several reasons for the rapid decrease in household size. First, young people prefer to leave parental homes at younger ages and establish their own homes. This tendency goes against the cultural tradition of having many generations under one roof. The choice to establish a new home is also made despite inconveniences associated with leaving the parental home, e.g., lack of child care and more household chores (Loucks et al., 2003). Second, the number of divorcees increased, while traditionally divorce was rare in Wolong (as in China

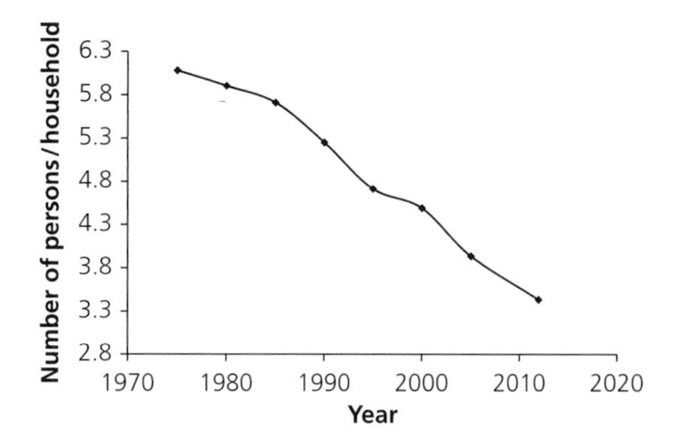

Figure 8.1 Dynamics of household size in Wolong Nature Reserve.

in general). Increasing divorces may lead to more households, even though the population size does not increase (Yu and Liu, 2007). Third, many people live longer than before due to improvements in living standards and medical care (see Figure 8.2). After a household member passes away, the number of people in the house decreases but the household still remains. Fourth, government policies also contribute to dwindling household size over time. For example, in 2001, government subsidies for the Natural Forest Conservation Program were distributed on a household basis. This arrangement provided a strong financial incentive for local households to split into smaller ones (Liu et al., 2005). Household dynamics in Wolong are similar to global patterns that also show faster

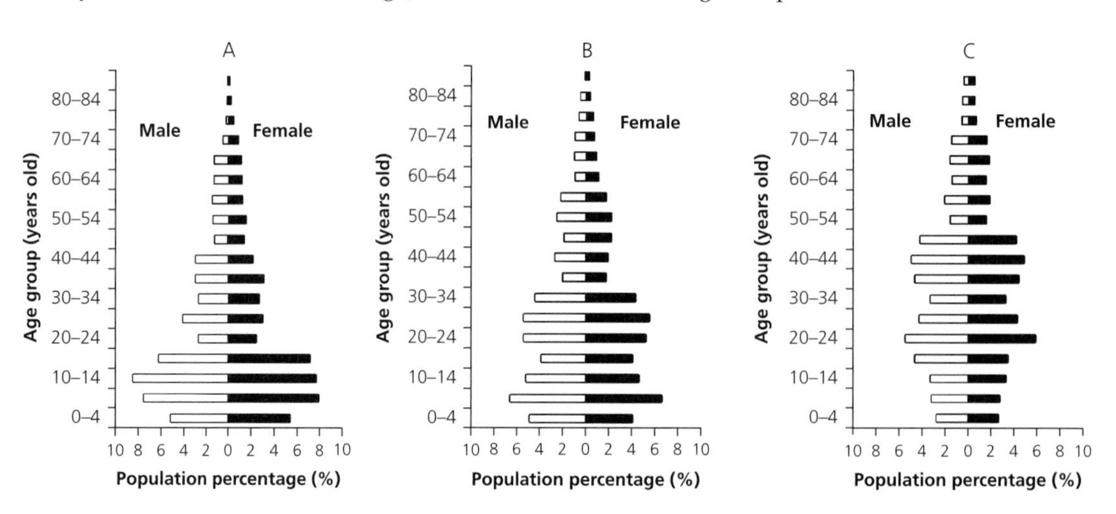

Figure 8.2 Population age/sex structure in (A) 1982, (B) 1996, and (C) 2006 in Wolong Nature Reserve.

increase in the number of households than population size, and continued decrease in household size (Bradbury et al., 2014, Liu et al., 2003).

The population structure in Wolong also changed dramatically between 1982 and 2006 (Figure 8.2). The large proportions of people in age groups 5–9, 10–14, and 15–19 as well as the decreased proportion of people in age group 0–4 (as compared to those in 1982) largely reflect the outcomes of the national family planning policy. This policy, introduced in 1978 and started in 1980, was initially known as a suggestion for "*wan* (later marriage), *xi* (prolonged time interval between births), and *shao* (fewer children)." The policy later developed into the stricter nationwide one-child policy. At the national level, China has witnessed curbed population growth due to its strict family planning policy. As a result, China has experienced drastic changes in its population composition. There has been an increase in the proportion of people of working age (15–64 years) and a decrease in the proportion of children (0–14 years) (Hussain, 2002). In Wolong, there was less control over family planning due to the large proportion of minority ethnic groups (e.g., Tibetan and Qiang). Minority ethnic groups enjoy special exemptions. Two to three children per couple were allowed for the 76% of people who belong to ethnic minorities (An et al., 2006, 2011).

The environmental implication of demographic dynamics could be substantial. The proportions and numbers of adults, particularly males, experienced steady increases in Wolong from 1996 to 2006, which expanded the available labor force that could be devoted to fuelwood collection and farming. According to a survey by Liu et al. in 1997 (Liu et al., 1999, 2005), there was a positive relationship between a local person's age and the number of days he or she spent collecting fuelwood. This relationship was non-linear with its peak in the age group of 25–59, followed by a sharp decrease after age 60.

To understand complex long-term human–nature interactions such as effects of demographic decisions, we developed multiple systems modeling approaches. In Section 8.3 we provide an overview of an ABM to characterize human–nature interactions in Wolong. Then, in Section 8.4, we use our model to simulate how demographic decisions affect long-term dynamics of population size, household number, and the environment (panda habitat in our context).

8.3 Agent-based systems model

The conceptual framework of our ABM "Integrative Model for Simulating Household and Ecosystem Dynamics" (IMSHED) is illustrated in Figure 8.3 (An et al., 2005). The left panel (T1) represents a symbolic snapshot of the households (represented by two households), the household members inside, and the forests on the landscape. The right panel (T2) represents all the households and forests at a later time. During the time between these two snapshots, many human and environmental processes take place. All local people (Person agents) go through individual life-history events (e.g., childbearing, in- or out-migration, or death). Person agents are heterogeneous in many dimensions such as demographic features (e.g., age or education), household locations, and fuelwood demand. For instance, Person A (female, the parent of B, C, D, and E) at the upper household dies. Persons C and D (both male, the children of A) move out of the household during this time interval for various reasons. Examples include going to college (i.e., education out-migration) or out-migrating to a spouse's home outside the reserve (i.e., marriage out-migration). Person E (female, the child of A) marries Person F (male, the child of J) at the lower household. They establish their own household. Person B (male, the child of A) remains in the parent's household. He marries a woman (Person K) who moves into the reserve through marriage (i.e., marriage in-migration). From T1 to T2, the lower household experiences structural changes but the household size remains stable. One Person (F) moves out as described above, and one Person (a baby) is born into the household.

The primary data sources for parameterizing and testing IMSHED are the 1996 Wolong agricultural census (Wolong Administration, 1996), the 2000 population census (Wolong Administration, 2000), and in-person surveys of 220 households that were conducted in 1999 (An et al., 2002, 2005). The forests (F1–F3, Figure 8.3) are modeled based on geographic information systems, remote sensing data, and literature (at two spatial resolutions of 90 m

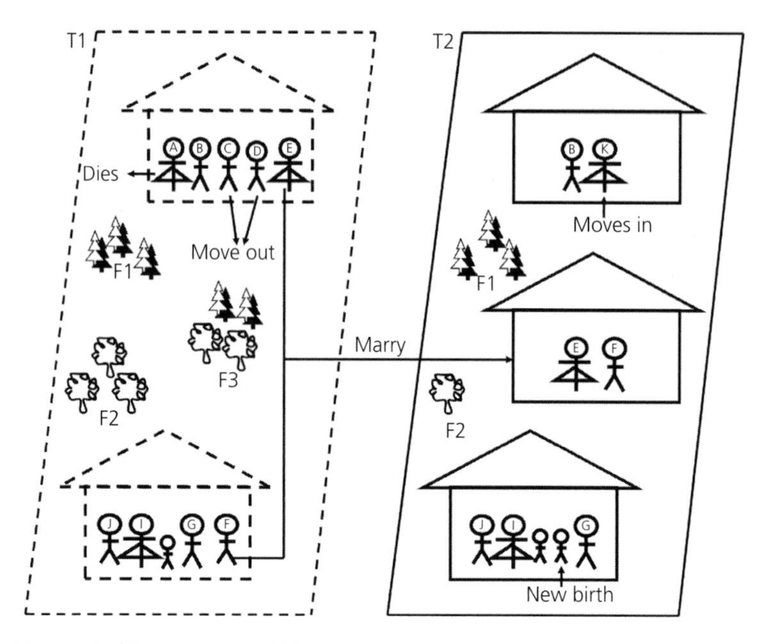

Figure 8.3 Conceptual framework of the Integrative Model for Simulating Household and Ecosystem Dynamics (IMSHED). Reprinted with permission from Springer Science and Business Media (originally printed in An and Liu 2010).

and 360 m). Land cover (vegetation type, volume, age, and growth rate), topography, and households are spatially explicitly represented in the ABM (An et al., 2005).

The ABM can simulate human–nature interactions over time. For example, the model simulates how individual decisions (e.g., about the age of marriage) made at a local site and a specific time affect dynamics of population size, household number, and panda habitat. The model captures the heterogeneous features of the individual persons (e.g., age and sex), households (e.g., location), and their local environment (e.g., available fuelwood collection sites). The impacts of these heterogeneous features are reflected in when (depending on the person's age and sex) and where (depending on the original household) a new household will be established. Once the new household is established, the model calculates the household's fuelwood demand in relation to the distance from the nearest fuelwood site according to the regression results from An et al. (2002).

After robust model verification and validation (An et al., 2005), we conducted a set of experiments concerning the effects of demographic decisions (An et al., 2006, 2011, An and Liu, 2010). For example, the model has been used to simulate fuelwood collection, a major interaction between the human and natural systems in Wolong. Effects include deforestation (e.g., F3, Figure 8.3) or volume reduction (e.g., F2, Figure 8.3) and habitat loss or degradation. Agents visit available forests within 1080 m, between 1080 and 2160 m, and over 2160 m from their households to collect fuelwood. There, they participate in fuelwood collection at observed probabilities of 0.48, 0.27, and 0.25, respectively (An et al., 2005). Using terrain data, all Person agents who collect fuelwood choose the path that has the least cost-distance between a household and its nearby forests so as to minimize energy consumption. All such agents choose the available forests with shortest cost-distance for fuelwood collection. Household agents may reduce their fuelwood demand as the remaining forests become less available and farther away (An et al., 2005). In this manner, the magnitude and location of household environmental impact due to fuelwood use may be modeled.

8.4 Effects of demographic decisions

In this section, we highlight several features of coupled systems that result from demographic decisions through running the ABM described above. Multiple relationships or feedback loops in coupled systems may sometimes strengthen or weaken (or even cancel out) one another. Due to complexities such as feedbacks, the system may feature emergent outcomes that cannot be explained or predicted by exploring individual system components alone.

8.4.1 Non-linearity and thresholds

Non-linearities are prevalent in Wolong, such as the non-linearity between population size and age at first marriage (MarryAge in Figure 8.4; An and Liu, 2010). When people postpone their marriages (MarryAge changes from 18 to 26, to 34, and finally to 38 years old), population sizes drop and show non-linear changes over time. The difference in population size over time becomes increasingly large. When people get married at 26 or later, population sizes start to decline at some time point (e.g., year 27 when MarryAge = 38; Figure 8.4A). This population inflection is a threshold (transition point between alternate states; Liu et al., 2007a).

The number of households also manifests complex non-linear relationships (Figure 8.4B). Even when population sizes decline, the numbers of households still increase. Also, the two curves for MarryAge = 34 and 38 resemble each other and display an "S"-shaped increase. This pattern shows that over the long term, the numbers of households would almost converge at the end of 50-year simulations despite the differences in age at first

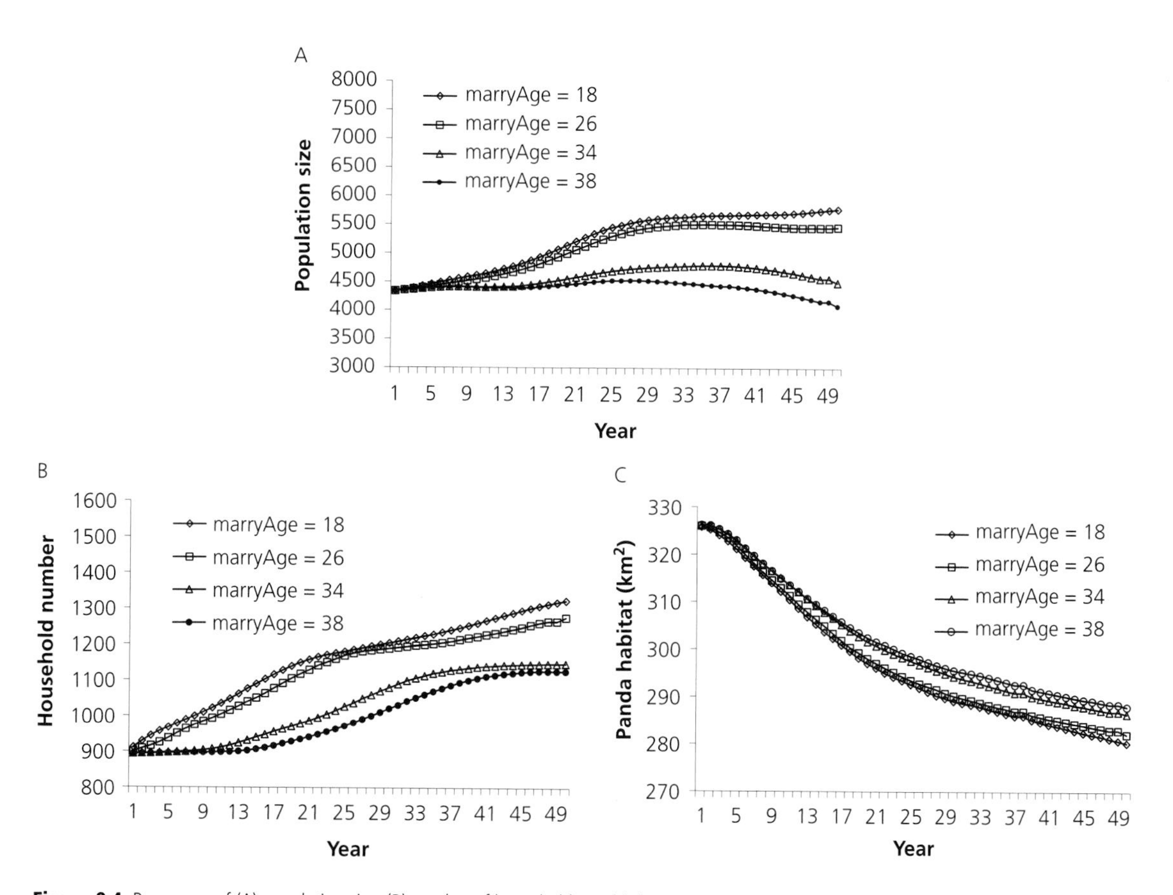

Figure 8.4 Responses of (A) population size, (B) number of households, and (C) amount of panda habitat to changes in age of first marriage (MarryAge). Reprinted with permission from Springer Science and Business Media (originally printed in An and Liu 2010).

marriage. Once the young people of this generation reach 34, they will begin to marry and most likely set up their new households and, therefore, the number of households will go up. However before their children (the next generation) grow up, the number of households will increase at a decreasing rate and finally level off. By that time, most people of this generation will have married and already established their households.

8.4.2 Time lags

Time lags commonly occur when key processes like population size, number of households, and panda habitat respond to changes in demographic decisions, e.g., age at first marriage, fertility, and the maximum age for childbearing (An and Liu, 2010). Consider the time interval between marriage and the first birth (BirthInterval in Figure 8.5) as an example. When this interval increases from one year to 11 years, population size responds quickly in about 3–5 years. But it takes nearly 20–25 years to see the corresponding change in the number of households, and 40–50 years for panda habitat to respond (Figure 8.5). Clearly when this interval is prolonged, children will be born later, which explains the nearly immediate response in population size. However only at the time that these birth-postponed babies grow up and establish their own households can the change in the number of households be observed. Panda habitat is the slowest to respond for many reasons, for example, forest volume may assimilate the increased demand for fuelwood associated with the incoming "new" households.

Another example is related to the time lags between the number of households and panda habitat (Figure 8.6). When fertility (numOfKids in Figure 8.6) changes from 1 to 6, the number of households and amount of panda habitat take about 17 years and 27 years to respond, respectively. Why does this average ten-year difference (over multiple simulation runs) in response time

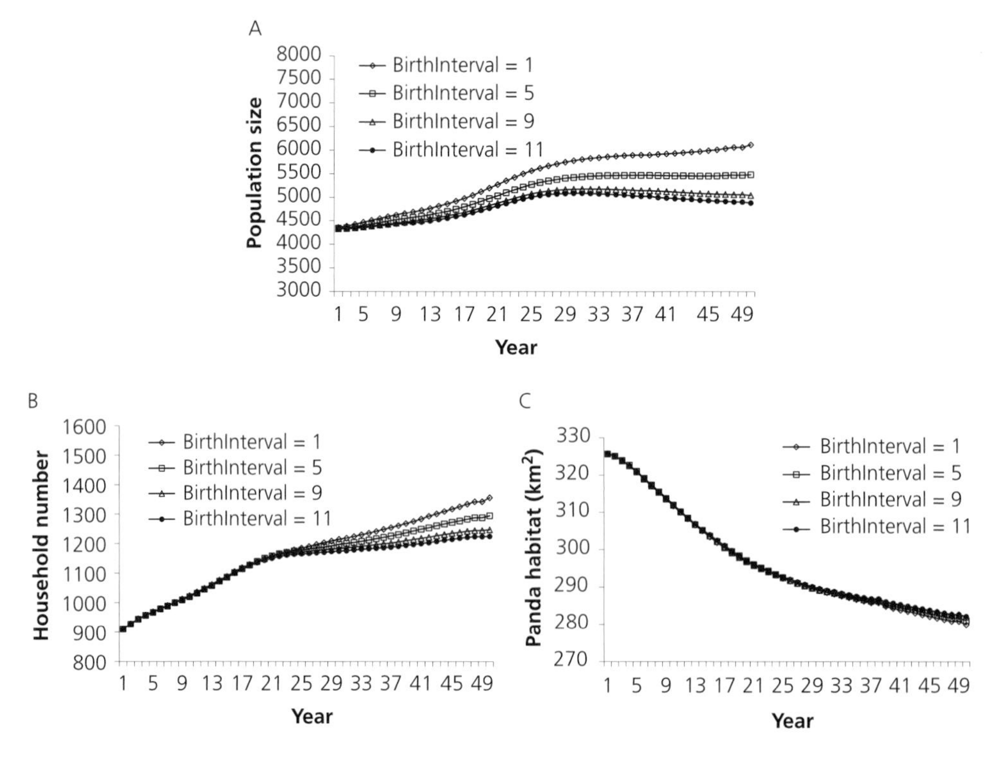

Figure 8.5 Responses of (A) population size, (B) number of households, and (C) amount of panda habitat to changes in the interval between marriage and first birth (BirthInterval). Reprinted with permission from Springer Science and Business Media (originally printed in An and Liu 2010).

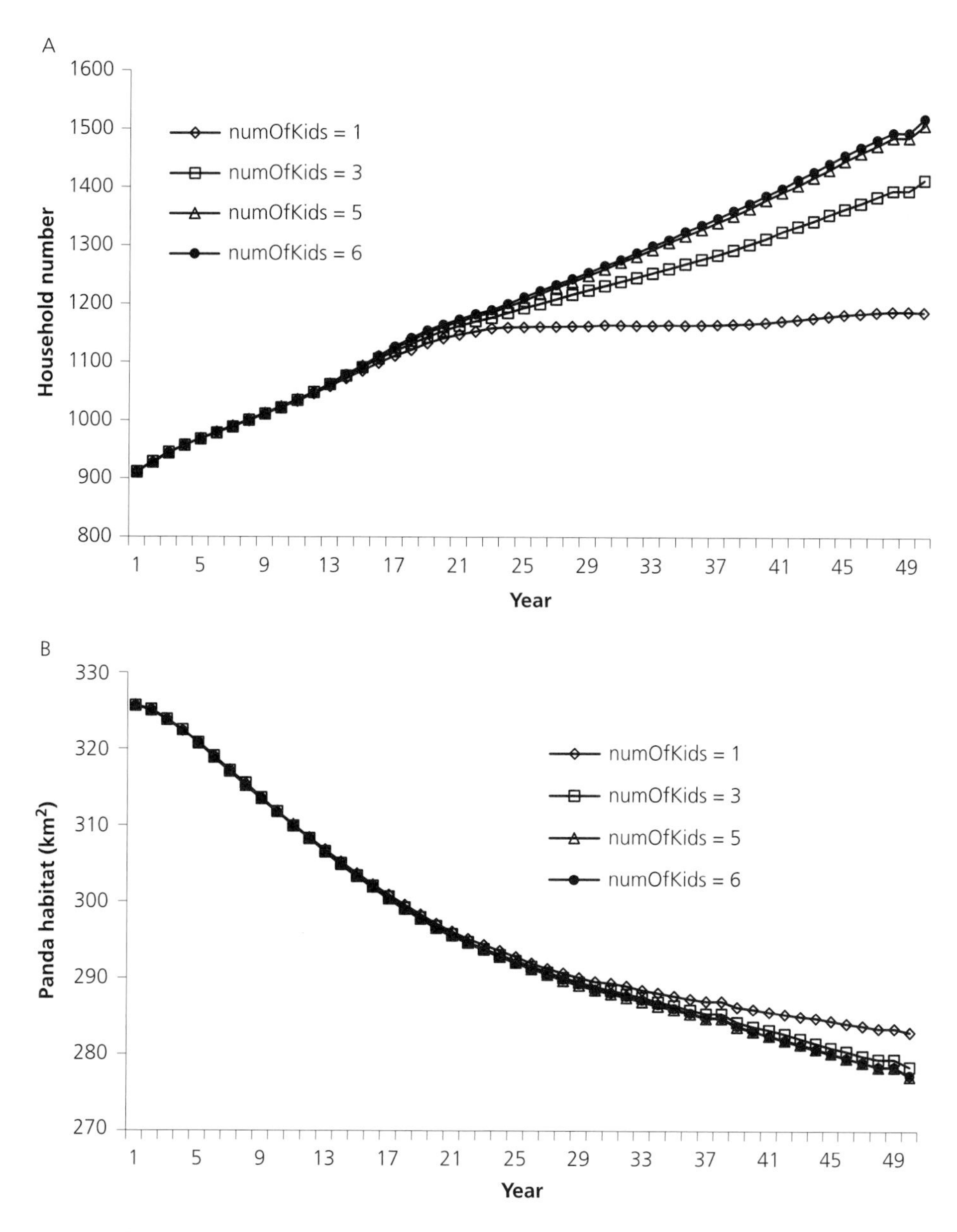

Figure 8.6 Responses of (A) number of households and (B) amount of panda habitat to changes in fertility (numOfKids). Reprinted with permission from Springer Science and Business Media (originally printed in An and Liu 2010).

occur? The increase in fuelwood demand that accounts for habitat loss is related more to the increase in number of households than to the number of people, and so human impact on habitat is most pronounced when young adults form their own households.

8.4.3 Resilience

Resilience is the ability of a system to maintain similar structure and functioning after disturbances (Folke et al., 2002, Liu et al., 2007a). It is prevalent in Wolong. For example, out-migration can be considered a disturbance to the coupled system. Wolong parents often take pride in their children's out-migration through education. They encourage their children to go to college and find permanent jobs outside Wolong, even though they are not personally willing to emigrate due to many reasons, such as lack of skills for working and living in cities (Liu et al., 2005). To understand the impact of this form of out-migration, we set the education out-migration rates (Edu Mig Rate in Figure 8.7; the probability that a person between 16 and 20 years old would go to college or technical school and migrate out) from 0.02 (baseline) to 0.05, 0.15, and 0.30. Surprisingly, this "good" disturbance does not cause an increase in the amount of panda habitat until 10–15 years later. In other words, the amount of panda habitat is resilient to out-migration for the first 10–15 years and only starts to respond after a time lag (Figure 8.7). The reason may be that when these young people move out, their original households still remain in Wolong and continue to use fuelwood, although a slightly lower amount. Only at the marriage of these young out-migrants can we see a substantial reduction in the number of households because out-migrants do not establish new households inside the reserve.

8.4.4 Heterogeneity

Coupled systems are characterized by multiple heterogeneous features. Under Wolong's specific spatial distribution of existing forests and households, demographic changes will not cause evenly distributed degradation or restoration among all forest or habitat types. We sought to analyze the

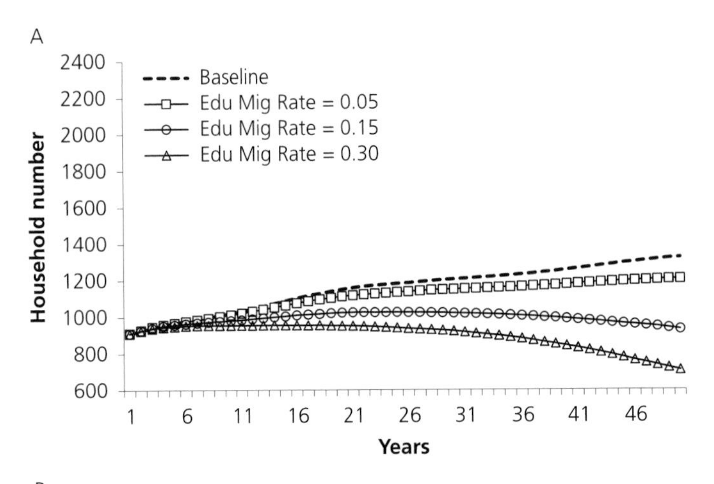

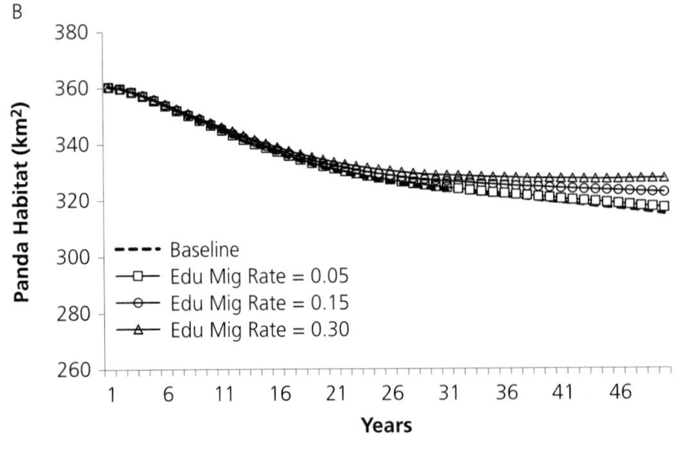

Figure 8.7 Responses of (A) number of households and (B) amount of panda habitat to changes in education out-migration rate (Edu Mig Rate).

effects of demographic expansion and contraction (due to factors that encourage or discourage population growth, respectively; see An and Liu, 2010). To do so, we mapped how panda habitat may respond to different policies in a spatially explicit manner. The simulation results show that, in year 50 (the end of the simulation), most of the saved habitat due to the demographic contraction policy is located in areas with coniferous forests. On the other hand, most of the lost habitat due to the demographic expansion is in areas with deciduous forests. These emergent phenomena may arise from the fact that in the demographic contraction scenario, people reduce or cease to collect fuelwood from places dominated by coniferous forests. Many such regions are at higher elevations and far from residential areas. In the demographic expansion scenario, these new households are often located in areas near their parental households. They collect fuelwood first in deciduous forests, which are at lower elevations and near residential areas. By explicitly modeling the spatial locations of new households and the least-cost fuelwood harvest sites, important differences in scenario outcomes can be identified.

8.5 Discussion and concluding remarks

CHANS researchers have paid much attention to data collection and analysis at the household level but many lack understanding of "intrahousehold processes" at the household member (individual) level (de Sherbinin et al., 2007). This chapter introduces an ABM that simulates individual-level demographic decisions and environmental interactions. The model can translate these dynamic household processes to explore emergent outcomes in populations and the environment. It has generated unique insights that could not be otherwise obtained. For example, we can see that the numbers of households evolve differently compared with population size. Population size already starts to decline or become stable while the number of households still keeps increasing (Figure 8.5). This difference can be attributed to reduced household size. Household sizes become smaller due to reasons such as reduction in multigenerational households, increase in divorce, increase in longevity, and division of households to receive more conservation subsidies. These and other simulation results (e.g., Figures 8.4 through 8.7) can be better understood from utilizing knowledge about individual decisions (such as fertility, time of marriage) and the associated driving forces. Demographic decisions play an essential role in affecting dynamics of coupled systems.

The ABM becomes more useful when there are a large number of links among the many demographic decisions, socioeconomic factors, and the environment (see also Chapter 14). For instance, the forest is affected by location-specific fuelwood demand. This demand is in turn a function of household age structure (consequence of earlier demographic decisions), distance from household to collection sites, and other factors (An et al., 2005). In the future, using the above socioeconomic, demographic, and environmental covariates to predict these demographic decisions (An et al., 2014, Zvoleff and An, 2014) could help researchers better understand complex relationships in coupled systems.

Coupled systems possess many complex features that deserve more attention, including those discussed in Section 8.4. Our research shows that the VCM (Brown, 2003, Marcoux, 1999) is not necessarily applicable in all instances. We provide an example in a small community surrounded by dense forests with high volume and fast forest growth. Here, increases in population size and number of households may not cause serious environmental degradation if the forest volume and regrowth surpass the demand from the community and elsewhere.

Spatial and temporal scales adopted in research affect what we observe and conclude. For instance, an increase in fertility is projected to cause observable changes in panda habitat after 30 years (Figure 8.6). If we only had data for a shorter period (e.g., 10–20 years), the conclusion would be that increasing fertility has no impact on panda habitat. However, this would be a misleading conclusion due to the time lag of impacts resulting from changes in fertility.

Our research in this chapter also has management and policy implications for promoting sustainability of coupled systems. For example, as shown in Section 8.4, different demographic decisions may cause major habitat change in either deciduous or coniferous forests. Coniferous forests, located in higher elevations compared to deciduous forests, are closer to the upper elevational boundary of panda habitat. If resources are limited, less priority can be placed on conserving habitat areas of coniferous forests due to their longer distances from residential areas and relative inaccessibility for local residents. Also worthy of mention is that individual-level demographic decisions, especially those less researched, such as marriage timing and birth interval, deserve more attention as they have long-term significant impacts on population and environmental dynamics. In addition, the complexity features in coupled systems may warrant data collection and analysis over long time frames (to account for time lags and feedback loops) and large geographic extents (to include spatial heterogeneity). Empirical research over several decades or larger spatial extents would in most instances be difficult, but modeling would be feasible.

8.6 Summary

Human populations in many places are undergoing demographic transitions involving complex shifts in marriage timing, fertility, mortality, and migration. But the interactions between individual characteristics (e.g., age and demographic decisions such as marriage) and the natural environment are not well understood. To fill this knowledge gap, we used an agent-based model to simulate interactions between demographic properties of individual households and environmental change in Wolong Nature Reserve in China, which has undergone profound shifts in demographics. Demographic characteristics such as time interval between marriage and age at first birth were shown to affect population size, household number, and fuelwood collection behaviors, which in turn impacted forests and panda habitat. Complex patterns arose from these demographic parameters such as non-linearities and thresholds, feedbacks, legacy effects, time lags, heterogeneity, and resilience. For example, when the time interval between marriage and first birth increased from one year to 11 years, population size responded quickly in about 3–5 years. But it took nearly 40–50 years for panda habitat to show less degradation with the increase in time interval between marriage and first birth. An example of heterogeneity can be seen in our simulation results showing improvements in coniferous forests far from households 50 years after population reduction. In contrast, declines were seen in deciduous forests closer to households after population expansion. Our work demonstrates the importance of considering and modeling individual-level differences and demographic decisions in coupled human and natural systems over long-term time frames.

References

An, L. (2012) Modeling human decisions in coupled human and natural systems: review of agent-based models. *Ecological Modelling*, **229**, 25–36.

An, L., He, G., Liang, Z., and Liu, J. (2006) Impacts of demographic and socioeconomic factors on spatio-temporal dynamics of panda habitat. *Biodiversity and Conservation*, **15**, 2343–63.

An, L., Linderman, M., He, G., et al. (2011) Long-term ecological effects of demographic and socioeconomic factors in Wolong Nature Reserve (China). In R.P. Cincotta and L.J. Gorenflo, eds, *Human Population: Its Influences on Biological Diversity*, pp. 179–95. Springer-Verlag, Berlin, Germany.

An, L., Linderman, M., Qi, J., et al. (2005) Exploring complexity in a human-environment system: an agent-based spatial model for multidisciplinary and multiscale integration. *Annals of the Association of American Geographers*, **95**, 54–79.

An, L. and Liu, J. (2010) Long-term effects of family planning and other determinants of fertility on population and environment: agent-based modeling evidence from Wolong Nature Reserve, China. *Population and Environment*, **31**, 427–59.

An, L., Lupi, F., Liu, J., et al. (2002) Modeling the choice to switch from fuelwood to electricity: implications for giant panda habitat conservation. *Ecological Economics*, **42**, 445–57.

An, L., Zvoleff, A., Liu, J., and Axinn, W. (2014) Agent-based modeling in coupled human and natural systems (CHANS): lessons from a comparative analysis. *Annals of the Association of American Geographers*, **104**, 723–45.

Bilsborrow, R.E. (2002) Migration, population change, and the rural environment. *Environmental Change and Security Report (ECSP)*. http://wilsoncenter.org/sites/default/files/Report_8_BIlsborrow_article.pdf.

Bilsborrow, R.E., Carr, D.L., Lee, D., and Barrett, C. (2001) Population, agricultural land use and the environment in developing countries. In D.R. Lee, and C.B. Barrett, eds, *Tradeoffs and Synergies: Agricultural Intensification, Economic Development and the Environment*, pp. 35–55. Commonwealth Agricultural Bureau International, London, UK.

Boserup, E. (1965) *The Conditions of Agricultural Growth*. Earthscan, London, UK.

Boserup, E. (2005) *The Conditions of Agricultural Growth: The Economics of Agrarian Change under Population Pressure*. Transaction Publishers, New Brunswick, NJ.

Bradbury, M., Peterson, M.N., and Liu, J. (2014) Long-term dynamics of household size and their environmental implications. *Population and Environment*, **36**, 73–84.

Brown, A.D. (2003) *Feed or Feedback: Agriculture, Population Dynamics and the State of the Planet*. International Books, Tuross Head, Australia.

Carter, N.H., Viña, A., Hull, V., et al. (2014) Coupled human and natural systems approach to wildlife research and conservation. *Ecology and Society*, **19**, 43.

Cohen, J.E. (2003) Human population: the next half century. *Science*, **302**, 1172–75.

Curran, L.M., Trigg, S.N., McDonald, A.K., et al. (2004) Lowland forest loss in protected areas of Indonesian Borneo. *Science*, **303**, 1000–1003.

de Sherbinin, A., Carr, D., Cassels, S., and Jiang, L. (2007) Population and environment. *Annual Review of Environment and Resources*, **32**, 345.

de Sherbinin, A., Vanwey, L.K., McSweeney, K., et al. (2008) Rural household demographics, livelihoods and the environment. *Global Environmental Change*, **18**, 38–53.

Ehrlich, P. (1968) *The Population Bomb*. Ballantine Books, New York.

Ehrlich, P. and Ehrlich, A. (2006) Enough already. *New Scientist*, **191**, 46–50.

Ehrlich, P.R. and Holdren, J.P. (1972) A bulletin dialogue on the "closing circle": Critique: One-dimensional ecology. *Bulletin of the Atomic Scientists*, **28**, 16–27.

Folke, C., Carpenter, S., Elmqvist, T., et al. (2002) Resilience and sustainable development: building adaptive capacity in a world of transformations. *Ambio*, **31**, 437–40.

Hussain, A. (2002) Demographic transition in China and its implications. *World Development*, **30**, 1823–34.

Liu, J., An, L., Batie, S.S., et al. (2005) Beyond population size: examining intricate interactions among population structure, land use, and environment in Wolong Nature Reserve, China. In B. Entwisle and P.C. Stern, eds, *Population, Land Use, and Environment: Research Directions*, pp. 217–37. National Academies Press, Washington, D.C.

Liu, J., Daily, G.C., Ehrlich, P.R., and Luck, G.W. (2003) Effects of household dynamics on resource consumption and biodiversity. *Nature*, **421**, 530–33.

Liu, J., Dietz, T., Carpenter, S.R., et al. (2007a) Complexity of coupled human and natural systems. *Science*, **317**, 1513–16.

Liu, J., Dietz, T., Carpenter, S.R., et al. (2007b) Coupled human and natural systems. *Ambio*, **36**, 639–49.

Liu, J., Linderman, M., Ouyang, Z., et al. (2001) Ecological degradation in protected areas: the case of Wolong Nature Reserve for giant pandas. *Science*, **292**, 98–101.

Liu, J., Mooney, H., Hull, V., et al. (2015) Systems integration for global sustainability. *Science*, **347**, 1258832.

Liu, J., Ouyang, Z., Tan, Y., et al. (1999) Changes in human population structure: implications for biodiversity conservation. *Population and Environment*, **21**, 45–58.

Loucks, C.J., Lü, Z., Dinerstein, E., et al. (2003) The giant pandas of the Qinling Mountains, China: a case study in designing conservation landscapes for elevational migrants. *Conservation Biology*, **17**, 558–65.

Malanson, G.P., Zeng, Y., and Walsh, S.J. (2006) Complexity at advancing ecotones and frontiers. *Environment and Planning A*, **38**, 619.

Malthus, T.R. (1798) *An Essay on the Principle of Population*. University of Michigan Press, Ann Arbor, MI.

Manson, S.M. (2005) Agent-based modeling and genetic programming for modeling land change in the Southern Yucatan Peninsular Region of Mexico. *Agriculture, Ecosystems & Environment*, **111**, 47–62.

Marcoux, A. (1999) *Population and Environmental Change: From Linkages to Policy Issues*. Sustainable Development Department (SD), FAO, Rome, Italy.

Meadows, D.H., Meadows, D.L., Randers, J., and Behrens, W.W. (1972) *The Limits to Growth*. Universe Books, New York, NY.

O'Neill, B.C., Dalton, M., Fuchs, R., et al. (2010) Global demographic trends and future carbon emissions. *Proceedings of the National Academy of Sciences of the United States of America*, **107**, 17521–26.

Population Reference Bureau (2014). *2014 World Population Data Sheet*. http://www.prb.org/pdf14/2014-world-population-data-sheet_eng.pdf.

Simon, J.L. (1990) *Population Matters: People, Resources, Environment, and Immigration*. Transaction Publishers, New Brunswick, NJ.

The World Bank (2015) *Population Growth (Annual%)*. http://data.worldbank.org/indicator/SP.POP.GROW.

U.S. Census Bureau (2009) *U.S. and World Population Clocks—POPClocks*. http://www.census.gov/ipc/www/popclockworld.html.

Vitousek, P.M. (1994) Beyond global warming: ecology and global change. *Ecology*, **75**, 1861–76.

Walsh, S.J., Messina, J.P., Mena, C.F., et al. (2008) Complexity theory, spatial simulation models, and land use dynamics in the Northern Ecuadorian Amazon. *Geoforum*, **39**, 867–78.

Wolong Administration (2000) *Nationwide Population Census Data*. Wolong Nature Reserve Administration, Wolong, Sichuan, China (in Chinese).

Wolong Administration (1996) *Wolong Agricultural Census*. Wolong Nature Reserve Administration, Wolong Nature Reserve, Sichuan, China (in Chinese).

Yu, E. and Liu, J. (2007) Environmental impacts of divorce. *Proceedings of the National Academy of Sciences of the United States of America*, **104**, 20629–34.

Zvoleff, A. and An, L. (2014) The effect of reciprocal connections between demographic decision making and land use on decadal dynamics of population and land-use change. *Ecology and Society*, **19**, 31.

CHAPTER 9

Dynamics of Economic Transformation

Wu Yang, Frank Lupi, Thomas Dietz, and Jianguo Liu

9.1 Introduction

Many places in China and other countries have experienced dramatic economic transformations (i.e., an abrupt or dramatic change in scale and/or structure of the economy) during the past few decades. Such economic transformations share some major drivers such as international and domestic trade, and conservation and development policies (DeFries et al., 2010, Liu et al., 2008, Scrieciu, 2007). They also led to both environmental and socioeconomic impacts at multiple scales (Bartolini and Bonatti, 2002, Liu and Diamond, 2005, Liu et al., 2008, 2013).

In many places, such as the Amazon basin, economic development has been achieved by converting forests into croplands. This transformation has led to massive losses of biodiversity and depletion or decline of ecosystem services (e.g., carbon sequestration, flood control, and soil retention). Similar cases are pervasive across the world, so that two-thirds of the Earth's ecosystems and provision of ecosystem services were negatively affected from the 1950s to the 2000s (Millennium Ecosystem Assessment, 2005). In some other places, however, policy interventions (e.g., payments for ecosystem services programs) have transformed local or regional economies (Yang et al., 2013a). Such measures have also safeguarded important natural capital to sustain ecosystem service provision. To achieve environmental and socioeconomic sustainability, it is crucial to understand the dynamics, causes, and impacts of such economic transformations.

While there are scattered pieces of evidence, systematic analyses of the dynamics and impacts of economic transformations are challenging. Barriers include the lack of long-term and multidisciplinary data and lack of consideration of interactions among important system components such as policies (Liu et al., 2013, Yang et al., 2013b). In this chapter, by taking advantage of our long-term research at our model coupled human and natural system of Wolong, we synthesize various quantitative and qualitative analyses to provide a picture of the dynamics, associated factors, and effects of economic transformation.

9.2 Economic transformation at Wolong Nature Reserve

Before the early 2000s, Wolong could be characterized as a subsistence-based agricultural economy. Most local people were farmers. They made a living primarily through subsistence activities such as crop cultivation (corn, potato, radish, and other vegetables), livestock husbandry (chickens, pigs, cattle, yaks, goats, and horses), traditional Chinese herbal plant collection, bee keeping, and timber and fuelwood collection (Liu et al., 1999, Yang et al., 2013c). There were occasionally some non-agricultural income sources through road construction, house building, transportation, and sporadic tourism businesses (Liu et al., 2012, Yang et al., 2013c). Due to the lack of alternative income sources, some local people resorted to poaching and illegal logging (Schaller et al., 1985, Wolong Nature Reserve, 2005).

An economic transformation began in the early 2000s, when the Wolong Administrative Bureau started to implement two national conservation policies (Natural Forest Conservation Program (NFCP)

Pandas and People. Edited by Jianguo Liu, Vanessa Hull, Wu Yang, Andrés Viña, Xiaodong Chen, Zhiyun Ouyang, and Hemin Zhang. Published 2016 by Oxford University Press.

and the Grain to Green Program (GTGP)). NFCP is designed to protect and restore natural forests via logging bans, afforestation measures, and forest monitoring (Liu et al., 2008, 2013). At Wolong, a payments for ecosystem services approach was adopted to motivate local people's participation and compliance (Yang et al., 2013a, b). GTGP is designed to convert cropland on steep slopes to forest or grassland. Mimicking GTGP, a local program called Grain to Bamboo Program (GTBP) was also implemented, which pays residents to convert croplands to bamboo plantations (Yang et al., 2013a). Considering the similarities of GTGP and GTBP in terms of aim and implementation method, here we regarded them as one integrated cropland conversion policy—Grain to Green/Bamboo Program (GTGB). It was also the time when large-scale tourism development was launched with the official approval of Wolong's first ecotourism development master plan by the central and provincial governments (Liu et al., 2012). To stimulate tourism development, the main road stretching across Wolong and connecting it to outside areas was upgraded and widened to a provincial road (or highway) in 2006. Subsequently many tourism facilities (a four-star hotel, private inns and restaurants, Giant Panda Museum, Panda Valley, and other scenic sites) were constructed (Liu et al., 2012, Wolong Nature Reserve, 2005). The number of tourists increased from approximately 20 000 in 1996 to a peak of 235 500 in 2006 (Liu et al., 2012). During the same period, overall tourism receipts increased forty-fold from approximately 1 million yuan to 42.4 million yuan. However, the tourism industry had a severe financial leakage problem with more than 96% of tourism-generated revenue being retained by the government and the tourism development company instead of going to local residents (He et al., 2009, Liu et al., 2012).

Long-term household survey data that we collected in Wolong effectively captures this pattern of economic transformation (Figure 9.1). Before the early 2000s, agricultural income was the major income source. However, the pattern started to change in the early 2000s with the amount of non-agricultural income exceeding agricultural income around 2001 (Figure 9.1). As is characteristic of most income distributions, median incomes are much smaller than average incomes (Figure 9.1). However, the temporal patterns for income sources were consistent whether we considered the net average income or net median income. Non-agricultural income was the primary driver of changes in total household income, the mean value of which increased from 6543 yuan in 1998 to 21 043 yuan in 2007. Agricultural income was relatively stable and total income steadily increased until the occurrence of the 2008 Wenchuan earthquake. After the earthquake, the mean values of total household income and agricultural income dropped sharply by 38.7% and 82.1%, respectively, in comparison to those in 2007. From 2008 to 2009, the mean and median values of total household income increased by 205% and 37.8%, respectively. Over 89.7% and almost 100% of this income was non-agricultural income in 2008 and 2009, respectively (Figure 9.1). But this was primarily because of temporary work opportunities triggered by post-earthquake reconstruction. It did not mean that the local economy rebounded quickly or the local households adapted rapidly after the earthquake. As soon as the post-earthquake reconstruction ends, both the local economy and households will face another transformation if they are to achieve long-term economic sustainability (see detailed discussion in Chapter 12).

9.3 Factors associated with the economic transformation

In this section, we provide quantitative evidence to help explain the economic transformation occurring in Wolong from 1998 to 2007. A number of factors may be associated with economic changes at the household level. The two main factors that we focus on here are macrosocioeconomic conditions (e.g., national and local conservation and development policies) and household-level characteristics (e.g., access to capital). With respect to the first factor, we focused on the four most important policies that potentially had large impacts on households' economic conditions. These are NFCP, GTGB, Electricity Subsidy Program (ESP), and Tourism Development Program (TDP). The first two are conservation policies and the second two are development policies. We briefly summarize the policies according to their implementation

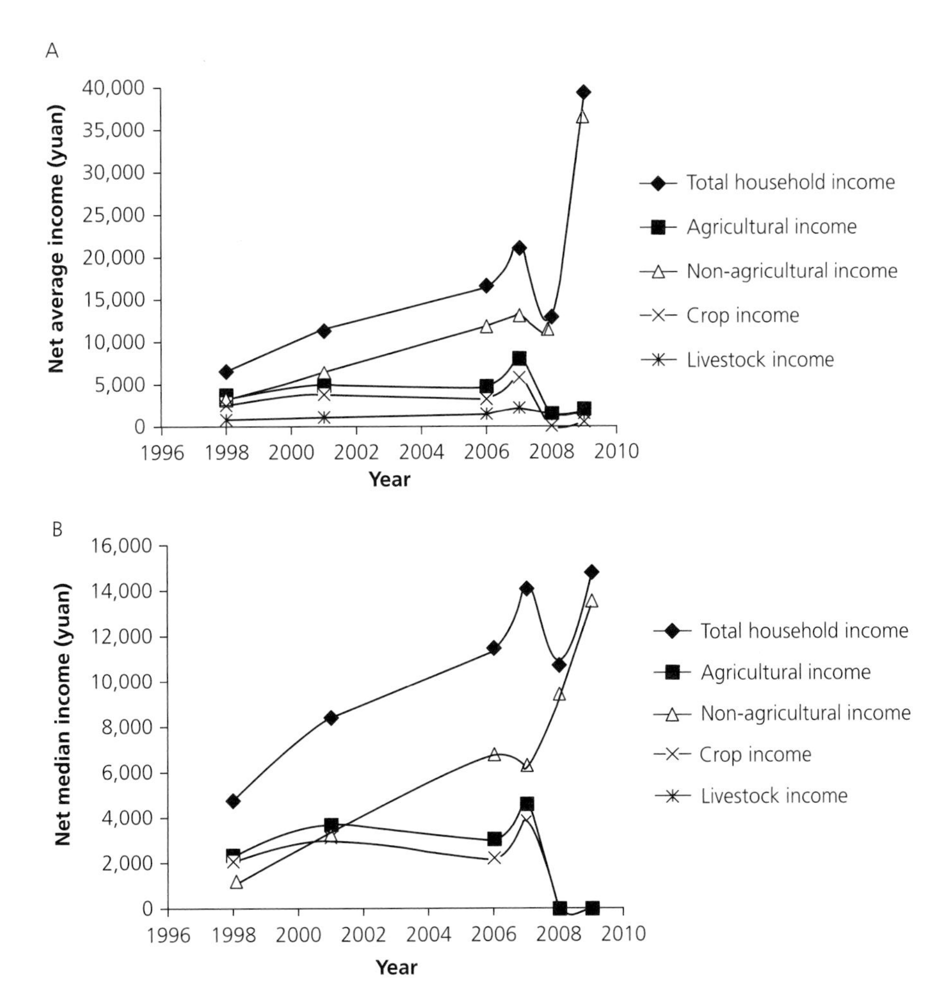

Figure 9.1 Dynamics of (A) net average income and (B) net median income based on household survey data in Wolong Nature Reserve. The number of observations was approximately 220, with small variations across years due to loss of some households and logistical difficulties of surveying certain households in some years. All monetary values are adjusted for inflation using the consumer price index with 2000 as the base year.

area, implementation period, and payments or subsidies to households in Table 9.1. With respect to the second factor, household-level characteristics, we were interested in examining households' access to different forms of capital. Access to capital may reflect heterogeneities in local households' environmental and socioeconomic behaviors (e.g., policy participation and compliance) and outcomes (e.g., forest-cover change and household income; Liu et al., 2012, Rustagi et al., 2010, Yang et al., 2013b).

We conducted regression analyses to examine relationships between income growth and these policy and capital factors (summarized in Table 9.2). The general function of the regression models is given in Equation 9.1:

$$Y = \beta_0 + P_1\beta_1 + P_2\beta_2 + C_f\beta_f + C_h\beta_h + C_n\beta_n + C_m\beta_m + C_s\beta_s + C_c\beta_c + \varepsilon \quad \text{(Equation 9.1)}$$

where Y is a vector of changes in total household income or changes in the percentage of agricultural income from 1998 to 2007; P_1 is a vector of variables indicating different policy interventions; P_2 is a vector of interaction terms of different policy intervention variables; C_f, C_h, C_n, C_m, and C_s are vectors of financial,

Table 9.1 Summary of key information on the four conservation and development policies at Wolong Nature Reserve (reproduced from Yang, 2013).

	Natural Forest Conservation Program (NFCP)	Grain to Green/Bamboo Program (GTGB)[b]	Electricity Subsidy Program (ESP)	Tourism Development Program (TDP)
Aim	Protect and restore natural forests through logging ban, afforestation, and payments	Convert cropland to timber or bamboo forests or shrublands with payments and alleviate poverty	Encourage the switch from use of fuelwood to electricity and improve local livelihoods	Provide alternative income sources to local households and generate additional funds for conservation
Type	Conservation policy with development goal	Conservation policy with development goal	Development policy with conservation goal	Development policy with conservation goal
Implementation method	Incentive-based mechanism	Incentive-based mechanism	Incentive-based mechanism	Partnership between local government and tourism companies
Implemented area	120 500 ha of land below tree line (i.e., 3600 m above sea level)	Grain to Green: 367.3 ha from cropland of 969 households; Grain to Bamboo: 81.9 ha from cropland of 530 households	Experimental zone of the reserve	Experimental zone of the reserve
Initial year	2001	Grain to Green: 2000; Grain to Bamboo: 2002	2002	2002[c]
Duration	10 years (renewed in 2010 for another 10 years to 2020)	8 years (Grain to Green: renewed in 2008 for another 8 years; Grain to Bamboo: ended in 2010)	Long term (unless terminated)	Long term (unless terminated)
Average payment rate	~900 yuan per household per year (almost tripled after 2010)[a]	Grain to Green: 3600 yuan per ha per year (halved after 2008); Grain to Bamboo: 13 500–18 000 yuan per ha per year	0.07 yuan per kilowatt-hour (discounted electricity price is 0.01 yuan per kilowatt-hour)	Not applicable

Notes:
[a]The Wolong Administrative Bureau changed the payment rate for NFCP from 900 yuan per household per year to 600 yuan per capita per year starting in 2011. The qualification for payment is any household member age seven years or above.
[b]The Grain to Bamboo Program refers to the project that was designed to convert cropland to bamboo for giant pandas. It was a local program that complemented the national Grain to Green Program. Given the similarities in terms of aim and implementation method, we regarded them as one integrated cropland conversion policy (i.e., GTGB).
[c]Dating back to the 1980s, Wolong Nature Reserve proposed and implemented the tourism development policy, but the State Forestry Administration of China did not officially approve the Ecotourism Development Plan at Wolong Nature Reserve until 2002.

human, natural, manufactured, and social capital, respectively; C_c is a vector of other contextual factors; β_s are vectors of coefficients; and ε is the error term assumed to be normally, identically, and independently distributed with mean of zero and constant variance. To avoid or reduce the collinearity between components and the interaction terms and facilitate interpretation, all continuous independent variables are mean-centered (Dalal and Zickar, 2012). We emphasize that we are modeling not income per se but *changes* in our income variables.

Our results show that different types of policy interventions and various forms of capital had different effects on household income growth (Table 9.3). The effects of policy interactions were statistically significant. The main effects of the two conservation policies (i.e. NFCP and GTGB) were both negative while their interaction effect was positive and higher than any of the main effects (Table 9.3). When controlling for all other variables at their mean values, households with one percentage point more NFCP or GTGB payment percentages had smaller increases in household income by 5.9% of the mean value (1288 yuan) and 0.7% of the mean value (155 yuan), respectively. The main effects of the two development policies (i.e., ESP and TDP) were both positive. While the main

Table 9.2 Summary of descriptive statistics of variables used for regression analyses. All monetary values in 1998 are discounted to values in 2007. The number of observations is 179, which is smaller than the 220 shown in Figure 9.1 due to missing data for some factors. Reproduced from Yang (2013).

	Variables	**Description**	**Mean (S.D.)**
Dependent variable	Changes in total household income	The difference of total household income in 2007 subtracting total household income in 1998 (thousand yuan)	21.988 (27.286)
	Changes in the percentage of agricultural income	The difference of percentage of agricultural income in 1998 subtracting the percentage of agricultural income in 2007	0.218 (0.389)
Policy variable	NFCP payment	The payment each household received annually from the NFCP (thousand yuan)	0.948 (0.183)
	NFCP payment percentage	NFCP payment divided by total household income in 2007	6.5% (6.0%)
	GTGB payment	The payment each household received annually from the cropland conversion policies, i.e., GTGP and GTBP (thousand yuan)	2.888 (2.320)
	GTGB payment percentage	GTGB payment divided by total household income in 2007	16.0% (15.3%)
	Tourism participation	Household participation in tourism business (1: participated; 0: did not participate)	0.274 (0.447)
	Electricity subsidy	Initial subsidy received for electricity consumption (thousand yuan)	0.086 (0.104)
	Electricity subsidy percentage	Electricity subsidy divided by total household income in 1998	2.7% (4.1%)
Financial capital	Initial total household income	Total household income in 1998 (thousand yuan)	6.285 (4.932)
	Initial percentage of agricultural income	The percentage of agricultural income in 1998	0.630 (0.313)
	Changes in total agricultural income	The difference of total agricultural income in 2007 subtracting total agricultural income in 1998 (thousand yuan)	7.817 (15.393)
Human capital	Number of laborers	The number of laborers in each household	2.820 (1.455)
	Changes in number of laborers	The difference in number of laborers in 2007 subtracting the number of laborers in 1998	−0.727 (1.795)
	Education	Education level of the most educated non-student adult in 2007 (year)	7.120 (3.432)
Natural capital	Cropland area	The total area of cropland for each household in 2007 (mu, 1 mu = 1/15 ha)	10.450 (4.163)
Manufactured capital	Distance to the main road	The Euclidean distance from each household location to the main road (km)	0.431 (0.629)
Social capital	Social ties to local governments	Whether the household had an immediate relative member working in local governments or government enterprises: 1: Yes; 0: No.	0.120 (0.326)

Notes: Dependent variable is the change of total household income from 1998 to 2007 calculated by the total household income in 2007 subtracting total household income in 1998.

effect of the tourism policy was positive, its interaction effect with NFCP was negative. Controlling for all other variables at their mean values, households participating in tourism businesses had income growth that was on average 24% higher (mean value of 5274 yuan). In addition, the effects of initial electricity subsidy percentage, increase of agricultural income, number of laborers in 1998, and increase in the number of laborers were all positive. But the effects of both the initial cropland area and distance from each household to the main road were negative (Table 9.3). Additional analyses show that households with more cropland or located farther away from the main road were less likely to participate in tourism businesses (Spearman's $p = -0.242$, $p = 0.001$ and Spearman's $p = -0.218$, $p = 0.003$, respectively). These households thus had lower income increases.

Our results also show varied effects on income structure for different policies and forms of capital (Table 9.4). As in the case with total income, there

Table 9.3 Effects of conservation and development policies on changes in total household income (reproduced from Yang, 2013).

	Variables	Unstandardized coefficients	Standardized coefficients	Robust S.E.
Policy variable	NFCP payment percentage	−128.811***	−0.286***	35.194
	GTGB payment percentage	−15.535*	−0.087*	7.707
	Initial electricity subsidy percentage	63.921*	0.097*	28.004
	Tourism participation	5.274†	0.086†	3.188
	NFCP payment percentage × GTGB payment percentage	387.458***	0.263***	89.473
	NFCP payment percentage × Initial electricity subsidy percentage	−188.653	−0.042	133.730
	NFCP payment percentage × Tourism participation	−228.758*	−0.201*	98.906
Financial capital	Initial total household income	−0.175	−0.032	0.294
	Changes in total agricultural income	1.114***	0.633***	0.251
Human capital	Number of laborers	2.767*	0.148*	1.283
	Changes in number of laborers	2.161*	0.143*	0.977
	Education	0.119	0.015	0.407
Natural capital	Cropland area	−1.451*	−0.130*	0.578
Manufactured capital	Distance to the main road, log-transformed	−1.783**	−0.132**	0.563
Social capital	Social ties to local governments (1: yes; 0: no)	0.319	0.004	3.919
Township	1: Wolong; 0: Gengda	4.253	0.078	2.671
	Constant	14.601***	—	2.788

Notes:
S.E., standard error. Dependent variable is the change of total household income from 1998 to 2007 calculated by the total household income in 2007 subtracting total household income in 1998. We mean-centered all continuous independent variables before entering them into the model. There are two townships (i.e., Wolong and Gengda) at Wolong Nature Reserve and thus we also controlled the township variable in our model. $N = 179$. $R^2 = 0.728$. All variance inflation factors were less than 5, demonstrating low multicollinearity. To avoid multicollinearity, we excluded the interaction terms between GTGB payment percentage and initial electricity subsidy percentage, and between GTGB payment percentage and tourism participation, which did not change our conclusions. †$p < 0.1$; *$p < 0.05$; **$p < 0.01$; ***$p < 0.001$.

Table 9.4 Effects of conservation and development policies on changes in agricultural income percentage (reproduced from Yang, 2013).

	Variables	Unstandardized coefficients	Standardized coefficients	Robust S.E.
Policy variable	NFCP payment percentage	−0.904	−0.141	0.663
	GTGB payment percentage	0.347*	0.136*	0.135
	Initial electricity subsidy percentage	1.218**	0.129**	0.347
	Tourism participation	0.037	0.043	0.041
	NFCP payment percentage × GTGB payment percentage	3.660**	0.174**	1.078
	NFCP payment percentage × Initial electricity subsidy percentage	−4.182	−0.066	4.052
	NFCP payment percentage × Tourism participation	−2.398**	−0.148**	0.870
	GTGB payment percentage × Initial electricity subsidy percentage	1.204	0.020	3.615
	GTGB payment percentage × Tourism participation	0.218	0.038	0.263
Financial capital	Initial total household income	−1.14e−3	−0.014	3.60e−3
	Initial agricultural income percentage	0.859***	0.685***	0.055
	Changes in total agricultural income	−6.86e−3**	−0.274**	2.51e−3
Human capital	Number of laborers	1.75e−4	6.59e−4	0.021
	Changes in number of laborers	0.018	0.083	0.015
	Education	0.005	0.041	0.005
Natural capital	Cropland area	−0.028***	−0.179***	0.007
Manufactured capital	Distance to the main road, log	−0.022*	−0.114*	0.009
Social capital	Social ties to local governments (1: yes; 0: no)	0.116*	0.096*	0.046
Township	1: Wolong; 0: Gengda	0.038	0.049	0.032
	Constant	0.112**	—	0.042

Notes:
S.E., standard error. Dependent variable is the change of agricultural income percentage from 1998 to 2007 calculated by agricultural income percentage in 1998 subtracting agricultural income percentage in 2007. Since the overall trend of agricultural income percentage was decreasing, we used the value in 1998 and subtracted the value in 2007 to make the interpretation of coefficients easier. We mean-centered all continuous independent variables before entering the model. $N = 179$. $R^2 = 0.771$. All variance inflation factors were less than 5, showing low multicollinearity. $^{*}p < 0.05$; $^{**}p < 0.01$; $^{***}p < 0.001$.

were also some statistically significant interaction effects among policies. The main effects of NFCP and tourism participation were not significant, while GTGB and the electricity subsidy significantly reduced the percentage of agricultural income. The interaction effect between NFCP and GTGB was positive, while it was negative between NFCP and tourism participation. The signs of these interaction effects were consistent with what we found for total household income. When controlling for all other variables at their mean values, an increase of one percentage point of the GTGB payment percentage reduced the percentage of agricultural income by 1.6% of the mean value (i.e., absolute change of about 0.3%). In addition, the initial electricity subsidy percentage, initial percentage of agricultural income in 1998, and social ties to local governments all significantly reduced the percentage of agricultural income from 1998 to 2007. But increased agricultural income from the base period, total area of cropland in 1998, and distance from each household to the main road all increased the percentage of agricultural income from 1998 to 2007.

9.4 Effects of economic transformation

While macrosocioeconomic conditions and household characteristics drive economic changes, changes in economic conditions at the reserve and household levels in return alter local residents' behaviors. Examples include logging, fuelwood collection, and land conversion, all of which have environmental and socioeconomic impacts. Below we provide a synthesis of these impacts based on our long-term research at Wolong.

Forest cover and panda habitat at Wolong experienced continuous deterioration between the 1960s and the 1990s, and then started to recover from the early 2000s. For example, the total amount of forest cover was 106 000 ha in 1965. Forest cover declined to 70 000 ha in 2001 and then recovered to 79 000 ha in 2007 (Yang et al., 2013a; Chapter 6). The previous deforestation and habitat destruction were mostly caused by logging and fuelwood collection (Bearer et al., 2008, Liu et al., 1999, 2001). The recent recovery is largely due to the implementation of NFCP and associated economic development (Yang et al., 2013a). Economic changes at the household level had a wide range of impacts on local households' activities that in turn shaped the dynamics of vegetation cover and wildlife habitat. Examples include fuelwood collection, land conversion, and tourism participation. Below is some empirical evidence about the processes underpinning these changes.

GTGB reduced the labor force engaged in farming, and much of the released labor was channeled into the labor force markets locally and far away in cities. The income generated from such labor directly and indirectly encouraged the use of electricity instead of fuelwood in Wolong (see Chapter 11). This indirect effect occurred via the lack of labor available to collect fuelwood due to out-migration and via a shift in lifestyle that encouraged the purchase of electronic appliances. As reported in Chapter 11 and Chen et al. (2012), when non-migration income and other factors were controlled, households with migrant workers on average consumed 1827 kg less fuelwood (29% of average fuelwood consumption by all households) than those without migrant workers.

Economic development is predicted to reduce the potential for land conversion after the anticipated expiration of GTGB and thus is likely to sustain the conservation benefits that have been achieved. When other demographic, biophysical, and socioeconomic factors are controlled, 1000 yuan increases in agricultural and non-agricultural income reduced the reconversion of GTGB land to cropland by 0.001 and 0.002 ha per household, respectively (Chen et al., 2009).

Economic development also influenced local households' participation in tourism businesses. Households with higher total income were more likely to participate in tourism businesses, which further expanded the income gap between participating households and those not participating. From 1998 to 2006, this gap in per capita income almost doubled (Liu et al., 2012).

9.5 Discussion and concluding remarks

Our long-term study in Wolong allowed us to document a striking pattern of economic transformation. With respect to drivers of this transformation, we found significant interaction effects among various conservation and development policies in the form of both antagonistic and synergistic effects. These findings raise critical concerns about traditional policy studies that tend to ignore policy interactions and possibly lead to biased results and even unanticipated consequences of policy decisions informed by such results. For example, in our case, NFCP and GTGB separately had a negative effect on the changes in total household income, but together led to a positive effect. This may appear to be a surprise, but makes sense after considering how people respond to policy interventions, how each policy takes effect, and how policies together interact with each other.

Generally, there are three types of adaptation strategies in response to policies or environmental impacts. For farmers, one is an incremental adaptation strategy. This strategy may involve incremental adjustments of agricultural activities, including growing more cash income crops (e.g., cabbage and traditional Chinese medicinal plants) than subsistence crops (e.g., corn and potatoes). Alternatives could include intensifying agricultural practices (e.g., increasing the use of fertilizers and pesticides) or regulating labor force allocation between livestock and crops. The second type of adaptation is a transformational strategy (Kates et al., 2012). Adaptations of this type are at a much larger scale or intensity, are new or almost new to a system, or involve other measures that dramatically transform the system (e.g., massive relocation of people). In the Wolong case, transformational adaptations may include substantially reducing or abandoning agricultural activities and going to cities for work or business opportunities. The third type is to take no action if the impact is positive or tolerable.

In our case, in households intensively involved in only one policy, they might adopt the incremental adaptation strategy to offset the negative policy impacts or may not do anything if the negative impacts were tolerable. For instance, the implementation of NFCP and GTGB each led to some negative economic impacts on households due to restrictions on forest use and loss of income from converted cropland, respectively. But the impacts of NFCP itself might not be enough to motivate households to adopt transformational adaptations. Households could easily offset the negative impacts through incremental adaptation measures. Example measures might include collecting more traditional Chinese medicinal plants for sale or increasing the production per unit area of their cropland by using more pesticides and fertilizers (Yang et al., 2013a, c). However, for many households who were substantially affected by both NFCP and GTGB, the incremental adaptation strategy, or no action strategy, would not allow them to maintain their livelihoods. This outcome may arise because on average the GTGB converted more than 60% of cropland. In addition to the negative impacts of NFCP, the small amount of remaining cropland could not generate enough income for many households even if they had adopted incremental adaptations such as using more pesticides and fertilizers. Therefore, many households had to adopt the transformational adaptation strategy to earn more non-agricultural income through tourism businesses, transportation ventures, and migrant work (Chen et al., 2012, Liu et al., 2012, Yang et al., 2013c). But there are also tradeoffs of such a transformational adaptation strategy. For example, as young and middle-aged laborers rush into cities, many children and seniors are left behind. This shift in age structure causes many social problems such as truancy, increasing crime rate of teenagers, and loneliness in old age. Even for those farmers and their children who have moved to cities, they often face socio-economic and psychological pressures. Such problems stem from substantial living and educational expenses, discrimination from urban residents, high stress, and depression (Wong et al., 2007, Yang et al., 2013d). For those households who participated in both NFCP and tourism, the story was quite different. Tourism participation itself had a marginally positive significant contribution to household income (Table 9.2). But, as we discussed above, most tourism revenue was retained by the local government and tourism companies; very few households earned a large amount of money from tourism participation. Thus, households affected by NFCP and which adopted transformational adaptation to participate in tourism businesses had a negative interaction effect from the two policies.

Economic transformation profoundly shapes government policies and local households' environmental and socioeconomic activities, and thus changes the dynamics of vegetation cover and wildlife habitat. But it should also be noted that, in turn, changes in vegetation cover and wildlife habitat also affect households' activities and livelihoods and form feedbacks. The recovery of forests and wildlife habitat encouraged the central government to renew both NFCP and GTGP for another phase (ten years for NFCP and eight years for GTGP). The payment rates approximately doubled for NFCP and halved for GTGP (Liu et al., 2013)—the local GTBP ended in 2010 without renewal due to lack of sustainable funding. The recovery of forests and wildlife habitat also increased the provision of various types of ecosystem services (e.g., water conservation, soil erosion control, carbon sequestration, and air purification). Recovered forests and habitat also attracted more tourists and generated more revenue for governments and local households (Liu et al., 2012, Yang et al., 2013a). Nevertheless, there were also some negative outcomes of these feedbacks. For example, before the implementation of NFCP and GTGP, forests were declining. Local households had to go farther away to find fuelwood and expend more labor to collect fuelwood (He et al., 2009). After the implementation of NFCP and GTGP, forests gradually recovered and wildlife increased. This latter change led to an increase in human–wildlife conflicts, such as damage to crops from wildlife including wild boars (Yang et al., 2013a).

The economic transformation at Wolong and its associated drivers and impacts may be representative of thousands of other similar cases in or around other protected areas across the world. Although the specific context in another place will be different, the new conceptual approaches, methodological innovation, and lessons learned at Wolong could be helpful and applicable to other places. Eventually, by replicating our analyses in other places, multiple site comparisons might make it possible to identify a comprehensive set of drivers of economic transformation and associated impacts, as well as key elements of policy design and important contextual factors that affect policy interactions.

9.6 Summary

Economic transformations have occurred and will continue to occur in many places across the world. To achieve environmental and socioeconomic sustainability, it is crucial to understand the dynamics and effects of such economic transformations. In this chapter, we examined economic transformations in our model coupled human and natural system in Wolong Nature Reserve. Results indicate that local livelihoods shifted from being primarily based on subsistence agriculture to diversifying into varied non-agricultural income sources (e.g. transportation, tourism, and conservation programs) beginning in the early 2000s. Both household-level access to multiple forms of capital and macrolevel policies played important roles in the transformations. In addition, we found evidence for interaction effects of policies. An example is the interaction between the NFCP and GTGB programs, which had negative effects separately on change in household income but a positive effect together. This finding may be related to the adaptation strategy adopted by households in response to new policies. Households may seek to offset negative impacts of one policy by changing their participation level in another policy. We also highlighted several impacts of the economic transformation. These included changes in forest cover and wildlife habitat, a shift in the allocation of the labor force, land conversion, and a boom in tourism ventures. These impacts in turn fed back to affect the economic state of individual households and the reserve as a whole. It is our hope that the interaction effects among policies and feedbacks between economic transformation and environmental changes are also applicable to many other places.

References

Bartolini, S. and Bonatti, L. (2002) Environmental and social degradation as the engine of economic growth. *Ecological Economics*, **43**, 1–16.

Bearer, S., Linderman, M., Huang, J., et al. (2008) Effects of fuelwood collection and timber harvesting on giant panda habitat use. *Biological Conservation*, **141**, 385–93.

Chen, X., Frank, K.A., Dietz, T., and Liu, J. (2012) Weak ties, labor migration, and environmental impacts:

toward a sociology of sustainability. *Organization & Environment*, **25**, 3–24.

Chen, X., Lupi, F., He, G., et al. (2009) Factors affecting land reconversion plans following a payment for ecosystem service program. *Biological Conservation*, **142**, 1740–47.

Dalal, D. K., Zickar, M. J. (2012) Some common myths about centering predictor variables in moderated multiple regression and polynomial regression. *Organizational Research Methods*, 15, **339**–62.

DeFries, R.S., Rudel, T., Uriarte, M., and Hansen, M. (2010) Deforestation driven by urban population growth and agricultural trade in the twenty-first century. *Nature Geoscience*, **3**, 178–81.

He, G., Chen, X., Bearer, S., et al. (2009) Spatial and temporal patterns of fuelwood collection in Wolong Nature Reserve: implications for panda conservation. *Landscape and Urban Planning*, **92**, 1–9.

Kates, R.W., Travis, W.R., and Wilbanks, T.J. (2012) Transformational adaptation when incremental adaptations to climate change are insufficient. *Proceedings of the National Academy of Sciences of the United States of America*, **109**, 7156–61.

Liu, J. and Diamond, J. (2005) China's environment in a globalizing world. *Nature*, 435, 1179–86.

Liu, J., Li, S., Ouyang, Z., et al. (2008) Ecological and socioeconomic effects of China's policies for ecosystem services. *Proceedings of the National Academy of Sciences of the United States of America*, **105**, 9477–82.

Liu, J., Linderman, M., Ouyang, Z., et al. (2001) Ecological degradation in protected areas: the case of Wolong Nature Reserve for giant pandas. *Science*, **292**, 98–101.

Liu, J., Ouyang, Z., Taylor, W.W., et al. (1999) A framework for evaluating the effects of human factors on wildlife habitat: the case of giant pandas. *Conservation Biology*, **13**, 1360–70.

Liu, J., Ouyang, Z., Yang, W., et al. (2013) Evaluation of ecosystem service policies from biophysical and social perspectives: the case of China. In S.A. Levin, ed., *Encyclopedia of Biodiversity* (vol. 3), pp. 372–84. Academic Press, Waltham, MA.

Liu, W., Vogt, C.A., Luo, J., et al. (2012) Drivers and socioeconomic impacts of tourism participation in protected areas. *PLoS ONE*, **7**, e35420.

Millennium Ecosystem Assessment (2005) *Ecosystems & Human Well-Being: Synthesis*. Island Press, Washington, DC.

Rustagi, D., Engel, S., and Kosfeld, M. (2010) Conditional cooperation and costly monitoring explain success in forest commons management. *Science*, **330**, 961–65.

Schaller, G.B., Hu, J., Pan, W., and Zhu, J. (1985) *The Giant Pandas of Wolong*. University of Chicago Press, Chicago, IL.

Scrieciu, S.S. (2007) Can economic causes of tropical deforestation be identified at a global level? *Ecological Economics*, **62**, 603–12.

Wolong Nature Reserve (2005) *Development History of Wolong Nature Reserve*. Sichuan Science and Technology Press, Chengdu, China (in Chinese).

Wong, D.F.K., Li, C.Y., and Song, H.X. (2007) Rural migrant workers in urban China: living a marginalised life. *International Journal of Social Welfare*, **16**, 32–40.

Yang, W. (2013) *Ecosystem Services, Human Well-being, and Policies in Coupled Human and Natural Systems*. Doctoral Dissertation, Michigan State University, East Lansing, MI.

Yang, W., Dietz, T., Kramer, D.B., et al. (2013d) Going beyond the Millennium Ecosystem Assessment: an index system of human well-being. *PLoS ONE*, **8**, e64582.

Yang, W., Dietz, T., Liu, W., et al. (2013c) Going beyond the Millennium Ecosystem Assessment: an index system of human dependence on ecosystem services. *PLoS ONE*, **8**, e64581.

Yang, W., Liu, W., Viña, A., et al. (2013a) Performance and prospects on payments for ecosystem services programs: evidence from China. *Journal of Environmental Management*, **127**, 86–95.

Yang, W., Liu, W., Viña, A., et al. (2013b) Nonlinear effects of group size on collective action and resource outcomes. *Proceedings of the National Academy of Sciences of the United States of America*, 110, 10916–21.

CHAPTER 10

Energy Transition from Fuelwood to Electricity

Wei Liu, Andrés Viña, Wu Yang, Frank Lupi, Zhiyun Ouyang, Hemin Zhang, and Jianguo Liu

10.1 Introduction

More than 2.5 billion people around the world, particularly in developing countries, rely on biomass fuels such as fuelwood as a primary source of energy to meet livelihood needs (Global Energy Assessment, 2012). The environmental and health impacts of using fuelwood are enormous. Indoor air pollutants emitted through the combustion of these fuels are responsible for over 2 million deaths per year (World Health Organization, 2009). Their use is also associated with serious environmental threats such as deforestation (An et al., 2002, He et al., 2009, Heltberg et al., 2000) and global climate change (Bond et al., 2008, Clancy, 2008). Thus, the transition from energy sources such as fuelwood to safer and more efficient alternatives has the potential to improve not only the global environment but also human well-being.

Energy transition from biomass fuels (e.g., fuelwood, charcoal, animal droppings, and agricultural residue) to alternative energy sources is often depicted using two models: the energy ladder model and the leapfrogging model. The energy ladder model argues that as income increases, households cease using biomass fuels and adopt energy alternatives (Barnes and Floor, 1996, Holdren and Smith, 2000, Leach, 1992, Macht et al., 2012). Thus, the model contends that income is the most important factor influencing fuel choice (Figure 10.1). The energy ladder model distinguishes three essential phases. These include reliance on (1) biomass fuels; (2) fossil fuels such as kerosene and coal, also called transitional fuels; and (3) modern energy sources such as natural gas, liquefied petroleum gas, and electricity (Heltberg, 2004, Holdren and Smith, 2000). This model has been criticized because fuel shifts rarely follow a linear path from one fuel source to another. In reality, shifts follow a multiple fuel adoption strategy in which new fuels are added while biomass fuels are seldom completely abandoned (Masera and Navia, 1997, Masera et al., 2000). The model has also been criticized because it mainly focuses on economic processes and disregards the effects of social and cultural processes on fuel choice (Jebaraj and Iniyan, 2006). Energy transition occurs in part because modern energy sources achieve higher efficiency or reduce the exposure to indoor pollutants. Another important cause is that modern energy use signifies higher socioeconomic status (Masera et al., 2000).

The leapfrogging model, on the other hand, purports that developing countries learn from developed countries. Developing countries are thus expected to adopt the most recent technologies rather than going through different steps of technological innovation. Under leapfrogging, households are introduced to modern sources of energy in a fast and cost-effective way. Example mechanisms include infrastructure development (e.g., electricity grids) and distribution of modern appliances (e.g., energy-efficient stoves) through subsidy programs (Goldemberg, 1998). Under such conditions, the energy ladder is replaced by a two-step path going

Pandas and People. Edited by Jianguo Liu, Vanessa Hull, Wu Yang, Andrés Viña, Xiaodong Chen, Zhiyun Ouyang, and Hemin Zhang. Published 2016 by Oxford University Press.

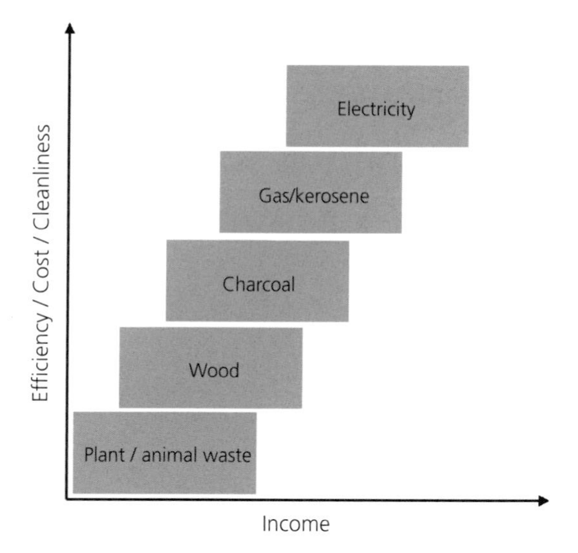

Figure 10.1 The "Energy Ladder" model, based on Holdren and Smith (2000).

from biomass fuels to modern energy sources, thus completely avoiding the transitional fuels. This model has been criticized on the basis that technology adoption will not occur without proper social, cultural, political, and economic conditions. Energy consumption patterns are important components of the lifestyles and worldviews affecting the behavior of rural people (Murphy, 2001). Thus, in places with access to multiple sources of energy, both income and the social organization of daily life constitute critical determinants of fuel choice (Foster, 1999).

China is a good example to illustrate the recent energy transition. The exceptional economic growth of China in the last three decades has intensified its demand for energy (Liu, 2010, Liu and Diamond, 2008, Liu et al., 2013). China's increasing energy consumption makes it one of the dominant players in the global energy consumption arena (International Energy Agency, 2013). China has experienced a continuous change in the consumption structure of different energy types. The proportion of biomass fuel has declined. Consumption of other forms of energy, particularly coal, oil, electricity, and piped natural gas has increased (Pachauri and Jiang, 2008). For instance, annual electricity consumption over the last decade has risen at annual rates between 5% and 7% (National Energy Administration of China, 2014). The rate of increase in 2013 was the highest so far, at about 7.5% (National Energy Administration of China, 2014). However, most of these increases are driven by the industrial sector, since household electricity consumption still remains comparatively low. At the household level, China tends to be following the energy ladder model more than the leapfrogging model. This ladder is observed more frequently in urban than in rural households. Rural households tend to diversify rather than completely switch their energy sources. Rural households thus end up utilizing a variety of energy sources simultaneously, namely the multiple fuel adoption strategy (Masera and Navia, 1997, Masera et al., 2000). As income level and accessibility to gridded energy sources increase, rural households "move upward" on the energy ladder. But they still keep biomass fuels and use them simultaneously with gridded energy sources, or as a supplement.

Evaluating energy transition in different contexts requires a holistic framework that incorporates not only socioeconomic and cultural domains but also environmental constraints and responses. As such, it requires analyzing the interactions, and perhaps more important, the feedbacks between different human and natural processes. Meeting such needs is the framework of Coupled Human and Natural Systems (CHANS; Liu et al., 2007a, b), such as socioecological systems (Anderson et al., 2008), population–environment systems (de Sherbinin et al., 2007, McNicoll, 2002, Mena et al., 2011), and human–environment systems (Fox and Vogler, 2005, Vitousek, 2006). This approach is founded on the explicit recognition that human and natural systems are coupled via flows of materials, energy, and information. The coupled systems can be treated as complex adaptive systems (Holland, 1992, Levin, 1998) with interacting feedbacks at multiple scales. Complex human–nature feedbacks at fine scales (e.g., households) are manifest in changes at broader scales (e.g., countries; An et al., 2014, Hull et al., 2015, Liu et al., 2015). This work has helped to differentiate the effects of endogenous (e.g., population growth) and exogenous (e.g., policies) factors (Laney, 2002), and to understand the interplay among them (Chowdhury and Turner, 2006). It also provides insights into the conditions under which households will adopt particular behaviors (including energy transition) in response to different social, economic, cultural, political, and environmental contexts (An et al., 2001, 2002, Chen et al., 2009, 2010, 2012, 2014).

Research on energy transition has been largely conducted at large scales (regional, national, and global). While these studies have provided useful broad insights, it is necessary to gain more understanding of energy transition at the local scale because ultimately all decisions are made and implemented at the local scale. In this chapter, we discuss the patterns and processes of energy transition in our model coupled system—Wolong Nature Reserve (Chapter 3). We begin with some background information about energy use in Wolong and general methods for energy analysis. We then present results on changes in quantity and structure of household energy use, changes in fuelwood harvesting activities, determinants of changes in energy use, and local attitudes toward energy transition. We conclude with reflections on the dynamic energy transition pathway that has taken place in Wolong and recommend topics for future research.

10.2 Background and methods

Until recently, fuelwood was the main source of energy in Wolong (Liu et al., 1999). The most preferred tree species sought for fuelwood include birch (*Betula sp.*), oak (*Cyclobalanopsis sp.*), maple (*Acer sp.*), and walnut (*Juglans cathayensis*). Similar to most other mountain regions around the world, the demand for fuelwood in Wolong is not restricted to cooking food, but is also used for heating, particularly during the winter season. But unlike many other regions, fuelwood is also used in Wolong to cook pig fodder and to smoke pork (An et al., 2001, 2002, Liu et al., 1999). Pigs were introduced into the reserve in the eighteenth century (Ghimire, 1997) and have since become the main supply of animal protein (An et al., 2001, Liu et al., 1999). As a result, most fuelwood demands in the reserve are driven by the need to cook pig fodder (An et al., 2002, Yang et al., 2013a).

Due to the lack of effective monitoring activities, overharvesting of wood by local residents used to be pervasive. Many of the trees harvested were not only used for fuelwood and house construction but also for illegal sale outside the reserve (Li et al., 1992). Therefore, during the last several decades of the twentieth century, forest harvest led to substantial deforestation and caused panda habitat degradation (Bearer et al., 2008, Li et al., 1992, Liu et al., 2001, Viña et al., 2007, Wolong Administration Bureau, 2004).

To reduce the local dependence on fuelwood, and thus the degradation of natural forests in the reserve, electricity was proposed as a clean alternative energy source during the 1970s (Li et al., 1992). A number of hydropower plants were built, most of which were run by state-owned companies from outside the reserve. While the plants generated a large quantity of electricity, the vast majority was sold to cities and only a small proportion was consumed by the local households. Barriers preventing the adoption of electricity by these households included the underdevelopment of the electricity grid system, which mostly only reached the households located along the main road. This shortcoming, coupled with the instability of electricity supply and lack of affordability, created a challenge for low-income households (An et al., 2002). As a result, by the late 1990s most local households with access to electricity used it only to power light bulbs and used fuelwood for cooking and heating.

To improve overall electricity supply and stability, and as part of a national rural electricity grid reform program, an upgrade of the local electricity grid in the reserve was initiated at the end of the 1990s. By 2003, the reserve-wide electricity system improvement was completed and a new hydropower plant was built (Wolong Administration Bureau, 2004). The electricity price for household use in the reserve increased from 0.155 yuan/kWh to 0.26 yuan/kWh during the winter months (due to lower river flows) and 0.24 yuan/kWh during the rest of the year. To improve the affordability of electricity and encourage the switch to its use from fuelwood, the Reserve Administration provided an electricity subsidy for households: a reduction in unit price of 0.08 yuan/kWh during winter and 0.06 yuan/kWh during the rest of the year.

To understand the energy transition in Wolong, we performed a longitudinal analysis using data acquired in two periods from a sample of households in the reserve. In 1999, we selected a stratified random sample of 220 households (corresponding to 23% of the total number of households inside the reserve at that time). We established the location of the households using differentially corrected global positioning system (GPS) receivers. We also conducted face-to-face interviews with the household heads (An et al., 2002). Information collected in the interviews included the choice to switch from fuelwood to electricity, household demographics, income structure, and energy consumption structure in the previous year. In 2007, we revisited 189 of these previously interviewed households, following the same face-to-face interview method and collecting the same information. During the 2007 survey, additional questions were asked pertaining to household energy use during both the late 1990s and in the prior two years (i.e., 2006/2007). These included household annual electricity expenses, amount of fuelwood consumption, frequency of fuelwood and electricity use for various purposes (including cooking, heating, and stewing pig fodder), location of fuelwood collection sites, preferred tree species, and harvesting methods (e.g., entire trees or tree branches). When collecting retrospective information, the life-history calendar method, widely used in social sciences (Axinn et al., 1999, Freeman et al., 1988), was used to improve respondents' recall accuracy. We also asked questions pertaining to their attitudes toward the conservation policies implemented after the late 1990s and their perceptions of relationships between these policies and household energy use.

Since the reduction in fuelwood use was considered the most important indicator of the energy transition in the reserve, we defined a household-level fuelwood consumption reduction rate, r, as:

$$r = \left(Fuelwood_{1998} - Fuelwood_{2007}\right) / \, Fuelwood_{1998} * 100\% \quad \text{(Equation 10.1)}$$

where $Fuelwood_{year}$ corresponds to the amount of fuelwood consumed per household in the particular year. From the late 1990s to 2006/2007, the mean household reduction rate (n = 184) was 73% (S.D. = 19%) and ranged from 17% to 100%. This wide range indicates a significant level of heterogeneity in the energy transition patterns among the sampled households. To understand this heterogeneity, we constructed the following generalized linear model:

$$r = \alpha + \beta_1 X_1 + \beta_2 X_2 + \ldots + \beta_k X_k + e \quad \text{(Equation 10.2)}$$

where X_k corresponds to different predictor variables. These included household-level demographic, economic, energy, geographic, and policy-related factors (Table 10.1), while β_k, α, and e represent the variable coefficients, the intercept, and the error term, respectively.

We controlled for the household energy consumption background level during the late 1990s by including two variables: annual fuelwood consumption and electricity use. Economic factors considered included non-agricultural income in 1998, the net change in household annual income between 1998 and 2007, and whether a household owned or operated a private hotel by 2007. Demographic factors included household size, number of laborers, number of seniors (60+ years old), and average education level of adults. Geographic variables included elevation of each household, the township in which a household was located, and the distance from a household to the main road. We also included the estimated number of households per unit area, based on a household density map at 30 m pixel resolution created using the Kernel Density Estimator (Beyer, 2004). This estimator uses the locations of all local households as input and calculates a fixed kernel density function using the quartic approximation for a Gaussian kernel. We used a kernel bandwidth of 1 km since it constitutes the average household dispersion distance within a local village group in the reserve. Finally, we included conservation policy factors based on measures of household participation in payments for ecosystem services (PES) programs (Table 10.1).

10.3 Changes in the quantity and structure of household energy consumption patterns

From 1975 to 1998, fuelwood consumption doubled (Liu et al., 1999), while the human population and

Table 10.1 Dependent and independent variables at household level used in the general linear model for determining household fuelwood consumption reduction rates ($n = 182$).

Variables	Description and Unit	Mean (SD)
Dependent variable		
r	Fuelwood consumption reduction rate	73% (19%)
Independent variables		
Energy baseline		
Fuelwood	Annual fuelwood consumption (in tons) in late 1990s	9.25 (4.72)
Electricity	Annual electricity consumption (in MWh) in 1998	1.31 (1.58)
Economic factors		
Δ Annual Income	Difference of total annual income (1000 yuan)	20.98 (27.27)
Ln(Non-ag income)	Log transformed household non-agricultural income in 1998	6.14 (2.86)
Tourism business	Whether the household owns or operates a private hotel by 2007: 1. Yes; 2. No	0.13 (0.33)
Demographic factors		
Δ Education	Net change of mean adult education (in years) in a household	0.26 (2.19)
Δ Household Size	Change in household size	−1.25 (1.84)
Δ Labor	Change in number of laborers	−0.42 (1.34)
Δ Senior	Change in number of seniors (60+ years old)	0.28 (0.70)
Geographic factors		
Township	1. Wolong Township; 0. Gengda Township	0.44 (0.50)
Elevation	Elevation of household location (in km)	1.82 (0.23)
Distance	Distance from household to the main road (in km)	0.43 (0.63)
Household density	Point kernel (30 m/pixel) density at household location in 2006	0.011 (0.005)
Policy factors		
GTGP/GTBP land	Total amount of cropland enrolled in GTGP and GTBP from 2000 to 2003 (mu)	6.44 (3.17)
NFCP subsidy	Ratio of NFCP subsidy to total household income in 1998#	43% (161%)

Note
using the income in 1998 as the benchmark.

the number of households in Wolong increased by 69% and 124%, respectively. The annual amount of fuelwood consumed by each household varied depending on household size, age structure, cropland area, and other socioeconomic conditions (An et al., 2001). In 1998, all households used fuelwood, ranging from approximately 1 to 30 tons. From 1998 to 2006/2007, the mean household annual fuelwood consumption decreased by almost 75%, from 9.2 tons to 2.4 tons. This relative value is between those of other studies which have reported reductions from as low as 40% (Chen et al., 2014, Yang et al., 2013a) to as high as 85% reported by the government (Wolong Administration Bureau, 2004). However, these different studies have been conducted using different methods during slightly different time periods. The mean household annual electricity use increased more than twofold, from 1.3 MWh to 4.2 MWh (Table 10.2). In addition, by 2007, 25 (13.6%) of the interviewed households reported a complete switch from fuelwood to electricity.

The pattern of household energy consumption also changed significantly. Fuelwood was the sole (75%) or primary (15%) source of energy for cooking food in about 90% of the sampled households during the late 1990s. But by 2006/2007, electricity had become the sole (39%) or primary (40%) source of energy for cooking food in about 80% of

Table 10.2 Changes in household energy use from the late 1990s to 2006/2007 ($n = 184$). Both mean and standard deviation values (in parentheses) are shown. Pairwise *t*-tests were conducted on all these changes and all *p*-values were less than 0.001.

	Late 1990s	2006/2007
Annual fuelwood consumption (tons)	9.2±4.7	2.4±2.0
Annual electricity consumption (MWh)	1.3±1.6	4.1±3.2
Ratio of household electricity expenses to total household income	3.6%	2.8%
Time spent on fuelwood collection activities (days)	61±30	19±19
Duration of cooking pig fodder (months)	11.6±2.1	9.2±3.3
Duration of heating (months)	5.5±1.6	4.9±1.6

the sampled households (Figure 10.2). For heating, during the late 1990s fuelwood was the sole energy source in more than 96% of the sampled households. This percentage decreased to about 13% in 2006/2007. About 55% of the sampled households completely (23%) or primarily (32%) relied on electricity for heating at this time (Figure 10.3). The mean annual duration of heating also decreased significantly from 5.5 months to 4.9 months (pairwise *t*-test, $p < 0.001$, Table 10.2). This change was probably due to the desire of households to reduce the electricity bill, as well as improvement in thermal insulation of houses in Wolong (W. Liu, pers. obs.).

In terms of cooking pig fodder, fuelwood was the sole source of energy during the late 1990s, and even in the period of 2006/2007 only two households reported using electricity for this purpose. However, the mean annual duration of cooking pig fodder among all the households decreased significantly, from 11.6 months to 9.2 months (pairwise *t*-test, $p < 0.001$, Table 10.2). This decrease is probably due to the decrease in the supply of fuelwood, an increase in the cost of labor, and also a change in the perception of the benefits of cooked vs raw fodder (W. Liu, pers. obs.).

Besides fuelwood and electricity, other energy sources, such as coal and kerosene, were rarely used by local residents. While many local restaurants purchased coal fines and made their own coal briquettes, few local households used these alternative energy sources as they are usually more expensive, less convenient, more polluting, and less accessible than electricity.

These results show a clear energy transition from fuelwood to electricity, occurring in about ten years. Such drastic and rapid transition seems to follow the leapfrogging model, in which fuelwood was effectively and rapidly replaced by electricity. However, because few households performed a complete switch and most continued to use fuelwood at a reduced rate, this leapfrogging model seemed to have followed a multiple fuel adoption strategy.

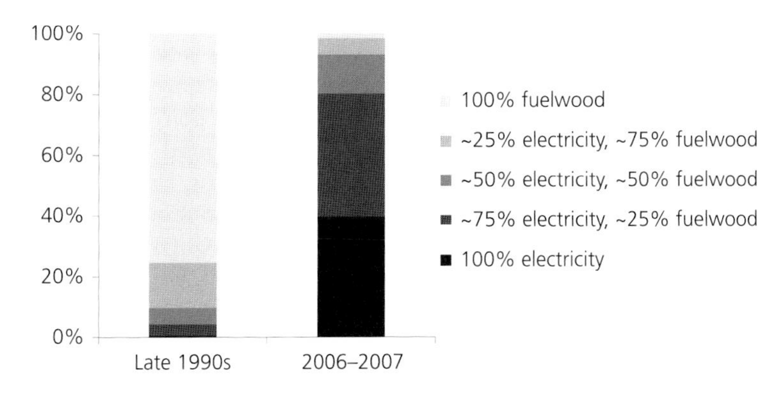

Figure 10.2 Changes in energy used in cooking food among local households from the late 1990s to 2006/2007. Fuelwood was the sole (75%) or primary (15%) source of energy for cooking food in about 90% of the sampled households in the late 1990s; in 2006/2007 electricity was the sole (39%) or primary (40%) source of energy for about 80% of the sampled households.

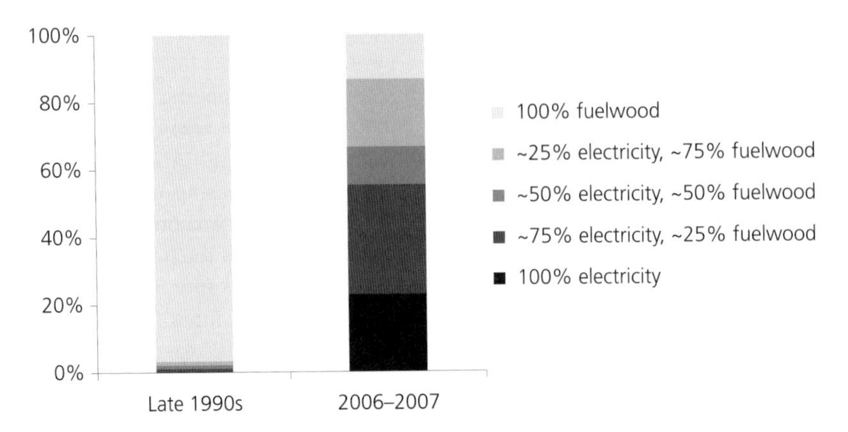

Figure 10.3 Changes in energy used for heating among local households from the late 1990s to 2006/2007. In the late 1990s fuelwood was the sole energy source for heating in more than 96% of the sampled households, but this percentage decreased to about 13% in 2006/2007. In 2006/2007, about 55% of the sampled households used electricity as the sole (23%) or primary (32%) source of energy for heating.

10.4 Changes in fuelwood harvesting activities

Together with the significant decline in fuelwood use, local residents' fuelwood collection activities also changed significantly. Local residents used to cut entire trees for both fuelwood and construction use. Harvesting of entire trees occurred more often in winter (between October and March), when most understory deciduous vegetation had defoliated. These conditions facilitated moving the wood through the forest (He et al., 2009). People from different households grouped together to carry out the extraction activities (e.g., transport the wood to their houses). This practice improved efficiency and minimized risks due to the complex topography (He et al., 2009). Instead of forming small groups, as was performed during the 1990s, by 2006/2007 most fuelwood collection activities were carried out by individual households and were not restricted to the winter months but were carried out year-round. Individual households made these changes in response to new policies that prevent harvesting entire trees and only allow collecting branches and small or dead trees (Chen et al., 2014, Yang et al., 2013a).

Over 75% of the interviewees reported that their households used to harvest fuelwood from natural forests during the late 1990s. In contrast, only about 10% of them reported that by 2006/2007 they still visited natural forests for fuelwood collection (Table 10.3). In contrast, plantation forests, which were planted mostly during the 1970s and 1980s near human settlements, became an important source of fuelwood (Table 10.3). A substantial increase in the use of driftwood and waste wood from various construction projects was also reported (Table 10.3), particularly by those households located along the main road. Mean annual labor spent in fuelwood collection activities was also reported to decrease between 1998 and 2006/2007 by ~70%, from 61 days to 19 days (Table 10.2). This reduction corresponded with the reduction in fuelwood consumption (Table 10.2).

The distances between households and fuelwood collection sites have increased and the areas for fuelwood collection have expanded. These changes may have occurred in response to the decline in fuelwood nearby and the implementation of the Natural Forest Conservation Program (NFCP) beginning in 2001 (Wolong Administration Bureau, 2004, Chen et al., 2010, 2012, 2014, Yang et al., 2013a, b; see Chapter 13). Under NFCP, all forms of logging were banned. In addition, a substantial portion of the forest in the reserve was divided into parcels and allocated to local households and household groups for monitoring (Yang et al., 2013b). Local residents were required to collect fuelwood only

Table 10.3 Changes in fuelwood collection sources between the late 1990s and 2006/2007 ($n = 184$). The numbers shown correspond to the ratio of households reporting collection of fuelwood from each source to total households sampled, in percent.

Fuelwood sources	Late 1990s	2006/2007
Natural forests	76.2%	10.3%
Replanted forests	2.2%	35.4%
Scrubland	80.0%	58.9%
Driftwood in the Pitiao River	1.1%	18.3%
Waste wood from construction areas	0.5%	2.9%

from within their own forest parcels. About 48% of the households that still use fuelwood reported that they travel longer distances than they used to in order to reach their NFCP parcels to collect fuelwood. About 80% reported that they had to cover larger areas to collect enough fuelwood, since by 2006/2007 they collected tree branches as opposed to entire trees, due to the logging ban imposed under NFCP.

These results show that the energy transition observed has not only reduced the dependency on fuelwood, but has also been accompanied by significant changes in the way fuelwood is procured. These changes are likely the result of the combination of conservation policy implementation (i.e., the NFCP logging ban) and the diversification of income sources. The former constrains the type of fuelwood harvested (since only branch collection is allowed under NFCP) and thus increases the harvesting area (and also the effort required) in order to collect enough fuelwood. The latter contributes to a reduction in the amount of time spent on fuelwood collection activities. Local people spend more time now on higher income-generating activities such as the production of cash crops and off-farm employment (e.g., tourism).

10.5 Determinants of fuelwood use reduction

Household fuelwood consumption reduction rates are significantly related to a variety of factors. Each of these factors was evaluated independently (models 1 through 5 in Table 10.4) and jointly (full model in Table 10.4). Results of the full model are discussed in the following paragraphs, while the results of models 1 through 5 are shown for completeness but are not discussed. Energy baselines, i.e., household fuelwood use ($p < 0.05$) and electricity consumption ($p < 0.01$) during the late 1990s, were used in all models as controlling factors. Both had significantly positive relations with fuelwood consumption reduction rates in all models in the case of electricity consumption, and in four out of six models in the case of fuelwood use (Table 10.4).

Among the economic factors evaluated, non-agricultural income and having a tourism business (i.e., households that owned or operated private hotels by 2007) exhibited negative and positive significant relationships with fuelwood consumption reduction rates, respectively (Table 10.4). The increase in income observed between the late 1990s and 2006/2007 was not related to the fuelwood-to-electricity switch observed in Wolong. The diversification of income sources may have been a primary factor. Therefore, the energy ladder model (Figure 10.1), as represented by an increase in income alone, may not fully support the energy transition in Wolong. However, other variables that did exhibit significant relationships with reduced fuelwood consumption (e.g., participation in tourism and educational attainment) could also be related to changes in income, and thus may implicitly provide some indirect support to the energy ladder model.

The completion of the main road in 1999, connecting Wolong to Chengdu, the capital city of Sichuan Province, also marked a major milestone in the energy transition in the reserve. This watershed event opened the local economy to outside markets. As a result, cash crops such as cabbages and radishes started to replace subsistence crops such as corn and potatoes, while tourism also started to boom (Liu et al., 2012). These processes generated a fourfold increase in the income of local households and reduced the poverty rates from 35.8% to 2.3% during the first decade of the twenty-first century (Liu et al., 2012). The human population increased by about 6% and the number of households by about 27% between 1998 and 2007 (Chapter 3).

Table 10.4 Model coefficients of general linear models on the determinants of household fuelwood consumption reduction rates ($n = 182$) between the late 1990s and 2006/2007. The mean variance inflation factor of the full model was 1.81. Values in parentheses correspond to standard errors. †, *, **, and *** represent significance levels at 0.10, 0.05, 0.01, and 0.001, respectively.

Independent variables	Model 1	Model 2	Model 3	Model 4	Model 5	Full model
Energy baseline						
Fuelwood	0.0044	0.0058†	0.0038	0.0059*	0.0051†	0.0071*
	(0.0030)	(0.0031)	(0.0030)	(0.0028)	(0.0029)	(0.0027)
Electricity	1.60***	1.26***	1.62***	1.45***	1.70***	1.23**
	(0.28)	(0.36)	(0.30)	(0.35)	(0.27)	(0.39)
Economic factors						
Δ Annual Income		−0.00068				−0.00062
		(0.00042)				(0.00046)
Ln(Non-ag income)		−0.0055				−0.0083*
		(0.0042)				(0.0040)
Tourism business		0.18***				0.14***
		(0.035)				(0.041)
Demographic factors						
Δ Education			0.011			0.020**
			(0.0073)			(0.0060)
Δ Household size			0.013			−0.0074
			(0.011)			(0.0095)
Δ Labor			−0.0054			0.017
			(0.017)			(0.015)
Δ Senior			0.0077			0.025
			(0.022)			(0.021)
Geographic factors						
Elevation				0.11		0.22*
				(0.10)		(0.091)
Township				−0.11*		−0.13**
				(0.042)		(0.039)
Distance				−0.048*		−0.060*
				(0.023)		(0.023)
Household density				8.26*		7.27*
				(3.42)		(3.64)
Policy factors						
GTGP/GTBP land					−0.011**	−0.0116**
					(0.0041)	(0.0039)
NFCP subsidy					−0.0082**	−0.0061†
					(0.0030)	(0.0033)
Constant	0.64***	0.66***	0.65***	0.40*	0.70***	0.33†
	(0.034)	(0.045)	(0.034)	(0.19)	(0.044)	(0.18)
Adjusted R^2	0.10	0.21	0.14	0.24	0.14	0.38

As stated above, reports of reductions in fuelwood use per household during this same period vary greatly, from average values as low as 40% (Chen et al., 2014, Yang et al., 2013a), to as high as 85% reported by the government (Wolong Administration Bureau, 2004). However, these reports support the finding that the reduction in fuelwood use is related to an increase in electricity consumption. This shift initiated a reserve-wide energy transition.

Among the demographic factors evaluated, only the change in adult educational attainment exhibited a significant positive relationship with reduced fuelwood consumption (Table 10.4). Educational attainment is assumed to be correlated with income, but the change in income alone was not a significant factor. This result suggests that not only changes in income but other metrics of a household's economic status need to be considered on their own, as well as in conjunction with other factors, to evaluate the energy transition. This result also suggests that the energy ladder model should be expanded by not treating income as the sole determinant of energy transition, particularly in rural settings like ours. Some other authors have reported similar results. For example, some researchers found a significantly positive relationship between a family's educational level and an ascension in the energy ladder (Gebreegziabher et al., 2009, Heltberg, 2004).

Regarding geographic factors, all variables evaluated exhibited a significant relationship with the fuelwood consumption reduction rate. Elevation and household density exhibited a positive relationship. Township and distance to the main road exhibited a negative relationship (Table 10.4). The regression results obtained for the dummy variable "Township" correspond with those for elevation, since households in Wolong Township are, on average, located at higher elevations than those in the other township (Gengda Township). Therefore, households in Wolong Township are exposed to colder winters and lower temperatures year-round, supporting the lower reduction in fuelwood consumption that was observed.

The positive relationship observed between fuelwood consumption reduction rates and household density could be explained through the diffusion of the adoption of technology. This is understood as the process of conscious behavioral change that comprises the acceptance, selection, and use of technology by an individual (Straub, 2009). Higher household densities tend to facilitate the diffusion of technology adoption. This is because an individual's willingness to adopt a technology is related to the decisions of others, which requires close proximity for developing proper observation of the behavior of others (Rogers, 2003). Higher household density also increases the efficiency of monitoring fuelwood collection, storage, and consumption activities within communities.

The negative relationship observed between the proximity to the main road and fuelwood consumption reduction rate could be due, on the one hand, to the fact that households closer to the road were more likely to have benefited from the electricity grid for a longer time than those farther from the road. On the other hand, households farther away from the road may experience lower costs/efforts to collect fuelwood, given their closer proximity to forested areas. Therefore, easier access to the electricity supply and longer distances to forested areas may be supporting the switch from fuelwood to electricity. In addition, households closer to the road tend to exhibit higher income (Viña et al., 2013), which provides indirect support for the energy ladder model.

Finally, participation in the Grain to Green Program (GTGP) and the Grain to Bamboo Program (GTBP) since the early 2000s exhibited a significant negative relationship with fuelwood consumption reduction rates (Table 10.4). Under GTGP and GTBP, local households received a subsidy from the government for converting their marginal cropland into tree plantations (mainly Japanese Larch; *Larix kaempferi*) and bamboo plantations (mainly umbrella bamboo; *Fargesia robusta*), respectively. Under NFCP, each household also received a subsidy contingent on the status of the forest parcel under its protection (Yang et al., 2013b). While these subsidies have been paid mainly in cash, the administration of Wolong encouraged local households to use them for purchasing electricity (Chen et al., 2014), thus further contributing to the energy transition. Yet, households with more cropland enrolled in GTGP and GTBP and with higher ratios of NFCP subsidy to total income in 1998 tended to be related to lower fuelwood consumption reduction rates.

Although this finding seems counterintuitive, it is important to note that almost all households in Wolong participated in these programs and that fuelwood consumption has indeed been reduced in Wolong. Therefore, our results suggest that the level of participation in these programs, rather than participation per se, is inversely related with the rate of reduction in fuelwood consumption. This could be explained by the fact that higher subsidy incomes tend to free farm labor, which then could be used to collect fuelwood. Participation in these programs may also encourage the use of branches rather than entire trees for fuelwood, therefore contributing to lower rates of reduction in fuelwood consumption. On the other hand, since 2000 NFCP and GTGP/GTBP have been associated with more non-farm income opportunities (Yang et al., 2013c; Chapter 9), such as tourism-related activities. These opportunities have, in turn, been related to higher fuelwood consumption reduction rates (Table 10.4). Perhaps the diversification of income sources is associated with a diversification of household energy sources rather than with a complete switch from fuelwood to electricity. However, given that participation in GTGP, GTBP, and NFCP constitutes endogenous choices, our results and their interpretation should be taken with a grain of salt.

10.6 Local residents' perception toward energy-related issues

Local residents generally recognized the various advantages of switching from fuelwood to electricity in their daily lives (Table 10.5). For instance, convenience and cleanliness were recognized by the majority. About a quarter of the study participants reported that using electricity saved time and labor required to collect fuelwood, allowing them to participate in alternative leisure and income-generating activities. Surprisingly, only 12% perceived that switching from fuelwood to electricity improved indoor air quality. This creates an educational opportunity through which a switch from fuelwood to electricity could be further enhanced.

In terms of the disadvantages of electricity use (Table 10.5), about half of the households interviewed felt using electricity was not as safe as using fuelwood, and about one-third thought that electricity was still quite expensive. In addition, local residents reported an average of 18 days of electricity outage in any given year within their neighborhoods, with most outages happening during the winter (i.e., dry) season. Almost 80% of the interviewees considered that these outages "seriously impacted" (44%) or "impacted" (33%) their daily lives. Interestingly, when asked about the government's electricity subsidy, 54% of the residents indicated that they had not heard about it. Finally, five interviewees (2.6%) stated that local traditions such as cooking big meals during ethnic holidays and making smoked pork for daily consumption were affected, since they could not properly complete such activities without the use of fuelwood.

Table 10.5 Local residents' perceptions of the advantages and disadvantages of switching from fuelwood to electricity ($n = 189$).

Advantages	
More convenient and saving time for cooking	85%
Cleaner	54%
Saving time and labor previously used for collecting fuelwood	26%
Better indoor air quality	12%
Disadvantages	
Not as safe as using fuelwood	51%
Costs money	34%
Affecting local traditions	3%

These results suggest that local residents' perception of modern energy sources such as electricity may influence their adoption. Such perceptions may also explain why local households in Wolong have not completely switched from fuelwood to electricity. The lack of a complete switch to electricity is otherwise counterintuitive, considering the large increase in income, higher accessibility to electricity grids, more affordable electricity prices, and higher access to technology.

10.7 Concluding remarks

Our results suggest that Wolong Nature Reserve is experiencing a fast energy transition from fuelwood to electricity, with little use of transitional fuels such as coal and kerosene by local households. Multiple economic, demographic, geographic, and governance factors exhibited stronger relationships with the switch from fuelwood to electricity than the changes in income alone. Contrary to what seems to be occurring at the national level, the leapfrogging model constitutes a better theoretical construct than the energy ladder model for explaining the patterns of energy transition in Wolong. However, similar to many other rural areas in China and around the developing world, households in Wolong so far have ended up utilizing both fuelwood and electricity simultaneously to fulfill their energy needs. Therefore, energy transition in Wolong seems to be better represented by the multiple fuel adoption strategy (Masera and Navia, 1997, Masera et al., 2000).

Energy transition in Wolong has a positive effect on the terrestrial ecosystems of the reserve. Local residents have used less fuelwood, and thus both forests (Chapter 6) and giant panda habitat (Chapter 7) have experienced a conspicuous continuous recovery since the early 2000s. The overall effects on the human system appear to also be positive, although the impact of hydropower plants on aquatic ecosystems is not clear. However, despite the expected reduction in indoor pollution through the adoption of cleaner energy sources such as electricity, little is known about the health benefits of the energy transition that have accrued to local residents. Studies should, therefore, be performed to evaluate the effects of the switch from fuelwood to electricity on aquatic systems and human health in Wolong.

Finally, additional longitudinal studies, combined with modeling following a coupled human and natural system framework (Liu et al., 2007a, b), need to be conducted. Researchers should try to determine what would be a sustainable energy structure for communities in Wolong, and if and when Wolong will eventually experience a complete switch to electricity, together with the conditions required for such a transition and the impacts on human and natural systems. Such evaluations will be valuable for further development of policies that effectively incentivize energy transitions in rural areas without negative impacts on local culture and human well-being. These policies can have an impact not only across China but around the developing world for achieving both environmental sustainability and human well-being goals.

10.8 Summary

More than 2.5 billion people around the world, particularly in rural areas of developing countries, still rely on biomass fuels (e.g., fuelwood) to fulfill their household energy requirements. This chapter described the patterns of energy transition from fuelwood to electricity and their main drivers in Wolong Nature Reserve. We conducted a longitudinal survey on this topic in a random sample of 189 Wolong households during 1998 and 2007. Contrary to what seems to be happening at the national level, Wolong is experiencing an energy transition that is occurring at a fast pace. From 1998 to 2006/2007, the mean household annual fuelwood consumption decreased by almost 75%, while the mean household annual electricity use increased more than twofold. This transition is related to a series of economic, demographic, geographic, and governance factors. Factors positively related to fuelwood reduction included operating a tourism business, educational level, elevation, and household density. Factors negatively contributing to the fuelwood reduction included distance to the main road and participation in incentive-based conservation programs. The transition does not constitute a complete switch from biomass fuels to electricity, but rather a simultaneous use of these different energy sources. Therefore, the theoretical construct that better explains the patterns observed in Wolong is the leapfrogging model combined with a multiple fuel adoption strategy. Future work should evaluate if the energy transition path observed in Wolong could be used for better policy development to help reach both human well-being and environmental sustainability goals.

References

An, L., Liu, J., Ouyang, Z., et al. (2001) Simulating demographic and socioeconomic processes on household level and implications for giant panda habitats. *Ecological Modelling*, **140**, 31–49.

An, L., Lupi, F., Liu, J., et al. (2002) Modeling the choice to switch from fuelwood to electricity—implications for giant panda habitat conservation. *Ecological Economics*, **42**, 445–57.

An, L., Zvoleff, A., Liu, J., and Axinn, W.G. (2014) Agent-based modeling in coupled human and natural systems

(CHANS): lessons from a comparative analysis. *Annals of the Association of American Geographers*, **104**, 723–45.

Anderson, C.B., Likens, G.E., Rozzi, R., et al. (2008) Integrating science and society through long-term socio-ecological research. *Environmental Ethics*, **30**, 295–312.

Axinn, W.G., Pearce, L.D., and Ghimire, D. (1999) Innovations in life history calendar applications. *Social Science Research*, **28**, 243–64.

Barnes, D.F. and Floor, W.M. (1996) Rural energy in developing countries: a challenge for economic development. *Annual Reviews in Energy and the Environment*, **21**, 497–530.

Bearer, S., Linderman, M., Huang, J., et al. (2008) Effects of fuelwood collection and timber harvesting on giant panda habitat use. *Biological Conservation*, **141**, 385–93.

Beyer, H.L. (2004) *Hawth's Analysis Tools for ArcGIS*. http://www.spatialecology.com/htools.

Bond, T., MacCarty, N., Ogle, D., et al. (2008) A laboratory comparison of the global warming impact of five major types of biomass cooking stoves. *Energy for Sustainable Development*, **7**, 5–14.

Chen, X., Lupi, F., An, L., et al. (2012) Agent-based modeling of the effects of social norms on enrollment in payments for ecosystem services. *Ecological Modelling*, **229**, 16–24.

Chen, X., Lupi, F., He, G., and Liu, J. (2009) Linking social norms to efficient conservation investment in payments for ecosystem services. *Proceedings of the National Academy of Sciences of the United States of America*, **106**, 11812–17.

Chen, X., Lupi, F., Viña, A., et al. (2010) Using cost-effective targeting to enhance the efficiency of conservation investments in payments for ecosystem services. *Conservation Biology*, **24**, 1469–78.

Chen, X., Viña, A., Shortridge, A., et al. (2014) Assessing the effectiveness of payments for ecosystem services: an agent-based modeling approach. *Ecology and Society*, **19**, 7.

Chowdhury, R. and Turner, B. (2006) Reconciling agency and structure in empirical analysis: smallholder land use in the southern Yucatan, Mexico. *Annals of the Association of American Geographers*, **96**, 302–22.

Clancy, J.S. (2008) Urban ecological footprint in Africa. *African Journal of Ecology*, **46**, 463–70.

de Sherbinin, A., Carr, D., Cassels, S., and Jiang, L. (2007) Population and environment. *Annual Review of Environment and Resources*, **32**, 345–73.

Foster, J.B. (1999) Marx's theory of metabolic rift: classical foundations for environmental sociology. *The American Journal of Sociology*, **105**, 366–405.

Fox, J. and Vogler, J.B. (2005) Land-use and land-cover change in montane mainland southeast Asia. *Environmental Management*, **36**, 394–403.

Freeman, D., Thornton, A., Camburn, D., et al. (1988) The life history calendar: a technique for collecting retrospective data. In C. Clogg, ed., *Sociological Methodology*, pp. 37–68. American Sociological Association, Washington, DC.

Gebreegziabher, Z., Mekonnen, A., Kassie, M., and Kohlin, G. (2009) *Urban Energy Transition and Technology Adoption: The case of Tigrai, Northern Ethiopia*. Environmental Economics Policy Forum for Ethiopia (EEPFE), Ethiopian Development Research Institute (EDRI), Addis Ababa, Ethiopia.

Ghimire, K.B. (1997) Conservation and social development: an assessment of Wolong and other panda reserves in China. In K.B. Ghimire and M.P. Pimbert, eds, *Environmental Politics and Impacts of National Parks and Protected Areas*, pp.187–213. Earthscan Publications, London, UK.

Global Energy Assessment (2012) *Toward a Sustainable Future*. International Institute for Applied Systems Analysis, Vienna, Austria and Cambridge University Press, Cambridge, UK.

Goldemberg, J. (1998) Leapfrog energy technologies. *Energy Policy*, **26**, 729–41.

He, G., Chen, X., Bearer, S., et al. (2009) Spatial and temporal patterns of fuelwood collection in Wolong Nature Reserve: implications for panda conservation. *Landscape and Urban Planning*, **92**, 1–9.

Heltberg, R. (2004) Fuel switching: evidence from eight developing countries. *Energy Economics*, **26**, 869–87.

Heltberg, R., Arndt, T., and Sekhar, N. (2000) Fuelwood consumption and forest degradation: a household model for domestic energy substitution in rural India. *Land Economics*, **76**, 213–32.

Holdren, J.P. and Smith, K.R. (2000) Energy, the environment, and health. In J. Goldemberg, ed., *The World Energy Assessment: Energy and the Challenge of Sustainability*, pp. 61–110. United Nations Development Programme, New York, NY.

Holland, J.H. (1992) Complex adaptive systems. *Daedalus*, **121**, 17–30.

Hull, V., Tuanmu, M.N., and Liu, J. (2015) Synthesis of human-nature feedbacks. *Ecology and Society* **20**(3), 17.

International Energy Agency (2013) *World Energy Outlook 2013*. OECD/IEA, Paris, France.

Jebaraj, S. and Iniyan, S. (2006) A review of energy models. *Renewable and Sustainable Energy Reviews*, **10**, 281–311.

Laney, R. (2002) Disaggregating induced intensification for land change analysis: a case study from Madagascar. *Annals of the Association of American Geographers*, **92**, 702–26.

Leach, G. (1992) The energy transition. *Energy Policy*, **20**, 116–23.

Levin, S.A. (1998) Ecosystems and the biosphere as complex adaptive systems. *Ecosystems*, **1**, 431–36.

Li, C., Zhou, S., Xiao, D., et al. (1992) The history and status of Wolong Nature Reserve. Wolong Nature Reserve, Sichuan Normal College, eds, *The Animal and Plant Resources and Protection of Wolong Nature Reserve*, pp. 326–42. Sichuan Publishing House of Science and Technology, Chengdu, China.

Liu, J. (2010) China's road to sustainability. *Science*, **328**, 50.

Liu, J. and Diamond, J. (2008) Revolutionizing China's environmental protection. *Science*, **319**, 37–38.

Liu, J., Dietz, T., Carpenter, S.R., et al. (2007a) Complexity of coupled human and natural systems. *Science*, **317**, 1513–16.

Liu, J., Dietz, T., Carpenter, S.R., et al. (2007b) Coupled human and natural systems. *Ambio*, **36**, 639–49.

Liu, J., Linderman, M., Ouyang, Z., et al. (2001) Ecological degradation in protected areas: the case of Wolong Nature Reserve for giant pandas. *Science*, **292**, 98.

Liu, J., Mooney, H., Hull, V., et al. (2015) Systems integration for global sustainability. *Science*, **347**, 1258832.

Liu, J., Ouyang, Z., Taylor, W.W., et al. (1999) A framework for evaluating the effects of human factors on wildlife habitats: the case of giant pandas. *Conservation Biology*, **13**, 1360–70.

Liu, W., Vogt, C.A., Luo, J., et al. (2012) Drivers and socioeconomic impacts of tourism participation in protected areas. *PLoS ONE*, **7**, e35420.

Liu, Z., Guan, D., Crawford-Brown, D., et al. (2013) A low-carbon road map for China. *Nature*, **500**, 143–45.

Macht, C., Axinn, W.G., and Ghimire, D. (2012) Household energy consumption: community context and the fuelwood transition. *Social Science Research*, **41**, 598–611.

Masera, O. and Navia, J. (1997) Fuel switching or multiple cooking fuels? Understanding inter-fuel substitution patterns in rural Mexican households. *Biomass and Bioenergy*, **12**, 347–61.

Masera, O., Saatkamp, B., and Kammen, D. (2000) From linear fuel switching to multiple cooking strategies: a critique and alternative to the energy ladder model. *World Development*, **28**, 2083–103.

McNicoll, G. (2002) Managing population-environment systems: problems of institutional design. *Population and Development Review*, **28**, 144–64.

Mena, C.F., Walsh, S.J., Frizzelle, B.G., et al. (2011) Land use change on household farms in the Ecuadorian Amazon: design and implementation of an agent-based model. *Applied Geography*, **31**, 210–22.

Murphy, J.T. (2001) Making the energy transition in rural East Africa: is leapfrogging an alternative? *Technological Forecasting & Social Change*, **68**, 173–93.

National Energy Administration of China (2014) *Total Electricity Consumption in 2013*. National Energy Administration, Beijing, China. http://www.gov.cn/gzdt/2014-01/14/content_2566377.htm.

Pachauri, S. and Jiang, L. (2008) *The Household Energy Transition in India and China*. International Institute for Applied Systems Analysis (Interim Report IR-08-009), Luxemburg, Austria.

Rogers, E.M. (2003) *Diffusion of Innovations*. Free Press, New York, NY.

Straub, E.T. (2009) Understanding technology adoption: theory and future directions for informal learning. *Review of Educational Research*, **79**, 625–49.

Viña, A., Bearer, S., Chen, X., et al. (2007) Temporal changes in giant panda habitat connectivity across boundaries of Wolong Nature Reserve, China. *Ecological Applications*, **17**, 1019–30.

Viña, A., Chen, X., Yang, W., et al. (2013) Improving the efficiency of conservation policies with the use of surrogates derived from remotely sensed and ancillary data. *Ecological Indicators*, **26**, 103–11.

Vitousek, P. (2006) Ecosystem science and human-environment interactions in the Hawaiian archipelago. *Journal of Ecology*, **94**, 510–21.

Wolong Administration Bureau (2004) *The History of Wolong Nature Reserve*. Sichuan Science and Technology Press, Chengdu, China (in Chinese).

World Health Organization (2009), *Global Health Risks: mortality and burden of disease attributable to selected major risks*. World Health Organization, Geneva, Switzerland. www.who.int/healthinfo/globalburdendisease/GlobalHealthRisksreportFront.pdf.

Yang, W., Dietz, T., Liu, W., et al. (2013c) Going beyond the Millennium Ecosystem Assessment: an index system of human dependence on ecosystem services. *PLoS ONE*, **8**, e64581.

Yang, W., Liu, W., Viña, A., et al. (2013a) Performance and prospects of payments for ecosystem services programs: evidence from China. *Journal of Environmental Management*, **127** 86–95.

Yang, W., Liu, W., Viña, A., et al. (2013b) The nonlinear effects of group size on collective action and resource outcomes. *Proceedings of the National Academy of Sciences of the United States of America*, **110**, 10916–21.

CHAPTER 11

Social Capital and Social Norms Shape Human–Nature Interactions

Xiaodong Chen, Wu Yang, Vanessa Hull, Li An, Thomas Dietz, Ken Frank, Frank Lupi, and Jianguo Liu

11.1 Introduction

Humans affect the natural environment through a variety of activities, including the overuse of natural resources, which results in ecosystem degradation worldwide (Millennium Ecosystem Assessment, 2005, Vitousek et al., 1997). Conservation efforts such as payments for ecosystem services (PES) have been implemented to recover some of these degraded ecosystems (Chen et al., 2009a, OECD, 1997). However, current conservation resources are still far less than what is required globally (James et al., 1999, 2001). Therefore, it is important to understand complex interactions in coupled human and natural systems (CHANS) that affect the consumption of environmentally significant resources and human responses to conservation policies. One emerging area of research in coupled systems involves appreciation for the fact that human actions are embedded in a social context (Dietz and Henry, 2008). Such research recognizes that human activities are often affected by social capital and social norms (Dietz et al., 2003, Ostrom, 2000).

One key debate in the sustainability of coupled systems is the degree of substitutability between natural capital and human capital (Daly and Cobb, 1989, Dasgupta, 2010, Neumayer, 2010). Natural capital is defined as the stock of goods and services that humans derive from the natural environment (Costanza et al., 1997, Daily et al., 2000). Human capital refers to the stock of knowledge, education, and abilities humans possess to produce economic values. In environmental economics, advocates of weak sustainability assume that long-term sustainability can still be achieved with a loss of natural capital, as long as human capital is available as a substitute. In contrast, advocates of strong sustainability assume limited or no substitutability between natural and human capitals. Thus, they emphasize that the amount of each type of capital should be maintained for future generations in order to achieve sustainability (Dasgupta, 2010, Solow, 1993). This debate has been expanded to include social capital, which is defined as the resources that people access through social ties and relations (Arrow et al., 2004, Lin, 2001). [We are aware that there are various definitions of social capital and some scholars define social networks and norms together as social capital (Lyon, 2000, Pretty and Ward, 2001).]

In addition to social capital, social norms are also important factors that may affect human–nature interactions. Social norms are shared understanding in a community about what actions are proper or correct, or improper or incorrect, and actors enforce norms to perpetuate the social systems to which they are committed (Bendor and Swistak, 2001, Coleman, 1990). Social norms may be sustained through internalized social–psychological values such as reputation, fairness, and self-esteem. Members in the community reward or punish people who follow or break the norms (Elster, 1989, Fehr and Gintis, 2007, Goldstein et al., 2008). More generally, social norms can also be simply what

Pandas and People. Edited by Jianguo Liu, Vanessa Hull, Wu Yang, Andrés Viña, Xiaodong Chen, Zhiyun Ouyang, and Hemin Zhang. Published 2016 by Oxford University Press.

most people do in a given situation. Social norms may even result from an equilibrium in which economically rational agents choose actions based on the expected behaviors of others (Young, 1996). Past studies have shown that social norms can substantially influence human behavior (Elster, 1989, Fehr and Gintis, 2007, Goldstein et al., 2008), such as collective actions in natural resources management (Dietz et al., 2003, Ostrom, 2000, Pretty, 2003) or decisions about environmentally significant consumption (Schultz et al., 2007). Under the influence of social norms, trust and reciprocity among community members encourage the engagement of collective activities in natural resources extraction. Formal or informal sanctions prevent the detachment from norms.

Despite the advances in research on both social norms and social capital, most studies on coupled systems do not account for either. We hypothesize that it is essential to incorporate social norms and social capital with demographic, economic, and environmental factors in order to fully understand the complexity of coupled systems. In this chapter, we outline our research efforts aimed at testing this hypothesis. We first provide a brief overview of how social norms and social capital affect human–nature interactions in our model coupled system—Wolong Nature Reserve (Chapter 3). Next, we discuss our research on the intersection of social capital, labor migration, and environmental impacts. We then discuss our findings on the impact of social norms on participation in PES programs, including both the Grain to Green Program (GTGP) and the Natural Forest Conservation Program (NFCP). We close the chapter with some implications of using social norms and social capital to shape human activities for sustainability.

11.2 Overview of social norms, social capital, and human–nature interactions in Wolong

The human system in Wolong is a dynamic rural community with nuanced cultural traditions and a rich history (Chapter 3). People living in Wolong interact with their environment in many complex ways. Examples include fuelwood collection (Chapters 7 and 10), farming and tourism (Chapters 4 and 13), and participation in PES programs such as NFCP and GTGP (Chapters 5 and 13). During these activities, inhabitants in Wolong have developed social capital and established social norms. For instance, during planting and harvesting seasons, groups of households often work together supporting different households at different times in order to improve the efficiency of labor use. Exchanges of labor are also common in fuelwood collection and raising livestock, for when groups of households work together and share facilities.

Two human activities driven by social norms and social capital that have particular relevance for sustainability in Wolong are labor migration and PES programs (both GTGP and NFCP). Below we explore each in detail with the goal of better understanding the role of social norms and social capital in mediating human–nature interactions and how this knowledge can improve management measures for sustainability.

11.3 The intersection of social capital, labor migration, and environmental impacts

Labor migration is a global phenomenon that is particularly prevalent in developing countries, where employment opportunities in distant areas spur individuals to move away from their homes in search for jobs. The impacts of social capital on labor migration are well understood. Social capital is important in gaining access to employment information and influential persons for employment (Bian, 1997, Granovetter, 1995, Lin et al., 1981, Yakubovich, 2005). It also plays a key role in making migration decisions (Hugo, 1998, Massey, 1990, Palloni et al., 2001) and reducing costs and risks of migration (Korinek et al., 2005). Studies on the impacts of social capital on labor migration often differentiate the strength of social ties because social ties with different strengths may affect labor migration in different ways. Among different types of social ties, relatives have stronger social ties than friends, and friends have stronger social ties than acquaintances (Bian, 1997, Granovetter, 1995). Strong social ties are often more reliable than weak social ties, and hence have stronger impacts on migration processes such as transportation and settlement (Massey and Espinosa, 1997, Wilson, 1998). Weak social ties are especially helpful for migrants in obtaining information about employment opportunities, and providing direct access to influential persons for employment (Bian, 1997, Granovetter, 1995, Yakubovich, 2005).

Many farmers in Wolong relocate to cities for short periods of time as labor migrants. This phenomenon is also very common in many other rural regions in China as a result of increasingly available employment opportunities from the rapid economic growth in cities (Johnson, 2000, Li and Zahniser, 2002). Most of the labor migrants from Wolong only work in cities temporarily and return to their home villages whenever needed. Labor migration is important to study in Wolong because it may affect the local environment (Chen et al., 2012a). First, remittances from labor migration may be used for electricity, hence reducing fuelwood use (see also Chapter 10). Second, reduced human population due to migration may reduce both the demand for fuelwood use and the labor supply for fuelwood collection. Third, other human activities negatively affecting the environment may also be reduced (e.g., medicinal herb collection or poaching). Moreover, labor migrants may help other local residents find employment opportunities and migrate to cities via social capital, so the cumulative effects of labor migration on the environment could be substantial in the long run.

Despite the well-established links between migration and social capital, their impact on the environment is not well understood. We set out to establish the relationship between social capital, labor migration, and the environment using our model system of Wolong. To do so, we followed the causal chain from social capital to labor migration by estimating the effects of social capital, human capital, and economic conditions on labor migration. We then followed the chain from labor migration to fuelwood consumption by estimating the effects of labor migration on fuelwood consumption. We conducted household interviews in 2005 with households that had (n = 129) and did not have (n = 215) labor migrants. During the interviews, we collected information on households' fuelwood consumption, human capital factors and economic status, social ties with people living or working in cities, and the amount of remittances that labor migrants sent back home (Chen et al., 2012a). We recorded the availability of different types of social ties. We considered social ties with relatives as strong ties, with acquaintances as weak ties, and with friends as ties of moderate strength.

Labor migration can be confounded with fuelwood consumption because labor migration is a process of self-selection rather than a process of controlled random assignment. Under such circumstances, any estimated causal effect can be spurious unless the process of self-selection is taken into account (Hirano and Imbens, 2002, Winship and Morgan, 1999). We adjusted for the confounding of labor migration with fuelwood consumption using the propensity score weighting method (Hirano and Imbens, 2002, Robins and Rotnitzky, 1995, Rosenbaum and Rubin, 1983). Propensity score techniques compare individuals in the treatment group (i.e., households with migrants) to individuals in the control group (i.e., households without migrants) with a similar propensity score (i.e., likelihood of being in the treatment group). This procedure allows the random assignment of treatment to be approximated (Rosenbaum and Rubin, 1983). The propensity score is defined as:

$$p(S) = \Pr(M = 1 \mid S) \qquad \text{(Equation 11.1)}$$

where M is a dummy variable of treatment, S is a group of covariates, and $p(S)$ is the probability of receiving the treatment, which can be estimated using a logistic regression model.

Assuming the relevant covariates were in the model, the average effect of labor migration on fuelwood consumption can be consistently estimated using a propensity score weighted general linear model (Hirano and Imbens, 2002, Hirano et al., 2003). The weights of the model are defined as:

$$\omega(M,S) = \frac{M}{p(S)} + \frac{1-M}{1-p(S)} \qquad \text{(Equation 11.2)}$$

That is, a household with labor migrants is weighted by $1/p(S)$ and a household without migrants is weighted by $1/(1-p(S))$. So the lower the propensity of having migrants for those households with migrants, the greater the weight would be.

Similarly, the higher the propensity of having migrants for households without migrants, the greater the weight would be. Therefore, the estimation of the average effect of labor migration on fuelwood use focuses mainly on the strongest overlap in propensity between the treatment group and the control group.

We first established a logistic regression model to estimate the propensity for labor migration based on people's human capital, economic status, and social capital. Then we estimated the average effect of labor migration on fuelwood consumption using a propensity score weighted general linear model (Chen et al., 2012a). All working-age people (18–60 years of age, 912 people) from the households that we interviewed were used to estimate the propensity for labor migration. Since the correlation in characteristics among people in the same household may result in heteroscedasticity in the regression, we used Huber's variance correction to obtain robust standard errors (Wooldridge, 2002).

We found that among social capital factors, the availability of acquaintances in cities was significantly positively correlated to labor migration, while the availabilities of relatives and friends were not statistically significant. The availability of acquaintances increased the odds of labor migration by 2.54 when other variables were held constant (Table 11.1). Our

Table 11.1 Determinants of labor migration (reproduced from Chen et al., 2012a).

Independent variables	Coefficient (adjusted standard error) [odds ratios]
Gender (male = 1; female = 0)	1.029*** (0.222) [2.798]
Age (years)	0.316** (0.117) [1.372]
Age squared	−0.005** (0.002) [0.995]
Marital status (married = 1; single = 0)	−1.725*** (0.344) [0.178]
Education (years)	0.186*** (0.043) [1.204]
Children (number of children younger than 15 years)	0.072 (0.182) [1.075]
Extended (have extended member = 1; no extended member = 0)	0.314 (0.312) [1.369]
Laborers (number of working-age people)	0.359** (0.126) [1.432]
Cropland (hectares)	−0.267 (0.870) [0.766]
Non-migration income (thousands of yuan)	−0.082*** (0.024) [0.921]
Township (Gengda Township = 1; Wolong Township = 0)	0.782*** (0.238) [2.186]
Relative (Have relatives in cities = 1; no relatives in cities = 0)	0.196 (0.217) [1.217]
Friend (Have friends in cities = 1; no friends in cities = 0)	0.197 (0.281) [1.218]
Acquaintance (Have acquaintances in cities = 1; no acquaintances in cities = 0)	0.930*** (0.244) [2.535]
Intercept	−8.003*** (1.959)
Pseudo-R^2	0.319

Notes:
$^{**}p \leq 0.01$; $^{***}p \leq 0.001$; $n = 912$

results suggest that households with weak social ties (acquaintances) were more likely to have labor migrants (Chen et al., 2012a).

Some human capital and economic conditions were also significantly correlated with labor migration (Table 11.1). Men were more likely to work as labor migrants than women, which is consistent with the observation that men are usually expected to assume more economic responsibilities for households in rural China. Significant effects of age and its quadratic term on migration suggested a quadratic relationship between age and labor migration. The estimated coefficients imply that as age increased, the probability of migration increased until 30 years old and then declined. Married people were less likely to migrate, while education increased the probability of migration. The number of working-age people in the household had a significant positive effect on labor migration. We also found a significant negative effect of non-migration income on labor migration (Table 11.1). The township indicator also had a significant effect on labor migration. People in Gengda Township, which is geographically closer to cities, were more likely to work as labor migrants than people in Wolong Township.

In order to estimate the average effect of labor migration on fuelwood consumption, we used the estimated propensities to weight a standard regression (Chen et al., 2012a). We found that households with labor migrants consumed significantly less fuelwood (1788 kg less) than households without migrants (Table 11.2). Household size had a significant positive effect on fuelwood consumption, as larger households usually had a higher demand for energy than smaller ones. Households with senior members consumed more fuelwood because senior people required more fuelwood for heating in winter (An et al., 2001). Non-migration income had a significant negative effect on fuelwood consumption because households with higher non-migration income tended to have better housing conditions and could afford more electricity (Table 11.2). The amount of cropland positively affected fuelwood consumption, presumably because households with more cropland tended to rely more on natural resources while households with less cropland relied more on off-farm opportunities to meet livelihood needs. Last, people in Gengda Township consumed less fuelwood than people in Wolong Township. One of the main differences between the two townships is that the elevation of Gengda Township is much lower than that of Wolong Township. The weather in Gengda Township is warmer than Wolong Township, and hence residents need less fuelwood for heating and cooking.

Table 11.2 Estimation of the average effect of labor migration on fuelwood consumption (reproduced from Chen et al., 2012a).

Independent variables	Coefficient (standard error)
Migration (Have labor migrant = 1; no labor migrant = 0)	−1788*** (405)
Household size (number of people)	573** (174)
Senior (have senior member = 1; no senior member = 0)	1253** (446)
Non-migration income	−114*** (24)
Cropland	7965*** (1414)
Township	−2506*** (429)
Relative	−290 (417)
Friend	−1067* (506)
Acquaintance	78 (479)
Intercept	4576*** (985)
Adjusted R^2	0.31

Notes:
$^*p \leq 0.05$; $^{**}p \leq 0.01$; $^{***}p \leq 0.001$. $n = 344$

While our results illuminate the importance of weak social ties for labor migration, we do not deny the role of strong social ties because some weak ties might develop through strong ties. However, our results support Granovetter's argument for "the strength of weak ties" (Granovetter, 1973, 1995) and suggest that strong ties alone may not be very helpful for labor migration. Compared to strong ties, weak ties are spread across wider social networks and are more likely to expand employment information in urban settings as well as gain access to influential persons in the job market directly. Furthermore, we found that having labor migrants substantially reduced household fuelwood consumption. Thus, weak social ties had a significant indirect effect on the environment in Wolong (Chen et al., 2012a). In our analysis, it is not reduced population through migration alone that mitigated human pressure on the environment. Rather, labor migration also reduced human pressures because there was a readily available, although costly, supply of electricity that could serve as a substitute for fuelwood. Through labor migration, local households were better able to afford to use electricity since labor migration complements rural agricultural income, a common phenomenon in transitional economies (Korinek et al., 2005). Our results provided evidence of partial substitutability among social capital, human capital, and natural capital.

11.4 Social capital and forest monitoring under the Natural Forest Conservation Program

Social capital also plays a crucial role in participation in PES programs. An example is NFCP, a program launched across China to protect natural forests through a series of measures, including a logging ban, artificial plantation and restoration, and forest monitoring (Liu et al., 2008, 2013; see also Chapters 5 and 13). In Wolong, NFCP officially started in 2001 and covered all the forested areas (120 500 ha). NFCP implementation in Wolong is unique compared to most other areas across China because it engaged households to participate in forest monitoring for payment incentives (Chen et al., 2014, Yang et al., 2013a). Approximately one-third of the forest areas (40 100 ha) near human settlements was assigned to a total of 1130 households for monitoring in the form of household groups. The average annual payment rate was around US$143 per household, with distant forest parcels assigned to large household groups with higher payments (Yang et al., 2013b).

Social capital comes into play via the interactions within and among monitoring groups. Group size (i.e., the number of households in each monitoring group) for each forest parcel varied from 1 to 16. The Wolong Administrative Bureau decided the household group composition, but groups had the autonomy to decide their monitoring strategies such as monitoring frequency, duration, and subdivision of group members to monitor in turns. The Bureau evaluated household-monitored forest parcels twice a year through random field assessments (e.g., inspection of illegal logging) and anonymous reporting (people who anonymously report illegal activities obtain cash rewards). If any illegal activities were detected in a household-monitored forest parcel, every group member would receive a penalty in terms of payment reduction, and legal sanctions if applicable (Yang et al., 2013b). The exception to this rule occurred if they could identify the person(s) who conducted the infraction (Yang et al., 2013b).

We expected that social capital might play an important role in affecting the collective monitoring behaviors among group members and across different monitoring groups under NFCP, and then influence the forest monitoring outcomes. On one hand, households with social ties to local leaders would have more influence on other households' monitoring and logging behaviors. Social learning literature (Henrich and Gil-White, 2001, Milinski et al., 2002, Rustagi et al., 2010) suggests that this may occur because leaders will often be disproportionally imitated by others. On the other hand, the social relationships among group members might affect their self-organization and monitoring strategies.

To examine our hypothesized effects of social capital on households' forest monitoring behaviors, we conducted quantitative and qualitative analyses based on our long-term empirical household survey data, and focus group and personal interviews (Yang et al., 2013b). Our dependent variable (i.e., the total laborer days of input per year by each household spent on monitoring) had a censored distribution (clustered with a minimum value of zero and maximum value of 20 in our case). Therefore, we used censored regression models (i.e., Tobit models) to estimate the parameters. The general form of a Tobit model (Wooldridge, 2002) is given by:

$$y_{1i} = \begin{cases} y_{1i}^{*} & a < y_{1i}^{*} < b \\ a & y_{1i}^{*} \le a \\ b & y_{1i}^{*} \ge b \end{cases} \quad \text{(Equation 11.3)}$$

$$y_{1i}^{*} = X_{1i}\beta_1 + u_i \quad \text{(Equation 11.4)}$$

where y_{1i} is the observed monitoring effort, y^*_{1i} is a latent variable that satisfies assumptions of a classic linear model, a is the minimum value, and b is the maximum value. X_{1i} is a vector of exogenous explanatory variables, β_1 is a parameter vector to be estimated, i is the ith observation, and u_i is an error term that has a normal distribution with a mean value of zero.

Our results suggest that households with strong social ties to local leaders on average spent 54% less labor on forest monitoring than those with weak social ties to local leaders (Table 11.3). We also found that households with strong social ties tended to be those whose adult members were more senior and educated (statistically significant with $p < 0.05$ and $p < 0.01$, respectively). Such findings were also confirmed by qualitative evidence from household interviewees, who admitted that they dared not and did not want to conduct illegal activities in forest parcels monitored by households with strong social ties. People behave in this manner because they do not want to impair the social relationships with those households possessing strong social ties since those are households whom they often need to ask for help. This opens the possibility that those with strong ties can leverage their informal networks to avoid sanctions. The staff members of the administrative bureau were mostly college graduates hired from outside (with few local ties), and all local people were encouraged to report illegal logging to receive a cash reward. We did not find a single case where staff members shielded households with strong ties so that those households could avoid or reduce monitoring. Therefore, this does not seem a plausible alternative explanation for our findings. We were left to conclude that in comparison to weak social ties to local leaders, strong social ties to local leaders reduced or even prevented illegal activities and saved actual monitoring efforts.

Additional causal analyses (Yang et al., 2013b) also confirmed previous claims in the literature about the conditions under which members in larger groups are more likely to free ride (i.e., do not contribute individual efforts for group benefits). In other words, people were less willing to contribute when

Table 11.3 Estimation of social ties to local leaders on households' forest monitoring efforts (reproduced from Yang et al., 2013b).

Variable	Description	Coefficients (robust S.E.)	Marginal effects
Social ties to local leaders	Binary: 0 for weak social ties; 1 for strong social ties	−5.377** (1.920)	−3.012
Quadratic term of group size	The quadratic term of group size (household^2)	−0.128** (0.041)	—
Group size	The number of households monitoring a single forest parcel (household)	1.331** (0.408)	0.767
Distance between each household and the main road	Euclidean distance between each household and the main road (km)	2.787* (1.216)	1.749
Laborers	Number of household laborers (individual)	0.296 (0.792)	0.186
Household size	Number of household members (individual)	−0.741 (0.630)	−0.465
Education	Average education of adult household members (year)	0.309 (0.369)	0.194
Household income (log)	Total household income in 2007 (yuan)	−0.011 (1.042)	−0.007
Percentage of agricultural income	Percentage of agricultural income to total household income (unitless)	−2.452 (2.760)	−1.539
Sampling weight	Sampling weight adjusting households sampled from the same monitoring groups	−1.432 (1.126)	−0.899
Intercept		8.921*** (2.360)	—

Notes:
*$p < 0.05$; **$p < 0.01$; ***$p < 0.001$. The analysis unit is the household. Dependent variable is total labor input for monitoring. The results are estimated by Tobit model. Total number of observations is 156. The left-censored and right-censored observations are 47 and 14, respectively. Standard errors are adjusted for clusters of households from the same monitoring groups. All independent variables were mean-centered before entering the model. All variance inflation factors were tested to be <5. Log pseudolikelihood is −390.962. Pseudo-R^2 is 0.035.

in larger groups compared to smaller groups. In contrast, our analyses also suggested that a household member would face higher social pressure in larger groups, which reduces free riding. This is because a member who did not participate in collective monitoring would be at the risk of damaging social relationships with only one other household in a two-member group but with nine other households if in a ten-member group. In this sense members of large groups may have quasi-ties (Frank, 2009) with each member of the group through their identification with the other members of a group. Violating a group norm may negate those quasi-ties. This argument is consistent with our finding that social capital influences a household's forest monitoring behaviors.

11.5 Social norms and enrollment in the Grain to Green Program

Another example of a PES program affected by social context is GTGP. GTGP has been implemented in Wolong since 2000 with the goal of increasing forest cover by converting cropland on steep slopes into forest (see also Chapters 5 and 13). Participants of the program receive an annual payment of 3450 yuan per ha for 8 years for planting and maintaining trees in their cropland. Almost all households in Wolong have participated in this program. Through GTGP, a total of 110 ha of cropland have been converted into forest. When most of the GTGP contracts started maturing at the end of 2008, the program was extended for another eight years at a reduced annual payment of 1875 yuan per ha.

GTGP has had a number of positive impacts on the natural environment in Wolong (Chen et al., 2009a). First, part of the labor force that was released from agriculture has been attracted to off-farm employment in more developed urban areas. This pattern was also found in other places that have adopted GTGP (Bao et al., 2005, Ge et al., 2006, Hu et al., 2006, Liu, 2005). As a result, human pressure on the natural environment has been reduced (Liu et al., 2007). Second, substantial fuelwood can be produced on GTGP land, which may alleviate the further degradation of panda habitat due to fuelwood collection in natural forests.

As with any policy or program, there is always uncertainty. We do not know whether the program will be continued in perpetuity and what will happen to the farmland that has already been converted to forest if the program were to end at some point. An important question that we have grappled with is, "If the program were to end, would farmers choose to convert the planted forestland back to agriculture?" The answer to this question may differ from household to household and may ultimately have a profound effect on the landscape and the sustainability of the coupled system. We hypothesized that social norms would affect people's decisions about whether to convert forest plantations back to agriculture. This prediction is based on our observations that local farmers made decisions that take substantially into account the behaviors of others in the community.

To test this hypothesis, we conducted interviews with 304 households in 2006. We collected information on households' socioeconomic conditions, characteristics of GTGP land plots, and land-use plans regarding GTGP land after the payment ends (Chen et al., 2009b). For those households that planned to convert some of their GTGP land back to agriculture, we further asked them to indicate their re-enrollment plan under different policy scenarios. The policy scenarios included payment level, duration of the program, and neighbors' behavior (i.e., percentage of neighbors reconverting at least one of their GTGP plots). We set the payment level at three possible values: 1500, 3000, and 4500 yuan per ha. The highest value was adjusted to 3750 yuan per ha after the first quarter of the survey. This revision allowed for more variation in responses. The three possible duration values were 3, 6, and 10 years. For neighbors' behavior, respondents were told that 25%, 50%, or 75% of their neighbors would convert at least part of their GTGP plots back to agriculture. We defined neighbors as those households who lived in the same group as the respondent. Group is an administrative unit within a village in rural China. Households in the same group tend to live closer geographically and have more interactions among one another.

Among a total of 735 GTGP plots that were enrolled by our interviewed households, only 166 plots were planned to be converted back to agriculture if the payment ended. We used stated-choice methods (Louviere et al., 2000) to evaluate the effects of social norms and other policy attributes on land re-enrollment. In order to more efficiently understand the main effect of each of the policy attributes, a main-effects design was used where policy attribute values were designed to be mutually orthogonal. We estimated the effects of policy attributes and the characteristics of households and GTGP plots on the land re-enrollment using a random-effects probit model (Wooldridge, 2002). The regression analyses were conducted based on those 166 GTGP plots that were planned to be reconverted after the payment ended.

We found that neighbors' behavior had a significant effect on respondents' plans of re-enrolling their GTGP plots in a new PES program (Table 11.4). We estimated that an additional 10% of neighbors' converting at least part of their GTGP plots back to agriculture reduced the respondents' re-enrollment intention by an average of 6.4%. Respondents indicated a higher likelihood of re-enrolling their GTGP plots when the program was to offer a higher payment. On average, an additional one yuan of payment increased the respondents' re-enrollment intention by 0.8%. These estimates contrast the effects of formal incentives versus informal norms. The effect of social norms on respondents' re-enrollment plans was non-linear at different levels of payments. The effect of social norms was largest at an intermediate payment and was smallest at the highest and lowest payments (Figure 11.1). In addition, the effect of program duration on respondents' re-enrollment plan was also non-linear. Respondents planned to re-enroll 23% more GTGP plots under a 6-year program than under a 3-year program. The planned re-enrollment under a 10-year program was not significantly different from a 6-year program (Table 11.4).

Some socioeconomic conditions and characteristics of GTGP plots also had significant effects on the planned re-enrollment of respondents (Table 11.4). We found that females and older respondents were more likely to re-enroll their GTGP plots than males and young respondents, probably due to less labor supply for crop production. Farming income had a significant negative effect on the planned re-enrollment, while off-farm income from employment outside of Wolong had a significant positive effect on re-enrollment. In comparison, income from off-farm employment within Wolong, including tourism employment, temporary off-farm employment, and permanent employment, did not have significant effects. Households with more cropland were more likely to re-enroll their GTGP plots. The respondents' expected fuelwood production from a GTGP plot had a significant positive effect on re-enrollment. The average distance from each household to its corresponding GTGP plots had a significant negative effect on re-enrollment. This result was probably because average distance was correlated to some unmeasured variables that negatively affect re-enrollment.

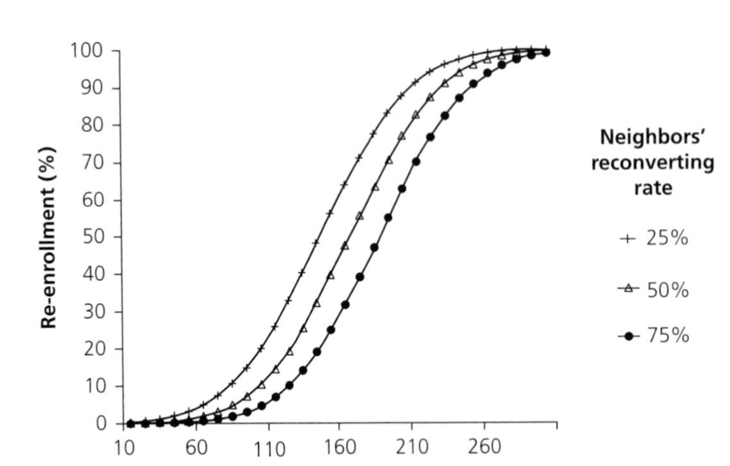

Figure 11.1 Estimated program re-enrollment under different levels of payment and neighbors' reconversion behavior (reproduced from Chen et al., 2009b).

Table 11.4 Estimation of policy attributes and other characteristics and their marginal effects on program re-enrollment (reproduced from Chen et al., 2009b).

Independent variables	Parameters (standard error)	Marginal effects
Neighbors' behavior	−1.662** (0.581)	−0.636
Conservation payment (yuan)	0.020** (0.003)	0.008
3-year duration (dummy, reference = 6 years)	−0.598* (0.277)	−0.230
10-year duration (dummy, reference = 6 years)	−0.270 (0.281)	−0.104
Gender (reference = female)	−0.841† (0.474)	−0.300
Age (years)	0.077** (0.023)	0.030
Education (years)	−0.049 (0.072)	−0.019
Farming income (1000 yuan)	−0.075† (0.042)	−0.029
Off-farm income from outside Wolong (1000 yuan)	0.253* (0.127)	0.097
Off-farm income from tourism employment in Wolong (1000 yuan)	0.046 (0.071)	0.018
Off-farm income from temporary employment in Wolong (1000 yuan)	0.063 (0.062)	0.024
Off-farm income from permanent employment in Wolong (1000 yuan)	0.054 (0.047)	0.021
Cropland after GTGP (mu)	0.361** (0.127)	0.138
Livestock (dummy)	0.406 (0.520)	0.157
Household size	−0.127 (0.176)	−0.049
Total land enrolled in GTGP (mu)	0.025 (0.085)	0.010
Area of land plot (mu)	−0.110 (0.246)	−0.042
Fuelwood production (kg)	0.003† (0.002)	0.001
Average walking distance from each household to its land plots (minutes)	−0.038** (0.012)	−0.015
Deviation of plot–household distance from the average distance (minutes)	0.015 (0.013)	0.006
Elevation (1000 m ASL)	0.050 (2.099)	0.019
Slope (degrees)	−0.038 (0.024)	−0.015
Aspect (180 = north-facing; 0 = south-facing)	−0.007 (0.006)	−0.003
Labor cost of reconversion (persons*days)	0.002 (0.003)	0.001
Geographic location (dummy)	0.498 (0.816)	0.185
Constant	−3.812 (4.902)	

Notes:
†$p \leq 0.1$; *$p \leq 0.05$; **$p \leq 0.01$.
Observations = 498; Number of plots = 166; Log likelihood = −219.209
Significant parameters for $\sigma_\mu = 1.836$ ($p < 0.01$) and $p = 0.771$ ($p < 0.01$) suggest the random-effects model is appropriate, and the test statistic $\chi^2 = 80.59$ ($p < 0.01$) indicates the random-effects model is preferred to the model without random effects.

11.6 Using social norms and social capital to shape human activities toward sustainability

The significance of social norms and social capital in influencing human–nature interactions in Wolong is informative not only for understanding coupled systems, but also for informing management of coupled systems worldwide for sustainability. If managers are aware of how human behavior is influenced by social context, this information can be harnessed for more efficient and effective policy making.

For instance, our study on social capital and labor migration demonstrated the significance of social ties for promoting rural–urban migration and in turn improving the rural environment. This phenomenon is common throughout the world, as the rapid economic development in urban regions has generated more employment opportunities. Greatly improved transportation and communications infrastructure facilitates the interactions between rural and urban areas and thus for building new social relations and the flow of rural labor to cities. While some parts of the world have formal government institutions to facilitate seeking employment in urban areas, these are absent in many rural areas such as our study area. As an alternative, rural migrants seek such information through their social ties to people in urban settings. Our study suggests an important opportunity for policy intervention lies in weak social ties in transitional economies. Compared to human capital, weak social ties to people in urban settings are relatively easy and inexpensive to develop. Policies that aim to shape labor migration for sustainability should provide employment information to rural people and facilitate communications between rural and urban settings.

We also found that social capital influences PES program behaviors. In our coupled system, social capital influenced households' forest monitoring behaviors under NFCP in many ways. Our findings provide empirical evidence that it is possible to manage collective action and resource outcomes through regulating social capital and factors interacting with social capital. For instance, policy makers may improve collective action and resource outcomes by improving social capital among group members and across different groups. This goal can be accomplished through education, enhancing interactions of local leaders from different groups, and helping groups with weak social ties to build social ties to local leaders. Policy makers can also allow members with good social relationships to self-organize (e.g., forming their own groups and deciding their own monitoring strategies). This approach may improve the efficiency and efficacy of the PES program.

We also found that social norms influenced PES program participation. In our coupled system, farmers' participation in GTGP was influenced by the participation decisions of their neighbors and people tended to conform to the majority. Under pro-PES social norms where most people would be willing to enroll their agricultural land in GTGP, the extra cost of enrolling an additional piece of land would be low due to the effects of social norms. In aggregate, the total costs of GTGP can be substantially reduced due to the effects of social norms. Even in regions where most people would initially be unwilling to participate, increased payments can move social norms toward participation. We also found that off-farm employment through labor migration increased people's willingness to participate in GTGP. As the economy continues to develop in urban China, the demand for labor also increases. This demand creates opportunities to reduce rural people's reliance on crop production and leverages the social norms of people in ecologically significant regions toward participation in GTGP.

As GTGP participation increases under pro-PES social norms, pro-PES social norms can be further enhanced via a positive feedback (Chen et al., 2012b). In order to take advantage of the effects of pro-PES social norms to obtain the targeted environmental benefits at the lowest cost, frequent social interactions among landholders are needed. Such interactions would facilitate the diffusion of information on social norms (Cialdini, 2003, Goldstein et al., 2008). However, most of the existing PES programs are implemented for a relatively long time frame, some over 20 years (Claassen et al., 2008, Liu et al., 2008, Pagiola, 2008). Therefore, landholders may not have frequent interactions regarding PES participation because they do not make PES enrollment decisions frequently. One approach to facilitating more frequent interactions about the decisions on PES participation is to divide landholders into multiple waves for enrollment. In this way, landholders who made the participation decisions at a later time can receive information on social norms from those who made participation decisions at earlier times. Our simulation study found that 11% additional land can be obtained at the same payment for enrollment in a PES program when this method is used (Chen et al., 2012b). Therefore, PES program designs that increase the diffusion of information on landholders' participation decisions can leverage pro-PES social norms to improve the efficiency of PES programs.

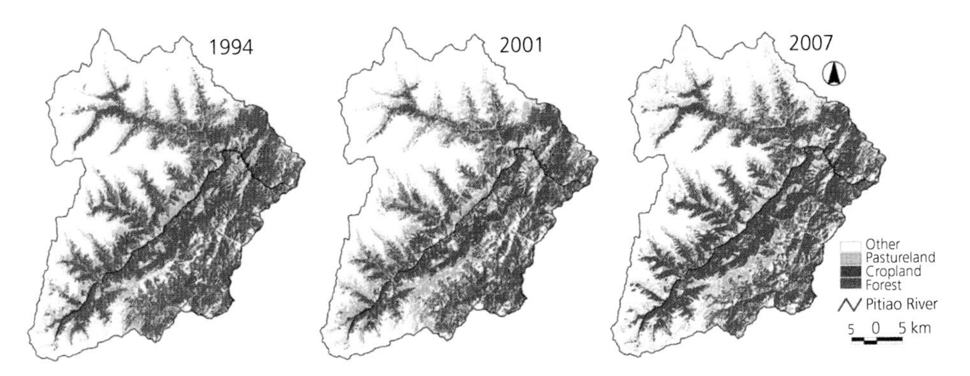

Plate 1: Maps of the spatial distribution of three major land-cover types in Wolong Nature Reserve between 1994 and 2007.

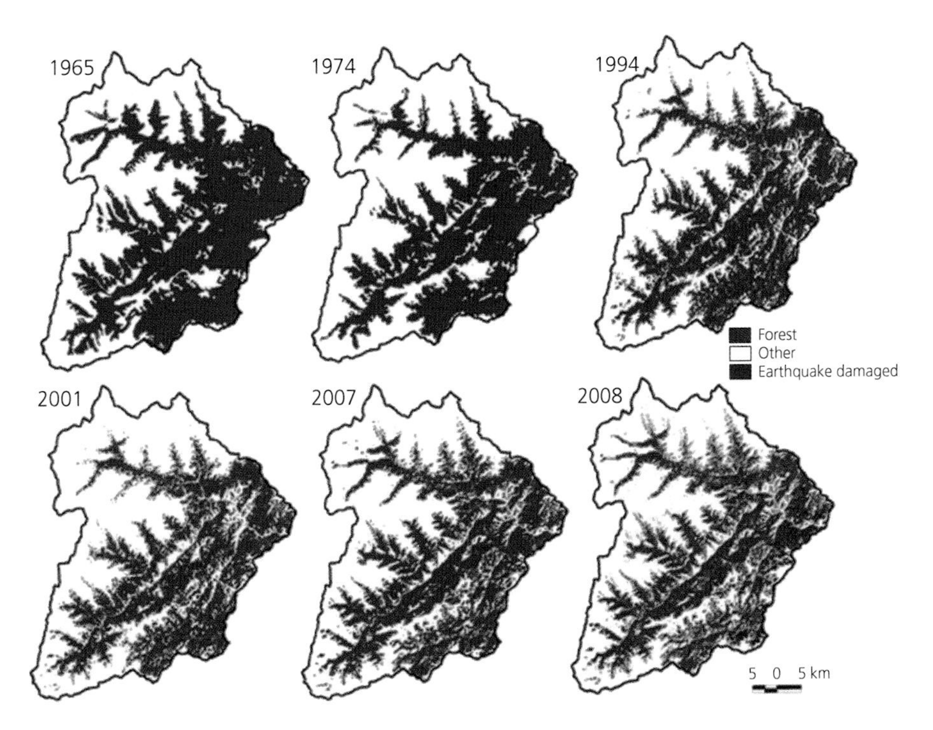

Plate 2: Time series of forest cover in Wolong Nature Reserve between 1965 and 2008 (based on Liu et al., 2001, Viña et al., 2007, 2011). Forest cover is shown in green, while areas of forest affected by landslides triggered by the May 12, 2008, Wenchuan earthquake are shown in red. Maps reprinted with permission from AAAS (originally published in Liu et al., 2001) and Springer Science and Business Media (originally published in Viña et al., 2011).

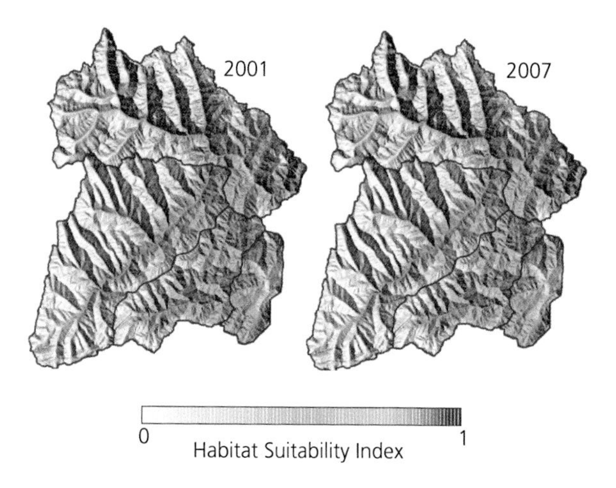

Plate 3: Topographic shaded relief maps showing the spatial distribution of panda habitat suitability index in the Wolong Nature Reserve in 2001 and 2007, based on phenological patterns derived from MODIS imagery (see Figure 7.3). Also shown are the locations of Wolong, Gengda, and Sanjiang townships inside and outside the reserve (see Figure 7.1).

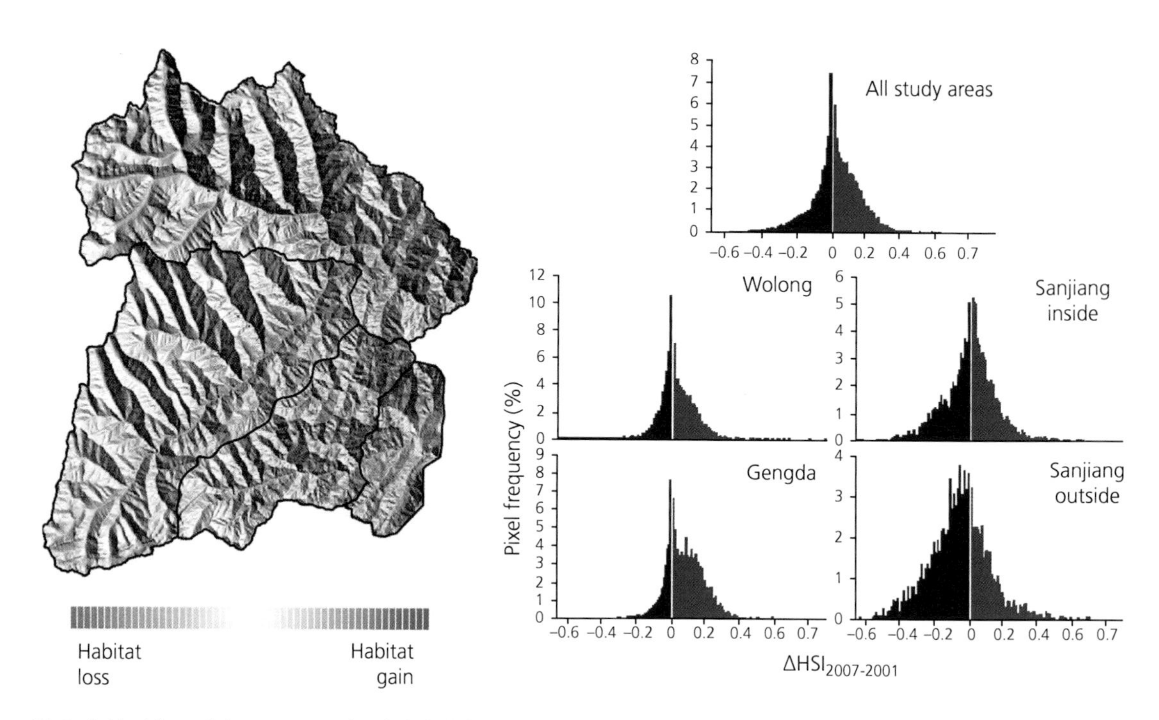

Plate 4: The left panel shows a topographic shaded relief map exhibiting the spatiotemporal dynamics of giant panda habitat suitability between 2001 and 2007 in the Wolong Nature Reserve. Also shown are the locations of Wolong, Gengda, and Sanjiang townships inside and outside the reserve (see Figure 7.1). Histograms in the right panel show the frequency distribution of the delta in habitat suitability index (HSI) pixel values across the entire study area and in each township. For color version see color plates. As with the colors shown in the map in the left panel, the red color represents negative deltas (i.e., loss of habitat suitability) while the blue color represents positive deltas (i.e., gains of habitat suitability). For clarity, the frequency of pixels exhibiting no change (i.e., HSI delta = 0) was omitted from the histograms.

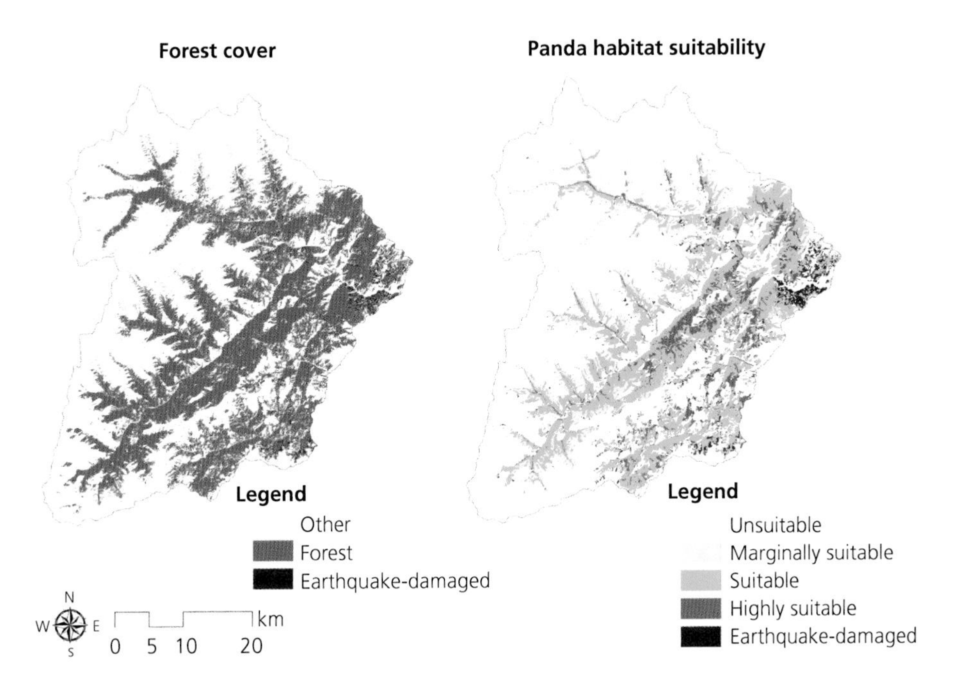

Plate 5: Wolong Nature Reserve and the impacts of the Wenchuan earthquake on forest cover and panda habitat. The panda habitat here is estimated only based on forest cover, elevation, and slope using the methods described in Liu et al. (2001) and thus is probably an overestimation of the actual habitat. Data are subsets of those appearing in Viña et al. (2011) and Liu and Viña (2014).

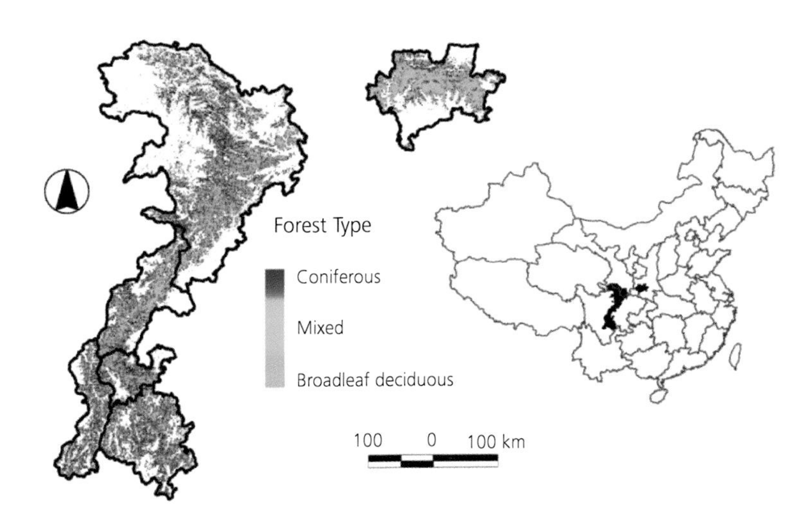

Plate 6: Spatial distribution in 2007 of coniferous, broadleaf deciduous, and mixed coniferous/deciduous forests in the mountain regions comprising the current geographic range of the giant panda. This map was obtained using the seasonal progression of vegetation in 2007, as measured by the Moderate Resolution Imaging Spectroradiometer (MODIS) sensor, for separating different forest types according to their different phenological signatures. Reprinted from Liu and Viña (2014) with permission from the Missouri Botanical Garden Press.

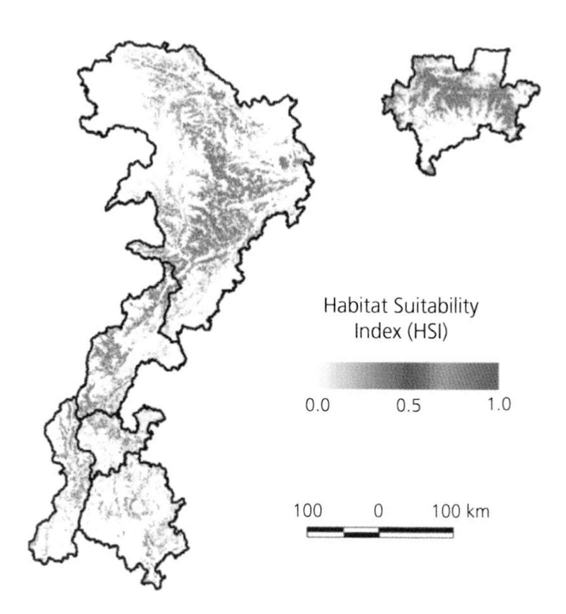

Plate 7: Spatial distribution in 2001 of giant panda habitat suitability index (HSI) values in the mountain regions comprising the current geographic range of the giant panda, obtained using 11 phenology metrics derived from an image time series of the Wide Dynamic Range Vegetation Index (WDRVI) calculated from data collected by the Moderate Resolution Imaging Spectroradiometer (MODIS) sensor. A higher HSI value indicates more suitable habitat.

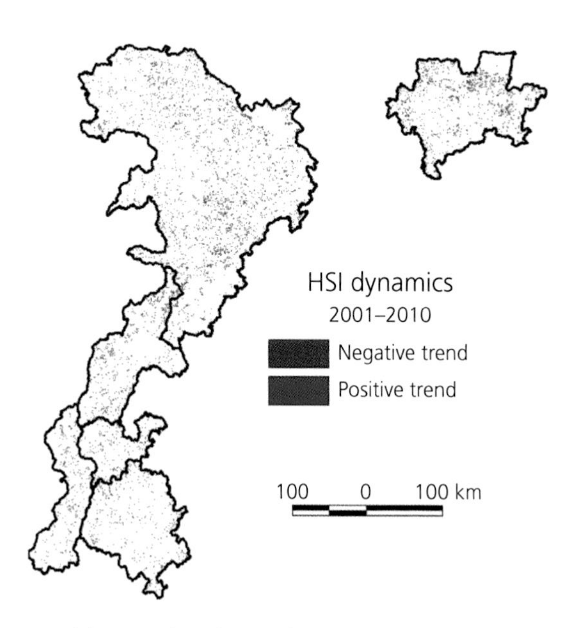

Plate 8: Spatial distribution of the areas exhibiting significant ($p < 0.05$) positive and negative trends in habitat suitability index (HSI) values between 2001 and 2010, in the mountain regions comprising the current geographic range of the giant panda.

11.7 Summary

Social factors can play a major role in human–nature interactions, but are understudied. In this chapter, we explored the role of social norms and social capital in human–nature interactions in Wolong Nature Reserve. We focused on labor migration and payments for ecosystem services (PES) programs. Results show that weak social ties of local residents to people in urban settings had a significant positive effect on labor migration. The availability of acquaintances increased the odds of labor migration by 2.54. Further, labor migration substantially reduced households' use of fuelwood. Results also indicate partial substitutability among social capital, human capital, and natural capital. Regarding PES programs, social capital played a role in households' collective action in forest monitoring of the Natural Forest Conservation Program. Social capital prevented or reduced illegal activities in one's forest parcel and thus saved monitoring efforts. Households with strong social ties to local leaders spent an average of 54% less labor on forest monitoring than those with weak social ties. Farmers' decisions about the enrollment of their land in the Grain to Green Program (GTGP) was significantly influenced by social norms (the decisions of other community members). An additional 10% of their neighbors intending to convert part of their GTGP plots back to agriculture reduced the respondents' intention to re-enroll by 6.4%. In addition, the effect of social norms was highest at an intermediate payment level. Our findings suggest that as economies develop and conservation policies are implemented, social factors should be leveraged to shape human–nature interactions toward sustainability.

References

An, L., Liu, J., Ouyang, Z., et al. (2001) Simulating demographic and socioeconomic processes on household level and implications for giant panda habitats. *Ecological Modelling*, **140**, 31–49.

Arrow, K., Dasgupta, P., Gouldner, L., et al. (2004) Are we consuming too much? *Journal of Economic Perspectives*, **18**, 147–72.

Bao, J., Tang, D., and Chen, B. (2005) Socioeconomic effects of Grain to Green Program in Sichuan Province. *Sichuan Forestry Exploration and Design*, **1**, 26–32 (in Chinese).

Bendor, J. and Swistak, P. (2001) The evolution of norms. *American Journal of Sociology*, **106**, 1493–545.

Bian, Y. (1997) Bringing strong ties back in: indirect ties, network bridges, and job searches in China. *American Sociological Review*, **62**, 366–85.

Chen, X., Frank, K.A., et al. (2012a) Weak ties, labor migration, and environmental impacts: toward a sociology of sustainability. *Organization & Environment*, **25**, 3–24.

Chen, X., Lupi, F., An, L., et al. (2012b) Agent-based modeling of the effects of social norms on enrollment in payments for ecosystem services. *Ecological Modelling*, **229**, 16–24.

Chen, X., Lupi, F., He, G., and Liu, J. (2009b) Linking social norms to efficient conservation investment in payments for ecosystem services. *Proceedings of the National Academy of Sciences of the United States of America*, **106**, 11812–17.

Chen, X., Lupi, F., He, G., et al. (2009a) Factors affecting land reconversion plans following a payment for ecosystem service program. *Biological Conservation*, **142**, 1740–47.

Chen, X., Viña, A., Shortridge, A., et al. (2014) Assessing the effectiveness of payments for ecosystem services: an agent-based modeling approach. *Ecology and Society*, **19**, 7.

Cialdini, R.B. (2003) Crafting normative messages to protect the environment. *Current Directions in Psychological Science*, **12**, 105–109.

Claassen, R., Cattaneo, A., and Johansson, R. (2008) Cost-effective design of agri-environmental payment programs: US experience in theory and practice. *Ecological Economics*, **65**, 737–52.

Coleman, J.S. (1990) *Foundations of Social Theory*. Harvard University Press, Cambridge, MA.

Costanza, R., Darge, R., Degroot, R., et al. (1997) The value of the world's ecosystem services and natural capital. *Nature*, **387**, 253–60.

Daily, G.C., Soderqvist, T., Aniyar, S., et al. (2000) The value of nature and the nature of value. *Science*, **289**, 395–96.

Daly, H.E. and Cobb, J.B., Jr. (1989) *For the Common Good: Redirecting the Economy toward Community, the Environment and a Sustainable Future*. Beacon Press, Boston, MA.

Dasgupta, P. (2010) The place of nature in economic development. In D. Rodrik and M. Rosenzweig, eds, *Handbook of Development Economics* (vol. **5**), pp.4039–5061. North Holland, Amsterdam.

Dietz, T. and Henry, A.D. (2008) Context and the commons. *Proceedings of the National Academy of Sciences of the United States of America*, **105**, 13189–90.

Dietz, T., Ostrom, E., and Stern, P.C. (2003) The struggle to govern the commons. *Science*, **302**, 1907–12.

Elster, J. (1989) Social norms and economic theory. *Journal of Economic Perspectives*, **3**, 99–117.

Fehr, E. and Gintis, H. (2007) Human motivation and social cooperation: experimental and analytical foundations. *Annual Review of Sociology*, **33**, 43–64.

Frank, K.A. (2009) Quasi-ties directing resources to members of a collective. *American Behavioral Scientist*, **52**, 1613–45.

Ge, W., Li, L., and Li, Y. (2006) Analysis on the sustainability about Grain to Green Program—a field survey of Grain to Green Program in Wuqi and Zhidan counties of Shaanxi. *Forestry Economics*, **11**, 33–49 (in Chinese).

Goldstein, N.J., Cialdini, R.B., and Griskevicius, V. (2008) A room with a viewpoint: using social norms to motivate environmental conservation in hotels. *Journal of Consumer Research*, **35**, 472–82.

Granovetter, M. (1995) *Getting a Job: A Study of Contacts and Careers*. (second edition). University of Chicago Press, Chicago, IL.

Granovetter, M.S. (1973) Strength of weak ties. *American Journal of Sociology*, **78**, 1360–80.

Henrich, J. and Gil-White, F.J. (2001) The evolution of prestige—freely conferred deference as a mechanism for enhancing the benefits of cultural transmission. *Evolution and Human Behavior*, **22**, 165–96.

Hirano, K. and Imbens, G.W. (2002) Estimation of causal effects using propensity score weighting: an application to data on right heart catheterization. *Health Services & Outcomes Research Methodology*, **2**, 259–78.

Hirano, K., Imbens, G.W., and Ridder, G. (2003) Efficient estimation of average treatment effects using the estimated propensity score. *Econometrica*, **71**, 1161–89.

Hu, C., Fu, B., and Chen, L. (2006) Impacts of "Grain for Green Project" on agriculture and rural economic development in the loess hilly and gully area—a case study in Ansai county *Journal of Arid Land Resources and Environment*, **20**, 67–72 (in Chinese).

Hugo, G. (1998) The demographic underpinnings of current and future international migration in Asia. *Asian and Pacific Migration Journal*, **7**, 1–25.

James, A., Gaston, K.J., and Balmford, A. (2001) Can we afford to conserve biodiversity? *Bioscience*, **51**, 43–52.

James, A.N., Gaston, K.J., and Balmford, A. (1999) Balancing the Earth's accounts. *Nature*, **401**, 323–24.

Johnson, D.G. (2000) Agricultural adjustment in China: problems and prospects. *Population and Development Review*, **26**, 319–34.

Korinek, K., Entwisle, B., and Jampaklay, A. (2005) Through thick and thin: layers of social ties and urban settlement among Thai migrants. *American Sociological Review*, **70**, 779–800.

Li, H. and Zahniser, S. (2002) The determinants of temporary rural-to-urban migration in China. *Urban Studies*, **39**, 2219–35.

Lin, N. (2001) *Social Capital: A Theory of Social Structure and Action*. Cambridge University Press, New York, NY.

Lin, N., Ensel, W.M., and Vaughn, J.C. (1981) Social resources and strength of ties—structural factors in occupational-status attainment. *American Sociological Review*, **46**, 393–403.

Liu, J., Dietz, T., Carpenter, S.R., et al. (2007) Complexity of coupled human and natural systems. *Science*, **317**, 1513–16.

Liu, J., Li, S., Ouyang, Z., et al. (2008) Ecological and socioeconomic effects of China's policies for ecosystem services. *Proceedings of the National Academy of Sciences of the United States of America*, **105**, 9477–82.

Liu, J., Ouyang, Z., Yang, W., et al. (2013) Evaluation of ecosystem service policies from biophysical and social perspectives: the case of China. In S.A. Levin, ed., *Encyclopedia of Biodiversity* (second edition), vol. **3**, pp.372–84. Academic Press, Waltham, MA.

Liu, K. (2005) Analysis on the prospect after Grain for Green Subsidy Policy. *Green China*, **4**, 30–31 (in Chinese).

Louviere, J.J., Hensher, D.A., and Swait, J.D. (2000) *Stated Choice Methods: Analysis and Applications*. Cambridge University Press, Cambridge, UK.

Lyon, F. (2000) Trust, networks and norms: the creation of social capital in agricultural economies in Ghana. *World Development*, **28**, 663–81.

Massey, D.S. (1990) Social structure, household strategies, and the cumulative causation of migration. *Population Index*, **56**, 3–26.

Massey, D.S. and Espinosa, K.E. (1997) What's driving Mexico–US migration? A theoretical, empirical, and policy analysis. *American Journal of Sociology*, **102**, 939–99.

Milinski, M., Semmann, D., and Krambeck, H.J. (2002) Reputation helps solve the "tragedy of the commons." *Nature*, **415**, 424–26.

Millennium Ecosystem Assessment (2005) *Ecosystems and Human Well-being: Synthesis*. Island Press, Washington, DC.

Neumayer, E. (2010) *Weak Versus Strong Sustainability: Exploring the Limits of Two Opposing Paradigms* (third edition). Edward Elgar Publishing, Cheltenham and Northampton, UK.

OECD (1997) *The Environmental Effects of Agricultural Land Diversion Schemes*. Organization for Economic Cooperation and Development (OECD), Paris, France.

Ostrom, E. (2000) Collective action and the evolution of social norms. *Journal of Economic Perspectives*, **14**, 137–58.

Pagiola, S. (2008) Payments for environmental services in Costa Rica. *Ecological Economics*, **65**, 712–24.

Palloni, A., Massey, D.S., Ceballos, M., et al. (2001) Social capital and international migration: a test using information on family networks. *American Journal of Sociology*, **106**, 1262–98.

Pretty, J. (2003) Social capital and the collective management of resources. *Science*, **302**, 1912–14.

Pretty, J. and Ward, H. (2001) Social capital and the environment. *World Development*, **29**, 209–27.

Robins, J.M. and Rotnitzky, A. (1995) Semiparametric efficiency in multivariate regression models with missing data. *Journal of the American Statistical Association*, **90**, 122–29.

Rosenbaum, P.R. and Rubin, D.B. (1983) The central role of the propensity score in observational studies for causal effects. *Biometrika*, **70**, 41–55.

Rustagi, D., Engel, S., and Kosfeld, M. (2010) Conditional cooperation and costly monitoring explain success in forest commons management. *Science*, **330**, 961–65.

Schultz, P.W., Nolan, J.M., Cialdini, R.B., et al. (2007) The constructive, destructive, and reconstructive power of social norms. *Psychological Science*, **18**, 429–34.

Solow, R.M. (1993) Sustainability: an economist's perspective. In N.S. Dorfman and R. Dorfman, eds, *Economics of the Environment: Selected Readings*, pp. 179–87. Norton, New York, NY.

Vitousek, P.M., Mooney, H.A., Lubchenco, J., and Melillo, J.M. (1997) Human domination of Earth's ecosystems. *Science*, **277**, 494–99.

Wilson, T.D. (1998) Weak ties, strong ties: network principles in Mexican migration. *Human Organization*, **57**, 394–403.

Winship, C. and Morgan, S.L. (1999) The estimation of causal effects from observational data. *Annual Review of Sociology*, **25**, 659–706.

Wooldridge, J.M. (2002) *Econometric Analysis of Cross Section and Panel Data*. MIT Press, Cambridge, MA.

Yakubovich, V. (2005) Weak ties, information, and influence: how workers find jobs in a local Russian labor market. *American Sociological Review*, **70**, 408–21.

Yang, W., Liu, W., Viña, A., et al. (2013a) Performance and prospects on payments for ecosystem services programs: evidence from China. *Journal of Environmental Management*, **127**, 86–95.

Yang, W., Liu, W., Viña, A., et al. (2013b) Nonlinear effects of group size on collective action and resource outcomes. *Proceedings of the National Academy of Sciences of the United States of America*, **110**, 10916–21.

Young, H.P. (1996) The economics of convention. *Journal of Economic Perspectives*, **10**, 105–22.

CHAPTER 12

Vulnerability and Adaptation to Natural Disasters

Wu Yang, Andrés Viña, Thomas Dietz, Vanessa Hull, Daniel Kramer, Zhiyun Ouyang, and Jianguo Liu

12.1 Introduction

12.1.1 General background

During recent decades, the number of natural disasters has increased globally. In particular, from 1980 to 2009, the number of weather-related disasters more than doubled (Neumayer and Barthel, 2011). The number and extent of natural disasters are likely to continue to increase in the future (IPCC, 2012). Therefore, it is increasingly important to understand how natural disasters affect ecosystems and people, and how ecosystems and people respond, recover, and adapt to natural disasters.

In this chapter, we illustrate the vulnerability of coupled human and natural systems to natural disasters. We use a broad definition of vulnerability as "the degree to which a system is likely to experience harm due to exposure to a hazard, either a perturbation or stress/stressor" (Turner et al., 2003). A related concept, resilience, is defined as the capacity or ability of a system to recover quickly from external hazards (Adger et al., 2005, Folke, 2006). Some scholars have conceptualized vulnerability in three components. That is, the vulnerability of a system is a function of (1) the exposure, (2) the sensitivity to hazardous conditions, and (3) the capacity (or resilience) to respond and recover (or adapt) to the effects of hazards (Adger, 2006, Gallopin, 2006, Smit and Wandel, 2006). An emerging vulnerability literature has thus tried to understand what kinds of conditions make a system exposed to hazards and what types of measures enhance the resilience of a system to hazards (Adger et al., 2005, Berke and Campanella, 2006, Folke, 2006, Gunderson, 2010, Klein and Zellmer, 2007). For instance, studies have found that a diverse economy (Berke and Campanella, 2006) and cross-scale interactions from state, federal, and international organizations (Adger et al., 2005) have helped human communities recover from disasters. Some flood prevention strategies, insurance programs, and regulatory policies accelerated the recovery of some communities from floods while other policies aggravated the vulnerability of other communities to floods (Klein and Zellmer, 2007). Some measures (e.g., building levees along a river to control floods) that are adaptive in the short term may become maladaptive in the long run. Ultimately, transformational adaptations (e.g., evacuation and relocation of human settlements) have to be adopted (Kates et al., 2012). Due to the lack of quantitative indicators, especially a lack of direct vulnerability indicators for human systems, there is little systematic and quantitative evidence on the vulnerability of coupled human and natural systems to disasters (Cutter et al., 2003, Yang et al., 2013a, 2015).

In response to the lack of quantitative indicators to measure vulnerability and factors affecting it, we conducted a systematic study in Wolong Nature Reserve based on years of data and numerous analyses of them. Specifically, we used the index system of human dependence on ecosystem services (IDES), as described in Chapter 5,

Pandas and People. Edited by Jianguo Liu, Vanessa Hull, Wu Yang, Andrés Viña, Xiaodong Chen, Zhiyun Ouyang, and Hemin Zhang. Published 2016 by Oxford University Press.

to assess the vulnerability of natural systems in terms of how their ability to deliver benefits to humans changes in response to natural disasters. This index includes a set of quantitative indicators to reflect the vulnerability of a system to the degradation or decline of ecosystem services (ES; Yang et al., 2013b). We also used the human well-being index (HWBI), a comprehensive and quantitative measure of outcomes of hazards, since the ultimate goal of human civilization is to improve human well-being (HWB; Yang et al., 2013a).

12.1.2 Natural disasters at Wolong Nature Reserve

Like many other biodiversity hotspots, Wolong Nature Reserve is an area susceptible to natural disasters. It is situated in the Longmen Mountain fault zone (see Figure 6.1), where the elevation climbs more than 6000 m from the Sichuan Basin to the eastern flank of the Tibetan Plateau in a distance of less than 100 km. In typical years, there are frequent storms at Wolong followed by mountain torrents, scattered landslides, and mud–rock flows during the rainy season from May to August. In winter, there is sometimes heavy snowfall and thus avalanches in the high-elevation regions (e.g., Balang Mountain and Siguniang Mountain) in the west. The affected locations are relatively far away from human settlements and the vegetation cover surrounding human settlements is dense. Thus, usually the impacts of these disasters to humans are small.

The Wenchuan earthquake (surface wave magnitude [M_s] 8.0 or moment magnitude [M_w] 7.9) that occurred on May 12, 2008, is an exception. This event was the most devastating earthquake in China since 1976, and its epicenter was along the eastern edge of Wolong. According to official statistics, across the entire earthquake-affected area (51 counties), there were 69 227 people killed and 374 643 injured (Xinhua News Agency, 2008a). A total of 1.5 million victims were evacuated and temporarily resettled as a result of the earthquake (Xinhua News Agency, 2008a). The rough estimate of economic losses was over one trillion yuan. This estimate includes damage to infrastructure, croplands, and tourism (Xinhua News Agency, 2008b). The earthquake also affected approximately 3.4% (1221 km^2) of the total area of natural ecosystems. Affected ecosystems included 977 km^2 of forest, 180 km^2 of shrub, 49 km^2 of meadow, 12 km^2 of barren land, 2 km^2 of water bodies, and 0.5 km^2 of snow-covered land (Ouyang et al., 2008). Approximately 5.9% of the total panda habitat (656 km^2) was damaged, of which 53% was inside nature reserves (Ouyang et al., 2008).

In this chapter, we (1) assess the short-term environmental and socioeconomic impacts of the Wenchuan earthquake on Wolong at the reserve and household levels using empirical work as well as IDES and HWBI; (2) examine what kinds of factors (including IDES) affect the vulnerability of local households to disasters, using the changes in overall HWBI as a measurement of vulnerability; and (3) analyze how the coupled system responded and adapted to the earthquake.

12.2 Impacts of the earthquake at the reserve level

12.2.1 Environmental impacts

The earthquake and aftershocks caused massive associated (secondary) disasters such as landslides, mud–rock flows, mountain torrents, and dammed lakes. Fortunately, the main seismic waves reverberated away from Wolong along the Longmen Mountain fault (see Figure 6.1) and did not cause more extensive and intensive effects inside the reserve compared to many areas outside. But the earthquake itself and associated disasters still led to relatively severe negative impacts on forests, habitat, and biodiversity.

The earthquake and associated disasters immediately caused the loss of 10% (79 km^2) of forest cover and 5% (35 km^2) of panda habitat inside Wolong (Figure 12.1; Plate 5). Most of the forest cover and panda habitat losses were concentrated in places near the epicenter in the eastern portion of the reserve and in regions with relatively steep slopes that pandas do not prefer. We found no evidence that the earthquake resulted in significant impacts on pandas directly as they avoided the immediate areas affected but still used surrounding areas (Zhang et al., 2011). However, a field investigation

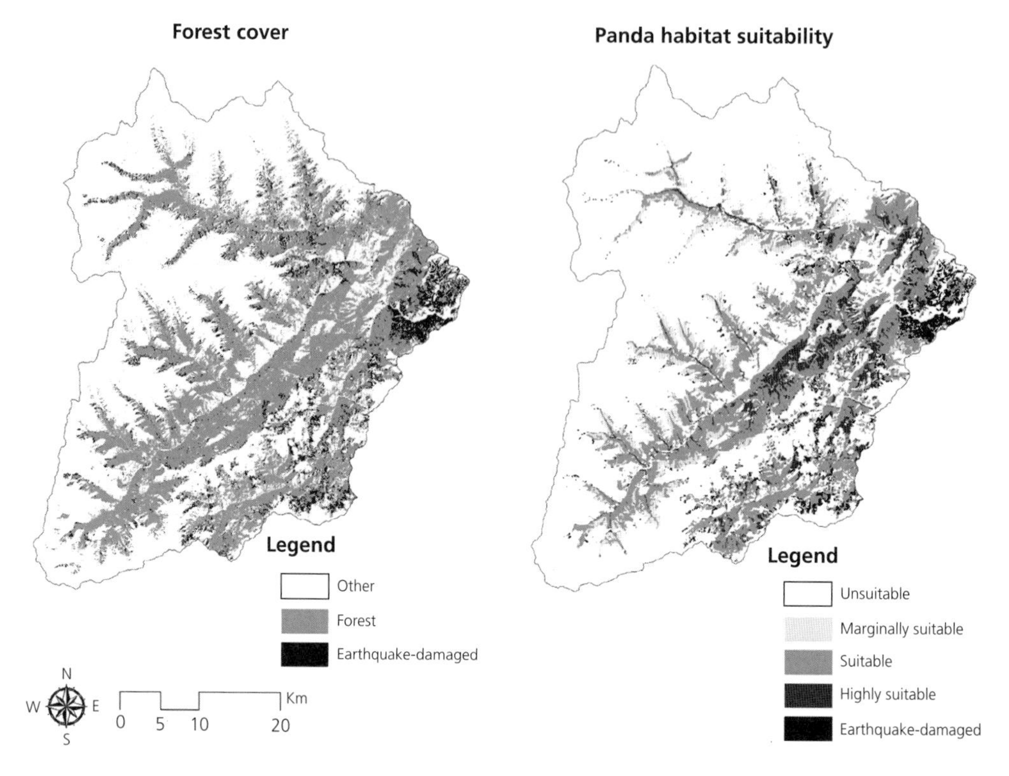

Figure 12.1 Wolong Nature Reserve and the impacts of the Wenchuan earthquake on forest cover and panda habitat. The panda habitat here is estimated only based on forest cover, elevation, and slope using the methods described in Liu et al. (2001) and thus is probably an overestimation of the actual habitat. Data are subsets of those appearing in Viña et al. (2011) and Liu and Viña (2014). For color version see color plates. [PLATE 5]

showed that the earthquake and associated disasters led to significant changes in biodiversity indicators. At the earthquake-damaged sampling-site level, species richness, the Shannon–Wiener index, the Simpson index, and species evenness for trees declined by 33.5%, 30.8%, 25.5%, and 16.3%, respectively. For shrubs, aside from an increase in species evenness by 3.3%, other indices including species richness, the Shannon–Wiener index, and the Simpson index decreased by 22.0%, 14.6%, and 7.6%, respectively (Zhang et al., 2011).

12.2.2 Socioeconomic impacts

The earthquake and associated disasters also caused enormous socioeconomic losses of residential houses, public infrastructure (e.g., roads, bridges, schools, and hospitals), cultural heritage, and agricultural, industrial, and tourism revenues. Approximately 98% of residential houses were damaged. Economic costs of lost infrastructure were estimated at 396 million yuan, of which hydropower facilities and tourism facilities accounted for 62% and 30%, respectively (Sichuan Department of Forestry, 2008). Total estimated loss of economic revenue was 207 million yuan. Industrial and service sectors accounted for 92.8% of this total, followed by agricultural plantations (3.6%), forestry (2.6%), and animal husbandry (1.0%). Some cultural heritage sites were damaged. One example is the centuries-old Lama Temple. Another is Wolongguan Old Street, a posting stage of the 2000-km, over 2300-year-old Ancient Tea-Horse Road connecting Sichuan Province via Yunnan Province to India (Sichuan Department of Forestry, 2008). In addition, within Wolong, 48 local residents and about a hundred workers and vehicle passengers were killed, and hundreds of people were injured (Yang et al., 2013a).

12.3 Impacts of the earthquake at the household level

Although the impacts of the earthquake at the reserve level have been well-documented, there is a lack of understanding of the impacts of the earthquake at the household level. We set out to fill this gap by investigating (1) the impact of the earthquake on households' dependence on ES, (2) the impact of the earthquake on household HWB, and (3) factors associated with change in HWB from pre-earthquake to post-earthquake.

12.3.1 Impacts on households' dependence on ES

To understand how the earthquake affected households' dependence on ES, we used the IDES derived in Chapter 5. We compared the values of the overall index and subindices immediately before and after the earthquake (2007 and 2009, respectively). The earthquake significantly reduced local households' overall IDES and subindices of provisioning and cultural services with only regulating services unchanged (Figure 12.2). Specifically, on average, from 2007 to 2009 the overall IDES, the subindex of provisioning services, and the subindex of cultural services decreased significantly by 48% (from 0.634 to 0.331), 86% (from 0.366 to 0.051), and by 57% (from 0.035 to 0.015), respectively (Figure 12.2). But our results show no significant change in the subindex of regulating services. To explain the differences in changes across subindices of IDES, we conducted some additional analyses (Table 12.1). We found that the earthquake on average reduced 70% and 43% of local households' net benefits from provisioning and cultural services, respectively. However, it did not affect net benefits from regulating services, which were realized from government payments. In the meantime, on average, net socioeconomic benefits (those not obtained from ecosystems services) increased by a factor of three, primarily due to the temporary increase in demand for laborers for reconstruction.

12.3.2 Impacts on households' well-being

A related question is how household-level HWB was affected by this disaster. Based on the HWBI developed in Yang et al. (2013a), we compared the overall index and subindices of HWB immediately

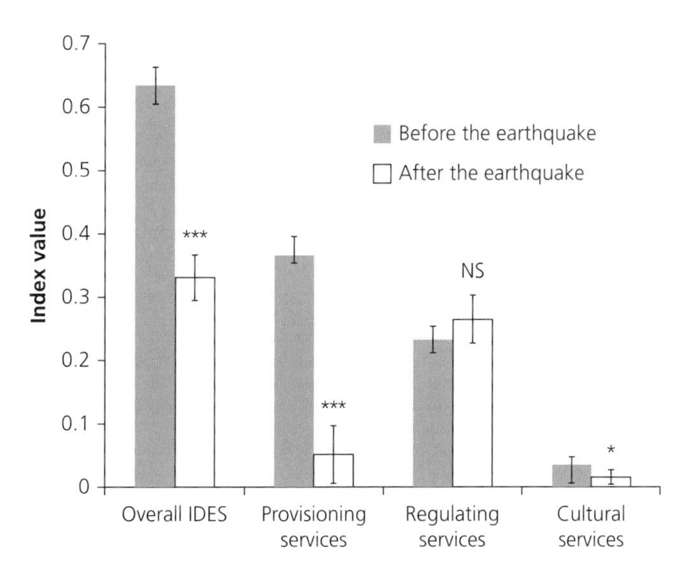

Figure 12.2 Impacts of the Wenchuan earthquake on households' dependence on ecosystem services. A higher value of the overall index or subindex indicates a higher dependence on the corresponding ecosystem services and thus a higher level of vulnerability to their damages or losses. Total number of observations: 101. $^{*}p < 0.05$, $^{***}p < 0.001$. Adapted from Yang et al. (2015).

Table 12.1 Impacts of the Wenchuan earthquake on households' net benefits.

Net benefit	Before the earthquake (2007)	After the earthquake (2009)	t statistics
Net benefit from provisioning services (thousand yuan)	10.279 (1.842)	3.063 (0.959)	4.821***
Net benefit from regulating services (thousand yuan)	3.519 (0.226)	3.468 (0.250)	−0.730
Net benefit from cultural services (thousand yuan)	0.992 (0.390)	0.424 (0.251)	1.686*
Net benefit from other socioeconomic activities (thousand yuan)	13.899 (2.612)	57.681 (9.964)	−4.434***

Notes: Numbers outside and inside parentheses are means and standard errors, respectively. Monetary values in 2009 were discounted to values in 2007. The same 101 households were sampled in 2007 and 2009. *$p < 0.05$, ***$p < 0.001$. Adapted from Yang et al. (2015).

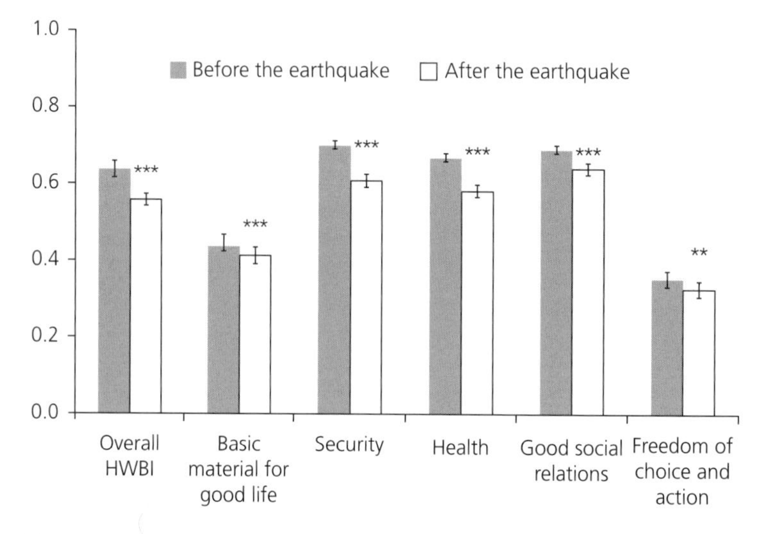

Figure 12.3 Impacts of the Wenchuan earthquake on households' well-being, based on household well-being index (HWBI) values. A higher value of the overall index or subindex indicates a higher state of well-being. Total number of observations: 101. **$p < 0.01$, ***$p < 0.001$. Adapted from Yang et al. (2015).

before and after the earthquake. The earthquake caused a significant reduction in local households' well-being, as indicated by both the overall index and subindices of HWB (Figure 12.3). The subindices of basic material for good life, security, health, good social relations, and freedom of choice and action decreased by 5.2%, 13.2%, 12.9%, 7.2%, and 7.4%, respectively (Figure 12.3). Together, these declines led to a 12.6% reduction in the overall index, despite the dramatic increase of net socioeconomic benefits as mentioned above. Both before and after the earthquake, the overall indices of affluent households were significantly higher than those of poor households (Table 12.2).

12.3.3 Factors associated with the change in HWBI

We used an ordinary least-squares regression model to determine factors associated with the change in HWBI for each household from before and after the earthquake. Explanatory variables in the model included pre-earthquake IDES, change in IDES from pre- to post-earthquake, and several

Table 12.2 Spearman's correlations between the human well-being index and affluence. $n = 101$. $*p < 0.05$, $**p < 0.01$.

	Household income in 2007 (yuan)	Household income in 2009 (yuan)
HWBI before the earthquake (in 2007)	0.309**	0.254*
HWBI after the earthquake (in 2009)	0.291**	0.309**

Source: Adapted from Yang et al. (2015).

Table 12.3 Summary of descriptive statistics of variables included in an ordinary least-squares model predicting change in human well-being from before (2007) to after (2009) the Wenchuan earthquake in May 2008.

Variable	Mean (S.D.)
Change in HWBI	−0.080 (0.010)
HWBI in 2007	0.638 (0.118)
IDES in 2007	0.634 (0.295)
Change in IDES	−0.303 (0.423)
Non-agricultural income share in 2007	0.581 (0.283)
Household income in 2009 (thousand yuan)	68.105 (104.715)
Change of household income per capita (thousand yuan)	11.160 (35.306)
Household size in 2007	3.317 (1.319)
Number of seniors (age ≥ 60 years)	0.653 (0.805)
Average education of adults in 2007 (years)	5.052 (2.760)
Percentage of female adults in 2007	0.469 (0.165)
House damage in the earthquake (0: low, 1: high)	0.624 (0.487)
Social ties to local leaders (0: weak, 1: strong)	0.109 (0.313)
Gender of interviewee (0: female, 1: male)	0.584 (0.495)
Age of interviewee (years)	53.604 (11.333)

Note: HWBI, human well-being index; IDES, index system of human dependence on ecosystem services. We calculated the change of HWBI, change of IDES, or change of household income per capita using the corresponding value in 2009 and subtracting the corresponding value in 2007. Local leaders refer to those who work for local governments and local government enterprises. Households that have at least one household member or one immediate relative (e.g., parents, children, or blood brothers) who is a local leader are regarded to have strong social ties to local leaders; otherwise, they are classified with those that have weak social ties to local leaders. $n = 101$. Adapted from Yang et al. (2015).

Table 12.4 Factors associated with the change in human well-being from before (2007) to after (2009) the earthquake.

Dependent variable: change in HWBI	Coefficient (robust S.E.)
HWBI in 2007	−0.111 (0.078)
IDES in 2007	0.096* (0.045)
Change of IDES	0.018 (0.026)
Agricultural income share in 2007	−0.118* (0.051)
Household income in 2009 (thousand yuan)	0.405 e−3** (0.122 e−3)
Change of per capita income (thousand yuan)	−0.980 e−3* (0.395 e−3)
Household size in 2007	0.016** (0.006)
Number of seniors (age ≥ 60)	−0.028 (0.018)
Average education of adults in 2007 (years)	0.007 (0.004)
Percentage of female adults in 2007	−0.113 (0.072)
House damage (0: low, 1: high)	−0.049** (0.018)
Social ties to local leaders (0: weak, 1: strong)	0.052* (0.025)
Gender of interviewee (0: female, 1: male)	0.031 (0.019)
Age of interviewee (years)	0.105 e−4 (0.001)
Intercept	−0.052 (0.077)

Notes: HWBI, human well-being index; IDES, index system of human dependence on ecosystem services. Dependent variable is the change in HWBI before and after the earthquake, which is HWBI in 2009 subtracting HWBI in 2007. R^2 of the ordinary least-squares regression is 0.272. The number of observations is 101. $*p < 0.05$, $**p < 0.01$. Variance inflation factors were tested to be less than 5.

socioeconomic variables (see descriptive statistics in Table 12.3). Our results show that pre-earthquake IDES was positively correlated with the change in HWBI, and that pre-earthquake agricultural income share was negatively associated with the change in HWBI (Table 12.4). Since HWBI was deteriorating due to the earthquake (Figure 12.3), the negative association indicates that households with higher dependence on agriculture suffered from a larger reduction in HWBI. In the meantime, households' dependence on ES pre-earthquake varied substantially across households for various types of ES besides agricultural sources (Yang et al., 2013b; see also Chapter 5). Thus, the positive association suggests that households more dependent on multiple types of ES experienced less reduction in HWBI. Nevertheless, the coefficient of change in IDES from pre- to post-earthquake was not significant (Table 12.4). One reason might be that the effect of pre-earthquake IDES overwhelmed the effect of its change. Another reason may be that the effect of change in IDES could take a longer time to appear. This time-lag effect has also been suggested in other studies (Liu et al., 2007, Raudsepp-Hearne et al., 2010, Yang et al., 2013a).

We next set out to examine whether the effect of IDES can be replaced by the separate indicators constituting it (i.e., the corresponding net benefits obtained from provisioning, regulating, or cultural services for households in 2007). To do so, we built an identical model to the one just described, but replaced pre-earthquake IDES with separate indicators corresponding to each ES category. We found that none of the separate indicators of IDES were significant predictors of HWBI (Table 12.5). These results

Table 12.5 Supplementary regressions for factors affecting the change in human well-being index.

Independent variable	Supplementary model #1	Supplementary model #2	Supplementary model #3
HWBI in 2007	−0.116 (0.082)	−0.119 (0.082)	−0.117 (0.081)
Net benefit from provisioning services in 2007 (thousand yuan)	2.13e−4 (5.29e−4)	—	—
Net benefit from regulating services in 2007 (thousand yuan)	—	−0.003 (0.005)	—
Net benefit from cultural services in 2007 (thousand yuan)	—	—	0.001 (0.002)
Change of IDES	−0.033 (0.029)	−0.027 (0.027)	−0.030 (0.028)
Agricultural income share in 2007	−0.065 (0.052)	−0.064 (0.044)	−0.053 (0.044)
Household income in 2009 (thousand yuan)	3.27e−4* (1.50e−4)	3.48e−4* (1.33e−4)	3.36e−4* (1.41e−4)
Change of per capita income (thousand yuan)	−8.38e−4 (5.42e−4)	−9.38e−4* (4.28e−4)	−8.94e−4* (4.35e−4)
Household size in 2007	0.012* (0.006)	0.014* (0.006)	0.012* (0.006)
Number of seniors (age > = 60 years)	−0.021 (0.017)	−0.023 (0.017)	−0.022 (0.017)
Average education of adults in 2007 (year)	0.006 (0.004)	0.006 (0.005)	0.006 (0.005)
Percentage of female adults in 2007	−0.090 (0.071)	−0.087 (0.070)	−0.094 (0.072)
House damage in the earthquake (0: low, 1: high)	−0.043* (0.018)	−0.041* (0.018)	−0.044* (0.018)
Social ties to local leaders (0: weak, 1: strong)	0.040 (0.025)	0.038 (0.026)	0.040 (0.026)
Gender of interviewee (0: female, 1: male)	0.033 (0.019)	0.032 (0.019)	0.033 (0.019)
Age of interviewee (year)	1.60e−5 (0.001)	5.36e−5 (0.001)	6.39e−5 (0.001)
Intercept	−0.018 (0.077)	−0.008 (0.081)	−0.018 (0.077)
F-statistic	2.01*	2.13*	2.08*
R^2	0.2485	0.2514	0.2490
n	101	101	101

Notes: HWBI, human well-being index; IDES, index system of human dependence on ecosystem services. Dependent variable is the change in HWBI calculated by subtracting HWBI in 2007 from HWBI in 2009. Numbers inside and outside parentheses are coefficients and robust standard errors, respectively. Variance inflation factors are tested to be less than 5. $*p < 0.05$. Adapted from Yang et al. (2015).

offer strong additional evidence for the validity and utility of the IDES as a composite index, which captures variation in HWB that cannot be detected through the use of separate indicators.

Our results also show that those who had larger decreases in HWBI were either disadvantaged households with lower access to multiple forms of capital or those subjected to more earthquake damage (Table 12.4). Specifically, there was a greater reduction in HWBI for households with fewer household members, less income after the earthquake, and weaker social ties to local leaders. Meanwhile, there was a greater reduction in HWBI for households with more property damage and income reduction due to the earthquake, a finding offering additional evidence of the model's validity. In addition, the coefficients of other controlled variables were not significant. These included HWBI before the earthquake, change of IDES from before to after the earthquake, the gender and age of the interviewee for each household, number of seniors, average education of adults, and percentage of female adults (Table 12.4).

12.4 How the system responded to the earthquake

12.4.1 Recovery and restoration plan at the reserve level

Shortly after the earthquake, the State Council of China developed a State Overall Planning for post-earthquake restoration and reconstruction. This measure aimed to restore conditions in all 51 counties affected by the earthquake to a level no worse than what they were immediately before the earthquake (SPGPWERR, 2008). This State Overall Planning was initially designed to be completed within three years but was later shortened to two years. Under this State Overall Planning, a detailed plan for Wolong was also designed for 2008 to 2015. A total budget of around two billion yuan was estimated necessary for the reserve-level reconstruction allocated for various purposes shown in Table 12.6. For environmental recovery and restoration, the general principle put forth was to focus on natural recovery of vegetation. But a set of artificial restoration projects (e.g., through manual seeding and plantation) were also implemented along the road and in important panda habitat regions (Zhang et al., 2014). Field investigation shows that artificial restoration accelerated vegetation recovery compared to natural recovery. Faster vegetation recovery also occurred in areas with more bare soil and gentler slope (Zhang et al., 2014). However, field assessment also indicates that artificial restoration projects could have been more effective with a better design for spatial targeting on forest parcels and panda habitat. The current projects only covered 12% and 7% of total damaged forest parcels and total damaged panda habitat, respectively (Zhang et al., 2014).

Table 12.6 Budget distribution for post-earthquake reconstruction funds provided by the Chinese government for Wolong Nature Reserve

Item	Budget (million yuan)
Conservation stations	454.6
Community infrastructure	381.6
Scientific research and facilities for giant pandas	336.9
Public service infrastructure	287.1
Post-earthquake disaster control	150.0
Schools	96.5
Administration buildings	52.0
Administrative, planning, and environmental assessment costs	263.8
Total	**2022.5**

Source: Sichuan Department of Forestry (2008).

Our informal interviews with local government officials in charge of the reconstruction indicated that local governments had set large-scale tourism development as a strategic priority. The hope was that it would rapidly recover and improve the local economy. However, our previous studies at Wolong (He et al., 2008, Liu et al., 2012) and synthesis of other studies across the world (Kiss, 2004) indicate that local people benefit minimally from large-scale tourism. Most of the revenue from large-scale tourism is captured by tourism companies and governments. Further, the local governments did not recognize that tourism is an unstable industry sector that is vulnerable to natural disasters (Wahab and Pigram, 1997). After the 2008 earthquake, the earthquake-associated disasters have often blocked the main road, and even seven years after the earthquake the tourism industry has not recovered. To find alternative income sources for local households, local governments also encouraged and compensated for raising livestock (e.g., pigs, chickens, cattle, and yaks). However, poorly managed livestock production can pose great threats and negative impacts on giant pandas and their habitat. For instance, our empirical study finds that the distribution of horses largely overlaps with that of giant pandas. Horses consume a substantial amount of bamboo and thus affect the food supply of giant pandas (Hull et al., 2014; see also Chapter 4).

12.4.2 Adaptation at the household level

Overall, our results show that almost all households suffered from the negative environmental and socioeconomic impacts of the earthquake and associated disasters after the earthquake, but the magnitudes of impacts were greater for disadvantaged households. Households' responses to the earthquake and associated disasters were both guided by the government recovery and restoration plan and restricted by the households' adaptive capacity.

Households received loans from the government-controlled banks (on average 26 190 yuan, 47% of the loans received) and from relatives and friends (on average 29 948 yuan, 53% of the loans received; Yang, 2013). These loans were used for repairing or reconstructing their houses. Overall, the quality of house and total constructed area per household improved after the earthquake (Yang, 2013). But households achieved such improvements at the cost of using their decades of savings and bearing an additional, substantial amount of debt. Many of them chose to do so because one of the government strategic priorities after the earthquake was tourism development. They expected that tourism would recover soon after the completion of reconstruction, and the spacious and high-quality houses would facilitate their participation in tourism businesses (e.g., lodging services and restaurants). Another reason for their investment in new housing is that the remaining cropland is unlikely to support many households' livelihoods, forcing them to switch from agriculture to tourism businesses or other non-agricultural activities. Our household survey data show that approximately 12% of the total cropland was destroyed by the earthquake and associated disasters and 24% by reconstruction after the earthquake (Yang, 2013). On average, only 0.2 ha of cropland per household remained in 2012. The bad road conditions have kept tourists away and shattered the hope for tourism-fueled economic recovery. In addition, the road's poor condition also threatened agricultural production because it made it very difficult and costly to rapidly transport some agricultural products (e.g., fresh vegetables) to outside markets.

Altogether, the government-led recovery and restoration plan and frequent associated disasters after the earthquake substantially limited the adaptive capacity of local households and shaped their adaptation measures. We observed that some households adopted incremental adaptation (e.g., switching crop types or investing more in animal husbandry). Others attempted transformational adaptation measures. Examples include abandoning agricultural production and traveling to cities for migrant work, going to their relatives, and permanently moving out of the reserve (Yang et al., 2013a). A few people also turned to the forest for poaching (Yang et al., 2013a). In Chapter 9, we showed that there was an abrupt increase of non-agricultural income and total household income after the earthquake, primarily due to temporary work opportunities provided by post-earthquake reconstruction. Nevertheless, this did not mean a rapid recovery or adaptation at the household level. As shown in Chapter 9 and this chapter, road transportation, agricultural production, and tourism businesses have not yet recovered. Once the post-earthquake reconstruction is completed, the temporary work income source will disappear or be substantially reduced, and local households will still have to find alternative long-term income sources through the incremental and transformational adaptation measures mentioned above.

12.5 Discussion and concluding remarks

The 2008 Wenchuan earthquake caused enormous socioeconomic and environmental losses in our coupled system of Wolong, which were further aggravated by long-lasting secondary disasters. Our findings suggest that disadvantaged people suffered more than others and were also more vulnerable, having less capacity to adapt and recover from the negative impacts. Their adaptation measures largely relied on the government recovery and restoration plan and were constrained by their limited adaptive capacity. Therefore, both short-term relief efforts and long-term reconstruction plans should target disadvantaged people who have less access to natural, human, manufactured, and/or social capitals. Some example measures include, but are not limited to, reducing or waiving their loans or interest rates from government-controlled banks, subsidizing their living costs (e.g., food, education, and health care expenditures), and providing training and employment opportunities.

In the long run, the priority of a reconstruction plan should be capacity building at multiple levels. In our case, local governments designed the reconstruction plan to attempt to restore and improve capacity at the reserve level but did not recognize the importance of enhancing capacity at the household level. Although the overall capacity at the reserve level was restored and even improved through massive investment, the adaptive capacity of households did not improve and might even have deteriorated. Such consequences are partly due to the massive and frequent associated disasters after the earthquake but are also largely due to poor reconstruction planning and inadequate governance. For example, the reconstruction plan underestimated the severity of secondary disasters occurring after the earthquake and did not ensure the accessibility of the main road. In addition, the reconstruction plan did not recognize the uncertainty of the tourism industry. The government also did not seriously consider the limited benefits that large-scale tourism development would provide for local people. Assuming that the tourism business will recover in the future, a potential complementary remedy that could be enacted is to redistribute a fair proportion of the revenue from large-scale tourism development to the local community. This goal could be accomplished by subsidizing tourism benefits and providing employment opportunities to local people. Future policy priorities may also promote switching to small-scale tourism development. This approach has been shown to provide more benefits to local people through greater opportunities to participate (He et al., 2008, Liu et al., 2012).

To enhance the resilience of local households to disasters, it is also important to expand income sources and diversify their dependence on ES according to the local context. Taking large-scale farms as an example, they might not be a good solution for Wolong due to pollution, poor and uncertain road conditions, and limited land resources. Instead, local governments might prioritize the cultivation of high-value local or eco-certified products (e.g., mushrooms, traditional Chinese herbs, and honey) that are not vulnerable to temporary road interruption and do not require much cropland. Considering the publicity given to Wolong and its high environmental quality, such products from the reserve could be highly competitive in outside markets. To cope with market fluctuations and other uncertainties (e.g., pests and diseases), as our results suggested, maintaining a balance of the dependence on multiple ES may reduce the risks and impacts of disasters.

The long-term socioeconomic and environmental impacts of the earthquake and associated disasters are difficult to predict. Impacts largely depend on reconstruction efforts guided by the central and local governments as well as the responses of local residents to earthquake impacts and government reconstruction efforts. Our previous studies (Hull et al., 2014, Yang et al., 2013a, 2015) suggest that changes in environmental conditions or HWB may affect both socioeconomic and environmental behaviors of humans, form feedbacks, and in turn affect the environment and people. Thus, to reduce the vulnerability and enhance the resilience of local communities, it is crucial to monitor, understand, and manage such feedbacks. This task can be accomplished by targeting certain population groups, enhancing their adaptive capacity at multiple levels, and diversifying their dependence on ES.

12.6 Summary

The number and scale of natural disasters have increased in the past few decades and may continue to rise in the future. Thus, it is crucial to understand quantitatively how to reduce the vulnerability, enhance the resilience, and increase the adaptive capacity of coupled human and natural systems to natural disasters. In this chapter, we demonstrated the utility of two quantitative indicators—the index system of human dependence on ecosystem services (IDES) and the human well-being index (HWBI)—in disaster research in our model coupled system of Wolong Nature Reserve. Models integrating the two indices helped to characterize the response of coupled systems to disasters. Results show that the 2008 Wenchuan earthquake caused a decline of 48% and 12.6% in dependence on ES and human well-being, respectively. Disadvantaged people who lacked access to different forms of capital suffered most from disasters both in terms of direct impacts and the post-disaster recovery. We found that the earthquake recovery and restoration plan helped to improve the overall adaptive capacity at the reserve level but did not restore and even deteriorated adaptive capacity at the household level. Our findings suggest that the design of disaster recovery and restoration plans should target capacity building at multiple levels, adapt to local contexts, and account for uncertainties. The methods and findings from our study may provide insights for such future research in Wolong and many other areas around the globe.

References

Adger, W.N. (2006) Vulnerability. *Global Environmental Change—Human and Policy Dimensions*, **16**, 268–81.

Adger, W.N., Hughes, T.P., Folke, C., et al. (2005) Social-ecological resilience to coastal disasters. *Science*, **309**, 1036–39.

Berke, P.R. and Campanella, T.J. (2006) Planning for post-disaster resiliency. *Annals of the American Academy of Political and Social Science*, **604**, 192–207.

Cutter, S.L., Boruff, B.J., and Shirley, W.L. (2003) Social vulnerability to environmental hazards. *Social Science Quarterly*, **84**, 242–61.

Folke, C. (2006) Resilience: the emergence of a perspective for social-ecological systems analyses. *Global Environmental Change—Human and Policy Dimensions*, **16**, 253–67.

Gallopin, G.C. (2006) Linkages between vulnerability, resilience, and adaptive capacity. *Global Environmental Change—Human and Policy Dimensions*, **16**, 293–303.

Gunderson, L. (2010) Ecological and human community resilience in response to natural disasters. *Ecology and Society*, **15**, 18.

He, G., Chen, X., Liu, W., et al. (2008) Distribution of economic benefits from ecotourism: a case study of Wolong Nature Reserve for Giant Pandas in China. *Environmental Management*, **42**, 1017–25.

Hull, V., Zhang, J., Zhou, S., et al. (2014) Impact of livestock on giant pandas and their habitat. *Journal for Nature Conservation*, **22**, 256–64.

IPCC (2012) *Managing the Risks of Extreme Events and Disasters to Advance Climate Change Adaption*. http://ipcc-wg2.gov/SREX/report,http://www.ipcc-wg2.gov/SREX/images/uploads/SREX-All_FINAL.pdf.

Kates, R.W., Travis, W.R., and Wilbanks, T.J. (2012) Transformational adaptation when incremental adaptations to climate change are insufficient. *Proceedings of the National Academy of Sciences of the United States of America*, **109**, 7156–61.

Kiss, A. (2004) Is community-based ecotourism a good use of biodiversity conservation funds? *Trends in Ecology & Evolution*, **19**, 232–37.

Klein, C.A. and Zellmer, S.B. (2007) Mississippi River stories: lessons from a century of unnatural disasters. *Southern Methodist University Law Review*, **60**, 1471–538.

Liu, J., Dietz, T., Carpenter, S.R., et al. (2007) Complexity of coupled human and natural systems. *Science*, **317**, 1513–16.

Liu, J., Linderman, M., Ouyang, Z., et al. (2001) Ecological degradation in protected areas: the case of Wolong Nature Reserve for giant pandas. *Science*, **292**, 98–101.

Liu, J. and Viña, A. (2014) Pandas, plants, and people. *Annals of the Missouri Botanical Garden*, **100**, 108–25.

Liu, W., Vogt, C.A., Luo, J., et al. (2012) Drivers and socioeconomic impacts of tourism participation in protected areas. *PLoS ONE*, **7**, e35420.

Neumayer, E. and Barthel, F. (2011) Normalizing economic loss from natural disasters: a global analysis. *Global Environmental Change*, **21**, 13–24.

Ouyang, Z., Xu, W., Wang, X., et al. (2008) Impact assessment of Wenchuan Earthquake on ecosystems. *Acta Ecologica Sinica*, **28**, 5801–09 (in Chinese).

Raudsepp-Hearne, C., Peterson, G.D., Tengö, M., et al. (2010) Untangling the environmentalist's paradox: why is human well-being increasing as ecosystem services degrade? *BioScience*, **60**, 576–89.

Sichuan Department of Forestry (2008) *Overall Planning for Post-Wenchuan Earthquake Restoration and Reconstruction at Wolong National Nature Reserve*, Chengdu, China (in Chinese).

Smit, B. and Wandel, J. (2006) Adaptation, adaptive capacity and vulnerability. *Global Environmental Change—Human and Policy Dimensions*, **16**, 282–92.

SPGPWERR (2008) *The State Overall Planning for Post-Wenchuan Earthquake Restoration and Reconstruction*. Report for State Planning Group of Post-Wenchuan Earthquake Restoration and Reconstruction (SPGPWERR), Beijing (in Chinese).

Turner, B.L., Kasperson, R.E., Matson, P.A., et al. (2003) A framework for vulnerability analysis in sustainability science. *Proceedings of the National Academy of Sciences of the United States of America*, **100**, 8074–79.

Viña, A., Chen, X., McConnell, W.J., et al. (2011) Effects of natural disasters on conservation policies: the case of the 2008 Wenchuan Earthquake, China. *Ambio*, **40**, 274–84.

Wahab, S. and Pigram, J.J. (1997) *Tourism, Development and Growth*. Routledge, London, UK.

Xinhua News Agency (2008a) Death Toll from China's May Earthquake Remains Unchanged at 69,227. http://news.xinhuanet.com/english/2008-09/25/content_10110269.htm.

Xinhua News Agency (2008b) Direct Economic Loss of the Wenchuan Earthquake is Estimated to be Above One Trillion. http://news.xinhuanet.com/local/2008-07/02/content_8488462.htm.

Yang, W. (2013) *Ecosystem Services, Human Well-being, and Policies in Coupled Human and Natural Systems*. Doctoral Dissertation, Michigan State University, East Lansing, MI.

Yang, W., Dietz, T., Kramer, D.B., et al. (2013a) Going beyond the Millennium Ecosystem Assessment: an index system of human well-being. *PLoS ONE*, **8**, e64582.

Yang, W., Dietz, T., Kramer, D.B., et al. (2015) An integrated approach to understanding the linkages between ecosystem services and human well-being. *Ecosystem Health and Sustainability*, **1**, 19.

Yang, W., Dietz, T., Liu, W., et al. (2013b) Going beyond the Millennium Ecosystem Assessment: an index system of human dependence on ecosystem services. *PLoS ONE*, **8**, e64581.

Zhang, J., Hull, V., Huang, J., et al. (2014) Natural recovery and restoration in giant panda habitat after the Wenchuan earthquake. *Forest Ecology and Management*, **319**, 1–9.

Zhang, J., Hull, V., Xu, W., et al. (2011) Impact of the 2008 Wenchuan earthquake on biodiversity and giant panda habitat in Wolong Nature Reserve, China. *Ecological Research*, **26**, 523–31.

CHAPTER 13

Human–Nature Interactions under Policy Interventions

Xiaodong Chen, Vanessa Hull, Wu Yang, and Jianguo Liu

13.1 Introduction

In today's globalizing world, few corners of the Earth have been untouched by humans. The world's natural systems are facing unprecedented threats due to human impacts, including loss of ecosystem services, degradation of habitats, increases in endangered species, and pollution (Crutzen, 2006). At the same time, the growing human population is facing complex challenges across the globe such as famine, natural disasters, political turmoil, and social inequities (Liu et al., 2015, Millennium Ecosystem Assessment, 2005). These far-reaching phenomena occurring in the human and natural realms are inherently linked to one another and are part of Coupled Human and Natural Systems (CHANS; Liu et al., 2007a, b, Alberti et al., 2011). The magnitude and severity of threats warrant effective policies to be put in place to manage the threats, prevent further losses to humans and nature, and strive to improve sustainability of these complex systems in the uncertain future (Dovers, 1996, Voss et al., 2006).

Policies for coupled systems can take on many forms and often act as feedbacks. They are usually implemented by institutions or governing bodies with the intention of shaping or changing human behaviors, such as those causing negative socio-economic and/or environmental impacts (Liu and Viña, 2014, Viña et al., 2011). Many policies are developed to target a specific sector or issue, such as agriculture, energy, family planning, or resource extraction. However, the effects of such policies are often overarching and impact multiple other sectors or issues. Policies also may be implemented at the local, regional, or global level, but their effects often penetrate all levels. For example, a local restriction on resource extraction can result in expansion of extractive activities to other areas of the globe (Liu and Raven, 2010).

Policies often incorporate a specific strategy for affecting human actions. Two broad categories often discussed in coupled systems research are exclusionary policies and incentive-based policies. Exclusionary policies involve prohibitions of human activities using punishments for violators. Incentive-based policies are geared toward rewarding people for their actions meant to promote sustainability. In the natural resource management realm, exclusionary policies are sometimes referred to as "fences and fines" (Brandon and Wells, 1992, Gockel and Gray, 2009) or "command-and-control" (Holling and Meffe, 1996, Nielsen et al., 2014) approaches. Such approaches involve, for example, fencing in a nature reserve and fining individuals who breach the restrictions on resource extraction. Such policies have come under fire in recent decades for being ineffective and socially unjust. This criticism is especially prominent in developing countries with widespread poverty and social unrest (Dietz et al., 2003, Holling and Meffe, 1996). In contrast, incentive-based policies have become popular in recent years with the emergence of creative solutions to long-standing conflicts between humans and nature. One example is payments for ecosystem services (PES) programs that provide

Pandas and People. Edited by Jianguo Liu, Vanessa Hull, Wu Yang, Andrés Viña, Xiaodong Chen, Zhiyun Ouyang, and Hemin Zhang. Published 2016 by Oxford University Press.

economic incentives to ecosystem services providers to undertake actions for desired environmental benefits (Chen et al., 2010, Wunder, 2007).

Although billions of dollars have been invested in conservation policies and programs globally, ecosystem degradation continues (Chen et al., 2009a, Ferraro and Kiss, 2002, James et al., 2001). Such failure has occurred in part because these conservation efforts are often not evaluated appropriately (Millennium Ecosystem Assessment, 2005). Evaluations of the effectiveness of conservation policies in coupled systems have been a great challenge due to the complexities of coupled systems. For instance, environmental conditions and human activities are often spatially heterogeneous and temporally dynamic (Chen et al., 2014). Further, unintended side effects may emerge from the implementation of a policy and might be harmful to the sustainability of the system (Liu et al., 2007b). Effective conservation makes it imperative to conduct in-depth studies on coupled systems, particularly with respect to characterizing and modeling the effects of policies on human–nature interactions in such systems.

In this chapter, we illustrate the effects of conservation policies on human–nature interactions in our model coupled system—Wolong Nature Reserve (Chapter 3). Specifically, we first provide an overview of the policy context in Wolong and then present some of our research findings on three policies in more detail. We highlight a couple of surprises and consequences that have arisen as a result of policy implementations, and then discuss potential ways to improve these policies. We conclude with reflections on lessons learned and ways forward in coupled systems research on policy interventions.

13.2 Policy interventions in Wolong Nature Reserve

Wolong Nature Reserve is an excellent system for understanding the effects of diverse policies on human–nature interactions. As described in Chapter 3, the reserve has a majestic, mountainous topographic template, rich biodiversity, a charismatic and world-famous endangered species—the giant panda—and a vibrant human community with ethnic minority groups. These characteristics set the stage for multifaceted and meaningful policy analyses that have national and global significance. As one of the first nature reserves designated in China, Wolong has long been held up as a model for nature reserve management and policy making across the country. By completing the first-ever reserve management plan in China (Ministry of Forestry, 1998), Wolong has been a laboratory for policy experimentation, whose results have far-reaching implications. In fact, policies directly affecting Wolong have also reached a global level. Examples include the designation of the reserve as part of the UNESCO Man and Biosphere Programme (MAB) network in 1980 and later as part of a World Heritage Site designated by UNESCO in 2006.

More than 95% of the inhabitants in Wolong are farmers. Much of their livelihood depends on farming and livestock breeding. As a nature reserve, Wolong prohibits timber harvesting. However, some of the inhabitants were involved in illegal timber harvesting, which was a major contributor to the deforestation after the establishment of the reserve (Liu et al., 2001). Another major human activity that affected forests was fuelwood collection, which provided the major energy source for cooking human food and pig fodder, and for heating in winter. Fuelwood collection had substantial impacts on forests. For instance, the average fuelwood consumption in 2004 alone was more than 6000 kg per household (Chen et al., 2012). Although most inhabitants preferred electricity over fuelwood, electricity was not reliable with its low voltage and was mostly used for electronic appliances and equipment only, e.g., television (An et al., 2002).

A prominent thread weaving through government policies since the reserve's inception has been a stated goal of promoting both biodiversity conservation and human development (Ghimire, 1997). Policies implemented in Wolong have been diverse and their effects varied. Many of the early policies were of the exclusionary "fences and fines" variety (Ghimire, 1997). For instance, early policies toward fuelwood collection aimed to restrict the amounts, times, and locations of fuelwood collection (He et al., 2009). These policies also prohibited cutting young trees or precious tree species. Because no economic incentives were provided through these policies, the management of fuelwood collection was difficult. With limited alternative energy sources, fuelwood collection and thus forest degradation continued (Viña et al., 2007). Early development policies involved road construction, which significantly improved access to services and markets for local residents. The first major road construction project was initiated in 1992, with a 35-million yuan investment from the central government with the goal of linking Wolong to the surrounding rural communities and markets (Liu, 2012). This project was integral to jumpstarting tourism in the region by improving the accessibility of this remote area to outsiders (Liu et al., 2012). Road construction has continued with additional improvement projects over the years. However, the accessibility and safety of the road are still not adequate. This is because the road is continually damaged by frequent landslides and rockslides due to the 2008 Wenchuan earthquake and its aftershocks (Viña et al., 2011; see also Chapter 12). The lack of reliable and safe road access has plagued the local economy and limited access to outside markets.

As time turned to the new millennium, large-scale incentive-based conservation mechanisms began to take shape. For instance, the government's investments in building an ecological hydroelectric power station greatly improved the reliability of electricity and inhabitants' accessibility to electricity (Chen et al., 2012). Thus, electricity became an alternative energy source to fuelwood (see Chapter 10). PES programs also came to the fore as a way of compensating local residents for participation in conservation programs, a policy shift that profoundly shaped human–nature interactions in Wolong.

13.3 Trials and tribulations of exclusionary policies

Exclusionary policies have a long history in Wolong. One of the early, failed exclusionary policies in Wolong was a relocation policy put forth in the early 1980s. The government sought to move 100 families living in areas near panda habitat down to the main road where their impacts on pandas and their habitats would be lower (Ghimire, 1997). A large apartment complex was constructed with support from the World Food Programme. But the families refused to relocate (Ghimire, 1997) because such housing was located far from their agricultural land and there were few other means to earn income. This shortcoming reveals the dangers of exclusionary methods that do not fully take into account the resource needs of the local people, and highlights the importance of considering spatial heterogeneity of resources on the landscape in policy making.

Complexities involving space are also evident in the exclusionary policy of Wolong's zoning scheme. Given the challenges of managing biodiversity conservation and human development within the same spatial extent, managers sought to develop a framework to spatially segregate the two activities across the landscape. This plan was part of a broader policy put forth by the central Chinese government that mandated all nature reserves in the country be divided into three zones—an experimental zone for human development, a core zone for biodiversity protection, and a buffer zone to soften the border between the two (Liu and Li, 2008). Most human activities are prohibited in the core zone, and only selected human activities are allowed in the buffer zone. The zoning scheme in Wolong (Figure 4.2), formally established in 1998, has the experimental zone centered on the main road running through the reserve and the existing housing structures (Ministry of Forestry, 1998). The core zone is mainly at higher elevations.

We conducted research to better understand the efficacy of this zoning scheme by performing different spatial overlays of various types of human and natural phenomena. We found that the zoning scheme could have been better designed to maximize protection of giant panda habitat. When the zoning was established, ~54% of highly suitable panda habitat was found outside the protected core zone, including 40% in the buffer zone and 14% in the experimental zone (Figure 13.1). Pandas inhabiting these areas may be more vulnerable under the lower levels of protection. Results also indicate that the zoning scheme was largely effective at containing large-scale infrastructural development (e.g., houses, roads, and tourist structures) to the experimental zone. But human activities such as raising livestock were not well contained and were found deep into the core zone (Hull et al., 2011; Figure 4.3). These findings reflect the difficulty of

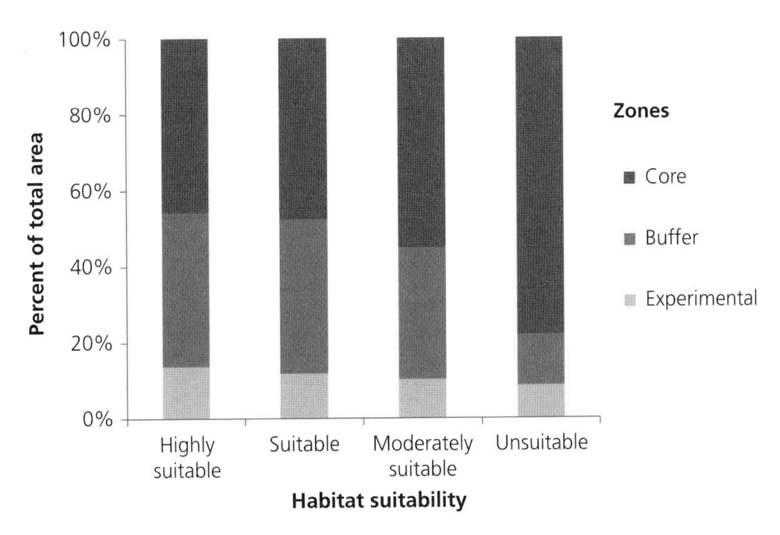

Figure 13.1 Distribution of giant panda habitat suitability classes across core, buffer, and experimental zones in Wolong Nature Reserve in 1997 (one year before zoning designation). Figure adapted from Hull et al. (2011). Reprinted with permission from Elsevier.

drawing "lines in the sand" with exclusionary zoning schemes. It is nearly impossible for local people to know where the boundaries lie on the rugged landscape and adjust activities accordingly (Hull et al., 2011). In addition, some recent revisions to the scheme involved an extension of the experimental zone to accommodate a new tourist venture (Hull et al., 2011). This area that underwent the zoning revision has regional importance for panda conservation because it is part of a key corridor connecting pandas in Wolong to neighboring panda habitats in this mountain range (Xu et al., 2006a).

13.4 Innovative solutions—payments for ecosystem services

The term PES is relatively new, but PES is not a new practice in Wolong although the scale was quite small in the past. The first PES program implemented in Wolong was a forest conversion program carried out in 1986 by the World Food Programme. This program paid some local people to convert 113 ha of cropland to forests (Ghimire, 1997). In recent years, PES programs have come to the fore in Wolong, with the implementation of two nationwide programs, the Natural Forest Conservation Program (NFCP) and the Grain to Green Program (GTGP), and one local program (the Grain to Bamboo Program, GTBP). The government initiated NFCP and GTGP because of the severe droughts in 1997 and the major floods in 1998, which had caused excessive deforestation across China (Liu and Raven, 2010, World Bank, 2001). Both programs aim to increase forest cover by providing payments to rural communities, especially in the middle-upper reaches of the Yangtze River and Yellow River basins (Liu et al., 2013, Xu et al., 2006b).

NFCP has been implemented in Wolong since 2001 (Chen et al., 2014; Figure 13.2). All households that existed in 2001 participated in the program. Through NFCP, each household, or groups of 2–16 households, are contracted to a natural forest parcel for monitoring to prevent illegal harvesting. Each participating household receives an annual payment of about 850 yuan, which accounted for ~5% of the average annual household income in 2005. People in Wolong are encouraged to purchase electricity with the NFCP payment in order to reduce their fuelwood consumption. Households who are found to be illegally harvesting in natural forests lose their NFCP contracts. Households also lose part or all of their NFCP payment for the year if illegal timber harvesting is found in the forest parcel that they are responsible for monitoring.

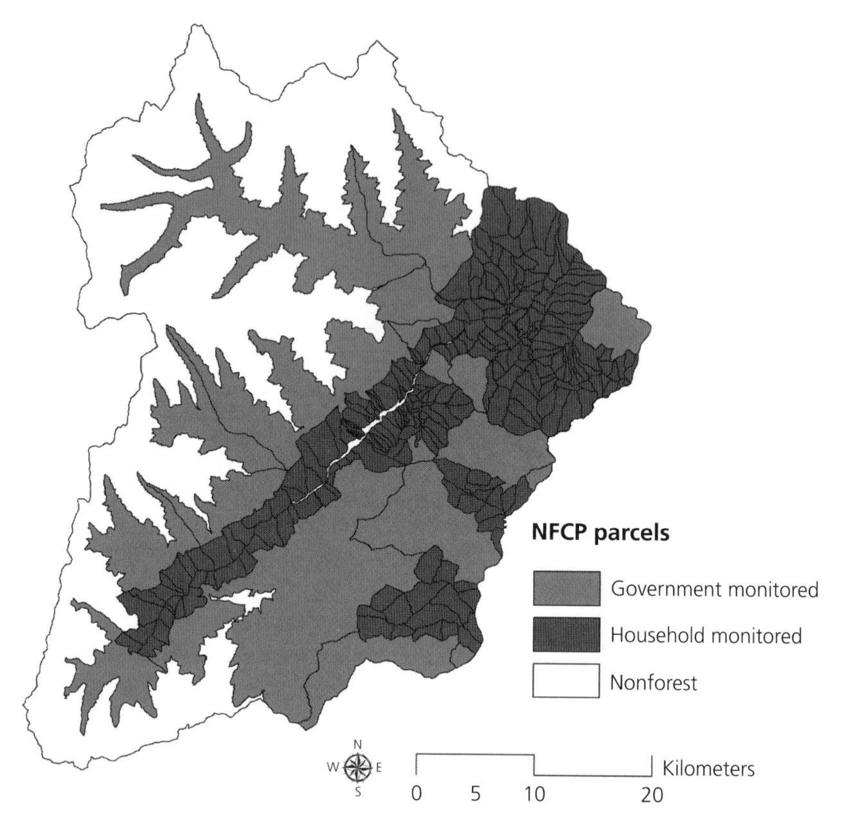

Figure 13.2 Distribution of household-monitored and government-monitored NFCP parcels in Wolong Nature Reserve.

NFCP has effectively prevented timber harvesting and has substantially reduced fuelwood collection. Since the implementation of NFCP, electricity consumption has doubled, and the labor spent on fuelwood collection has been reduced by about half (Yang et al., 2013a). After NFCP began in 2001, there was a net forest recovery of 11.4% by 2007, in contrast to 14.2% net forest loss (from 1072 km^2 to 920 km^2) between 1987 and 2001 (Yang et al., 2013a). Although many factors may contribute to the changes in forest cover, the abrupt changes observed before and after 2001 are mainly due to the implementation of NFCP (see also Chapter 6). We estimated a counterfactual baseline of what forest cover would have looked like without NFCP in 2007 based on historical changes in forest cover in the reserve from 1965 to 2001 (Yang et al., 2013a). We used this historical trend approach for counterfactual baseline estimation because this is the only feasible and most reliable approach in context. Blank control sites without NFCP are impossible here because NFCP is a national policy covering most of the country. This method was reliable because NFCP was the main driver of forest-cover change from 2001 to 2007 in the reserve. Our estimations suggest that the forest recovery trend is statistically significant ($p < 0.05$, Figure 13.3). This PES program might have increased forest cover by 11 000 ha between 2001 and 2007 (from 68 000 ha to 79 000 ha), roughly 5.5% of the total land in Wolong (Yang et al., 2013a; Figure 13.3). In addition, not all forest parcels contracted to local households under NFCP had equal forest recovery. Medium-sized groups of households were more effective in monitoring to prevent illegal harvesting than smaller or larger groups of households (Yang et al., 2013b; Figure 13.4). In terms of underlying mechanisms of the non-linear effect, there are two opposing forces from group size in affecting collective action (e.g., forest

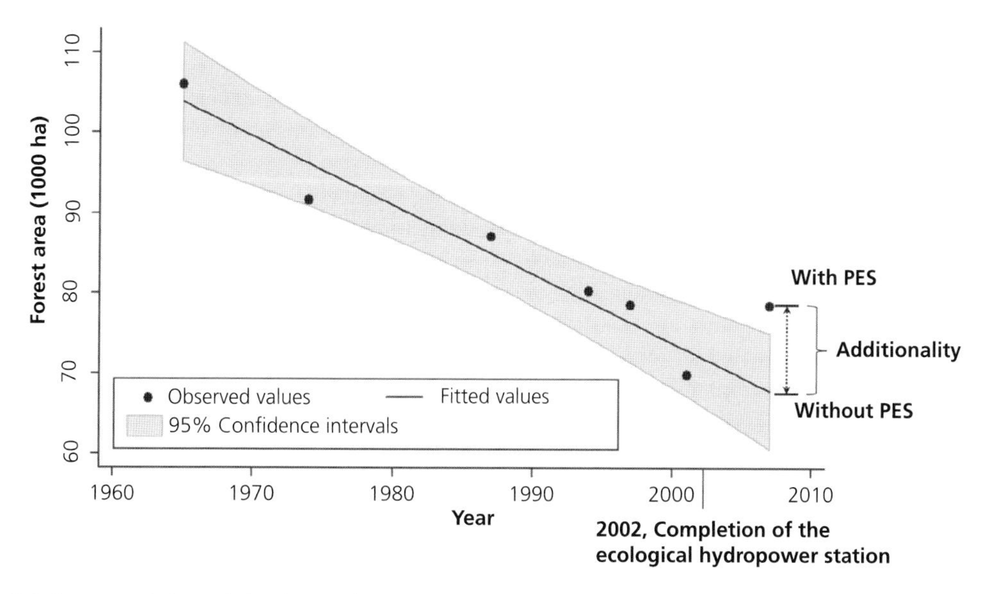

Figure 13.3 Forest cover before and after the Natural Forest Conservation Program (NFCP) was enacted in 2001 in Wolong Nature Reserve. Forest cover trends from 1965 to 2001 were used to estimate the counterfactual with and without NFCP in 2007. Figure reproduced from Yang et al. (2013a) with permission from Elsevier.

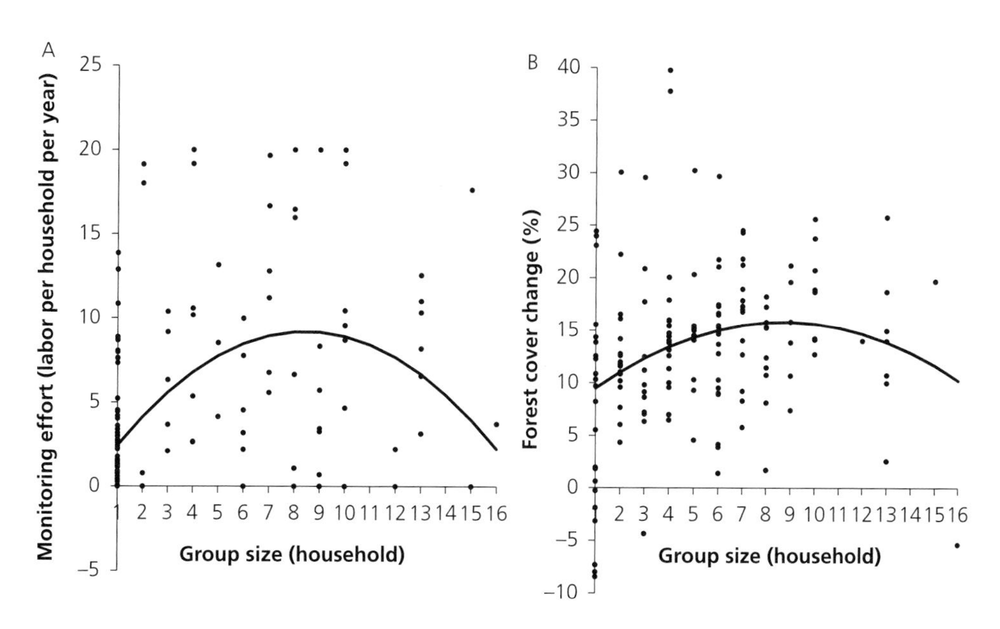

Figure 13.4 The effect of group size on collective action and forest outcomes. Predicted (A) monitoring effort and (B) forest cover change in Wolong from 2001 to 2007 under different group sizes for Natural Forest Conservation Program (NFCP) monitoring. The line shows predicted fit and dots show observations. Figure reprinted from Yang et al. (2013b).

monitoring here) and resource outcomes (e.g., forest cover) by simultaneously increasing free riding (i.e., obtaining group benefits without incurring costs) and enhancing enforcement to reduce free riding (Yang et al., 2013b). The confirmed non-linear effect and underlying mechanisms disentangle the long-lasting debate of the "group size paradox" (Esteban and Ray, 2001, Ostrom, 2005) and provide theoretical and empirical guidance for common-pool resource research and management.

GTGP has been implemented in Wolong since 2000 to convert sloping cropland into forests. Households that enroll in GTGP receive an annual payment of 3450 yuan per ha for eight years. GTGP was open for enrollment in 2000, 2001, and 2003 (Chen et al., 2009a). The main criterion is that cropland plots with slopes over 25 degrees had a higher priority for enrollment, although cropland plots with lower slopes were also allowed to enroll. Almost all the households in Wolong participated in GTGP, and a total of 367.3 hectares of cropland was enrolled by 2005 (Wolong Nature Reserve, 2005). As most GTGP contracts matured by the end of 2008, the government extended the program for another cycle of eight years at a lower payment of 1875 yuan per ha (Chen et al., 2009b).

GTGP has produced substantial impacts on the inhabitants. Our survey of 304 households in 2006 found that on average each household enrolled about 0.36 ha of land into the program, corresponding to ~56% of their land. The GTGP payment accounted for about 8% of the annual household income in 2005 (Chen et al., 2009b). If the program were to end, only about 22% of the land that had been enrolled in GTGP was planned to be converted back to crop production. In fact, ~68% of the households who participated in GTGP indicated that they will not reconvert any of their GTGP land to farming. The low reconversion rate was because GTGP participants tended to enroll marginal land into the program. In addition, households with more off-farm income were planning to reconvert less GTGP land after the payment ends (Chen et al., 2009a). Since the implementation of GTGP, much labor has been released from agriculture and invested in off-farm employment. Off-farm income has increased dramatically since the implementation of GTGP, and accounted for ~38% of the annual household income in 2005 (Chen et al., 2009b). As more off-farm opportunities become available from the rapid economic development in China, most GTGP participants may choose not to reconvert their GTGP land.

The Grain to Bamboo Program pays local residents to plant bamboo along a 15-m buffer on both sides of the main road in Wolong. Although the area involved is small (81.9 ha), the household payments were high, with 2143–2857 yuan per ha per year given to each participating household (Yang et al., 2013a). The program has been successful at increasing bamboo cover along the roadway and contributing to rural household incomes. Bamboo from these plots is harvested and fed to captive pandas in the China Conservation and Research Center for the Giant Panda at Wolong. But this bamboo has little direct value for the conservation of the wild pandas, considering giant pandas are usually distributed far away from roadways.

13.5 Surprises and unintended consequences

There have been many surprises as a result of policy implementation in Wolong. Perhaps the most surprising finding is the interaction effect of NFCP and GTGB (GTGP and GTBP together) on household income (see details in Chapter 9). Separately, NFCP and GTGB each had a negative impact on household income. But jointly their interaction led to a positive effect. These are due to the different adaptation strategies of households in response to policy implementation. When the negative impacts from one policy are tolerable, households may not react or adopt an incremental adaptation strategy (e.g., use more pesticides and fertilizers to increase crop production and thus earn more income) to offset the losses. When two policies have been simultaneously implemented and the negative impacts are not tolerable or cannot be easily offset, households will adopt transformational adaptation measures (e.g., switching from agriculture to migrant work or business activities) to maintain their livelihood.

Another example is the difference in efficacy between GTGP and NFCP on forest recovery. On average, each participating household received 36% higher payment through GTGP than NFCP as of 2008. However, the forest recovery between 2001 and 2007 detected via satellite data analysis has been attributed mainly to NFCP (Chapter 6). Young trees in the GTGP plots were difficult to detect from the satellite data, although they were evident on the ground. The land converted to forests through GTGP accounted for only ~1% of a total of 105 km^2 of increases in forested area during this period. Another related surprise is that although fuelwood collection has been substantially reduced since NFCP was initiated, Wolong inhabitants still rely on fuelwood for some of their energy needs (see also Chapter 10). Some residents have started cutting tree branches instead of whole trees, which may lead to substantial cumulative impacts on forests over time.

As ecosystems recover and forest cover increases, many wildlife populations have also increased. As a result, an unintended consequence of these programs is increased wildlife crop raiding (e.g., by deer and wild boar; Yang et al., 2013a). Crop raiding may affect local people's livelihood, their perception of PES programs, and land use in the future (Hill et al., 2002). Agricultural land plots that are close to the natural forests and GTGP plots are especially vulnerable to crop raiding. Therefore, many inhabitants indicated that they would like to enroll these agricultural lands into GTGP if the enrollment is offered again. The negative impacts of crop raiding on crop production has changed the originally perceived opportunity costs of GTGP enrollment when the program began, which may directly affect people's land-use decisions after the program ends.

Another surprise from policy making in Wolong has revolved around raising livestock, which is encouraged by development programs such as the Western China Development program as a way of improving the economic livelihoods of rural people without significantly harming the local ecosystem (Melick et al., 2007). In fact, after the timber harvesting ban, the government encouraged local residents to take up livestock. The belief was that the forest would recover and inhabitants would recover income lost as a result of the ban on timber harvesting (Melick et al., 2007). However, managers did not anticipate that livestock might in turn threaten the very conservation goals the timber-harvesting ban sought to achieve. This surprising effect occurred in Wolong when livestock such as horses overflowed from grazing areas and penetrated the panda habitat. There they consumed substantial amounts of bamboo, in some cases resulting in declines in panda use (see Chapter 4, Hull et al., 2014). In response to this threat, a ban on raising horses was put in place recently in Wolong. This ban also has had unintended consequences, because there has since been a resurgence of other livestock species such as sheep and yaks that have been observed penetrating the forest in greater numbers than before the horse ban.

13.6 Policy improvements

Results from our research indicate that the policies discussed in the previous sections could be improved. For example, the current zoning scheme could be enhanced by adjusting the core, buffer, and experimental zones to better protect panda habitat from human impacts in the future (see Figure 4.7). Currently, 40% of the panda distribution is outside the core zone. The core zone also includes a significant amount of alpine habitat that cannot support giant pandas (Hull et al., 2011). Several key areas of the experimental and buffer zones overlap with the panda distribution. Although the zoning scheme has been helpful for keeping developers from infiltrating habitat, zoning was not a solid "stand-alone" policy that could prevent all threats to core panda habitat. We recommend that managers view zoning as part of a larger toolbox for panda conservation.

With regard to GTGP, substantially more land can be obtained if cost-effective targeting can be adopted to improve the current GTGP scheme. GTGP was designed as a PES program with a flat payment, i.e., all participants are paid the same price. In order to maximize environmental benefits, PES programs must target the land that provides environmental benefits with the least cost, i.e., cost-effective targeting (Babcock et al., 1996). We conducted targeting by ranking all GTGP plots from high to low according to the benefit-to-cost ratio and enrolling plots with higher ratio first (Chen et al., 2010). It would cost 120 000 yuan vs 320 000 yuan to obtain 80% of GTGP land through cost-effective targeting vs the flat payment scheme, respectively. In addition, the difference between cost-effective targeting and the flat payment scheme increased as the amount of enrolled land increased (Figure 13.5). For instance, obtaining 50% and 90% of land through the flat payment scheme will cost 82 000 and 585 000 yuan, which are respectively about 1.9 and 3.4 times the costs through cost-effective targeting for enrollment.

Since the inception of NFCP, a substantial amount of fuelwood consumption has been replaced primarily with electricity use, which is at least partially because participants were better able to afford electricity given the NFCP payment. In fact, on average, the NFCP payments were much more than participants' electricity expenses. Policy designs that more effectively address people's energy needs (e.g., replace the cash payment with an electricity payment) may result in more conservation gains (Chen et al., 2014). Furthermore, medium-sized groups of households for monitoring forests were more effective in preventing illegal harvesting than smaller or larger groups of households (Yang et al., 2013b). Therefore, combining small groups and dividing large groups into optimal sizes may also enhance the conservation gains from NFCP. In addition, the forest management outcomes can also be improved through regulating factors that enhance collective action. Specific measures include punishing free riding, enhancing enforcement, improving social capital across monitoring groups and within each group, and encouraging self-selection of group members to allow members with good social relationships to form groups autonomously.

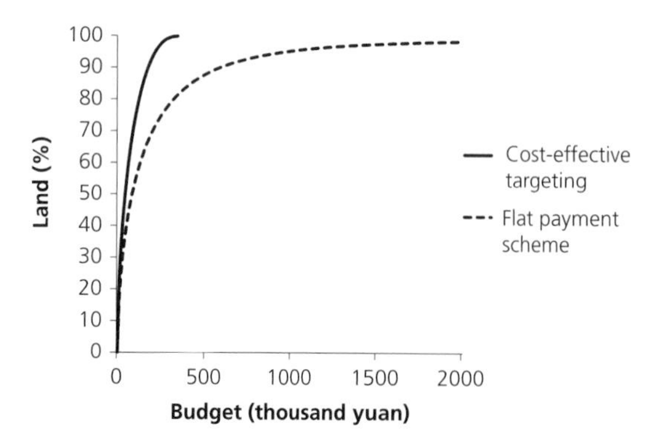

Figure 13.5 Percentage of land obtained through cost-effective targeting and the flat payment scheme for the Grain to Green Program (GTGP) in Wolong. Figure reproduced from Chen et al. (2010) with permissions from John Wiley and Sons.

13.7 Conclusions

Wolong has given us a unique window into understanding the complexities of policy implementation in CHANS. We learned that each policy has its own strengths and weaknesses. For instance, the zoning scheme is good for preventing infrastructural development but bad for monitoring individual human behaviors across space. The incentive-based programs were good for involving the local people and considering their livelihoods, but may not be designed in the most cost-effective manner. Research on the diverse and surprising effects of policies in coupled systems, such as what we have done here, can help to pinpoint the challenges of balancing competing conservation and development goals.

One important observation on policy implementation in Wolong is that each policy does not exist on its own, but rather interacts with many other policies being implemented simultaneously. Policies can complement one another. For example, core areas identified by the zoning scheme and areas demarcated for protection in NFCP overlap with one another across space, providing a double layer of protection. In addition, in Chapter 9 we described the surprising results of our modeling efforts. Results showed that NFCP and GTGP each had a negative effect of their own on household income, but a positive effect when implemented together. Policies also can compete with one another. For example, raising livestock in Wolong (and its associated evolving policy issues) came to the fore in part after timber-harvesting policies prevented people from gaining income from the timber trade. Another example is that GTGP has significantly decreased the labor force allocated to farming and the income earned from farming. This change has in turn affected all other policies, such as by providing more labor for tourism and off-farm activities. Further research is needed to tease apart these interaction effects in order to maximize policy effectiveness and prevent policies from offsetting one another.

Our work has also illustrated key differences in short-term and long-term effects of policies on human–nature interactions. For instance, positive effects of NFCP on forest cover took several years to occur. Effects of GTGP have not been fully realized, considering that the planted seedlings are still maturing. Effects of both policies on households are also changing over time, as initial positive attitudes toward the policies have shifted somewhat as time has gone on and wildlife raiding of crops has increased due to improved forest cover. These temporal effects highlight the mismatch between policies that are often static for a period and human–nature interactions that are constantly changing over time. Policies need to be dynamic and adaptive in order to stay ahead of emerging sustainability challenges. For example, our studies suggest that the zoning scheme should be changed as the distribution of panda habitat changes. Conservation incentive structuring also should be adapted to account for evolving household demographic and economic changes, such as new households appearing on the landscape.

Aside from the policies that we have chosen to focus on in this chapter, there are several other policies that deserve further study to determine their role in this complex coupled system. For example, population regulation policies (Chapter 8), post-earthquake reconstruction policies (Chapter 12), and tourism policies (Chapter 4) all play roles in this complex system. Each set of policies has its own challenges and opportunities to affect the sustainability of the system. Research that embraces complex interactions and synergies among the diverse policies will help to shed light on how to manage Wolong for a brighter future.

13.8 Summary

Many government policies are implemented simultaneously, but often evaluated separately with little attention to their interaction effects. In Wolong Nature Reserve, a number of policies have been implemented for conservation, including reserve zoning schemes and incentive-based programs such as Natural Forest Conservation Program (NFCP) and the Grain to Green Program (GTGP). In this chapter, we explored the role of these policies in mediating human–nature interactions and illustrated surprises resulting from their implementation. The zoning scheme has been largely effective for adjusting many human activities in different zones, but has not prevented all harmful human actions in the core zone. NFCP has substantially reduced illegal harvesting. GTGP has converted about half of local people's cropland into forest plantations. Separately, NFCP and GTGB (GTGP and the local Grain to Bamboo Program [GTBP] taken together as a single cropland conversion policy) each had a negative impact on household income. But jointly their interaction led to a positive effect. Even though households receive 36% higher payment from GTGP than NFCP, the latter is responsible for the majority of forest recovery. Also, there has been an unexpected increase in wildlife raiding of farmers' crops as a result of both GTGP and NFCP. The effectiveness of these policies can be improved by including more panda habitat in the core zone, cost-effective targeting in GTGP, and incorporating energy needs and optimal monitoring mechanisms into the design of NFCP. Our studies suggest that multiple policies should be designed and conservation resources be allocated together to maximize their effectiveness.

References

Alberti, M., Asbjornsen, H., Baker, L.A., et al. (2011) Research on coupled human and natural systems (CHANS): approach, challenges, and strategies. *Bulletin of the Ecological Society of America*, **92**, 218–28.

An, L., Lupi, F., Liu, J., et al. (2002) Modeling the choice to switch from fuelwood to electricity: implications for giant panda habitat conservation. *Ecological Economics*, **42**, 445–57.

Babcock, B.A., Lakshminarayan, P.G., Wu, J.J., and Zilberman, D. (1996) The economics of a public fund for environmental amenities: a study of CRP contracts. *American Journal of Agricultural Economics*, **78**, 961–71.

Brandon, K.E. and Wells, M. (1992) Planning for people and parks: design dilemmas. *World Development*, **20**, 557–70.

Chen, X., Frank, K.A., Dietz, T., and Liu, J. (2012) Weak ties, labor migration, and environmental impacts: toward a sociology of sustainability. *Organization & Environment*, **25**, 3–24.

Chen, X., Lupi, F., He, G., and Liu, J. (2009a) Linking social norms to efficient conservation investment in payments for ecosystem services. *Proceedings of the National Academy of Sciences of the United States of America*, **106**, 11812–17.

Chen, X., Lupi, F., He, G., et al. (2009b) Factors affecting land reconversion plans following a payment for ecosystem service program. *Biological Conservation*, **142**, 1740–47.

Chen, X., Lupi, F., Viña, A., et al. (2010) Using cost-effective targeting to enhance the efficiency of conservation investments in payments for ecosystem services. *Conservation Biology*, **24**, 1469–78.

Chen, X., Viña, A., Shortridge, A., et al. (2014) Assessing the effectiveness of payments for ecosystem services: an agent-based modeling approach. *Ecology and Society*, **19**, 7.

Crutzen, P.J. (2006) The "Anthropocene." In E. Ehlers and T. Krafft, eds, *Earth System Science in the Anthropocene*, pp. 13–18. Springer, Berlin, Germany.

Dietz, T., Ostrom, E., and Stern, P.C. (2003) The struggle to govern the commons. *Science*, **302**, 1907–12.

Dovers, S.R. (1996) Sustainability: demands on policy. *Journal of Public Policy*, **16**, 303–18.

Esteban, J. and Ray, D. (2001) Collective action and the group size paradox. *American Political Science Review*, **95**, 663–72.

Ferraro, P.J. and Kiss, A. (2002) Direct payments to conserve biodiversity. *Science*, **298**, 1718–19.

Ghimire, K.B. (1997) Conservation and social development: an assessment of Wolong and other panda reserves in China. In K.B. Ghimire and M.P. Pimbert, eds, *Environmental Politics and Impacts of National Parks and Protected Areas*, pp. 187–213. Earthscan Publications, London, UK.

Gockel, C.K. and Gray, L.C. (2009) Integrating conservation and development in the Peruvian Amazon. *Ecology and Society*, **14**, 11.

He, G., Chen, X., Bearer, S., et al. (2009) Spatial and temporal patterns of fuelwood collection in Wolong Nature Reserve: implications for panda conservation. *Landscape and Urban Planning*, **92**, 1–9.

Hill, C., Osborn, F., and Plumptre, A.J. (2002) *Human-Wildlife Conflict: identifying the problem and possible solutions. Albertine Rift Technical Report Series*, **vol. 1**. Wildlife Conservation Society, New York, NY.

Holling, C.S. and Meffe, G.K. (1996) Command and control and the pathology of natural resource management. *Conservation Biology*, **10**, 328–37.

Hull, V., Xu, W., Liu, W., et al. (2011) Evaluating the efficacy of zoning designations for protected area management. *Biological Conservation*, **144**, 3028–37.

Hull, V., Zhang, J., Zhou, S., et al. (2014) Impact of livestock on giant pandas and their habitat. *Journal for Nature Conservation*, **22**, 256–64.

James, A., Gaston, K.J., and Balmford, A. (2001) Can we afford to conserve biodiversity? *Bioscience*, **51**, 43–52.

Liu, J., Dietz, T., Carpenter, S.R., et al. (2007a) Complexity of coupled human and natural systems. *Science*, **317**, 1513–16.

Liu, J., Dietz, T., Carpenter, S.R., et al. (2007b) Coupled human and natural systems. *Ambio*, **36**, 639–49.

Liu, J., Linderman, M., Ouyang, Z., et al. (2001) Ecological degradation in protected areas: the case of Wolong Nature Reserve for giant pandas. *Science*, **292**, 98–101.

Liu, J., Mooney, H., Hull, V., et al. (2015) Systems integration for global sustainability. *Science*, **347**(6225), 1258832.

Liu, J., Ouyang, Z., Yang, W., et al. (2013) Evaluation of ecosystem service policies from biophysical and social perspectives: the case of China. In S.A. Levin, ed., *Encyclopedia of Biodiversity* (second edition), **vol. 3**, pp. 372–84. Academic Press, Waltham, MA.

Liu, J. and Raven, P.H. (2010) China's environmental challenges and implications for the world. *Critical Reviews in Environmental Science and Technology*, **40**, 823–51.

Liu, J. and Viña, A. (2014) Pandas, plants, and people. *Annals of the Missouri Botanical Garden*, **100**, 108–25.

Liu, W. (2012) *Patterns and Impacts of Tourism Development in a Coupled Human and Natural System*. Doctoral Dissertation, Michigan State University, East Lansing, MI.

Liu, W., Vogt, C.A., Luo, J., et al. (2012) Drivers and socioeconomic impacts of tourism participation in protected areas. *PLoS ONE*, **7**, e35420.

Liu, X. and Li, J. (2008) Scientific solutions for the functional zoning of nature reserves in China. *Ecological Modelling*, **215**, 237–46.

Melick, D., Yang, X., and Xu, J. (2007) Seeing the wood for the trees: how conservation policies can place greater pressure on village forests in southwest China. *Biodiversity and Conservation*, **16**, 1959–71.

Millennium Ecosystem Assessment (2005) *Ecosystems and Human Well-being: Synthesis*. Island Press, Washington, DC.

Ministry of Forestry (1998) *Wolong Nature Reserve Master Plan*. Beijing, China (in Chinese).

Nielsen, M.R., Jacobsen, J.B., and Thorsen, B.J. (2014) Factors determining the choice of hunting and trading bushmeat in the Kilombero Valley, Tanzania. *Conservation Biology*, **28**, 382–91.

Ostrom, E. (2005) *Understanding Institutional Diversity*. Princeton University Press, Princeton, NJ.

Viña, A., Bearer, S., Chen, X., et al. (2007) Temporal changes in giant panda habitat connectivity across boundaries of Wolong Nature Reserve, China. *Ecological Applications*, **17**, 1019–30.

Viña, A., Chen, X., McConnell, W.J., et al. (2011) Effects of natural disasters on conservation policies: the case of the 2008 Wenchuan Earthquake, China. *Ambio*, **40**, 274–84.

Voss, J.-P., Bauknecht, D., and Kemp, R. (2006) *Reflexive Governance for Sustainable Development*. Edward Elgar Publishing, Cheltenham, UK.

Wolong Nature Reserve (2005) *Development History of Wolong Nature Reserve*. Sichuan Science Publisher, Chengdu, China (in Chinese).

World Bank (2001) *China: Air, Land, and Water: Environmental Priorities for a New Millennium.* World Bank, Washington, DC.

Wunder, S. (2007) The efficiency of payments for environmental services in tropical conservation. *Conservation Biology*, **21**, 48–58.

Xu, J., Yin, R., Li, Z., and Liu, C. (2006b) China's ecological rehabilitation: unprecedented efforts, dramatic impacts, and requisite policies. *Ecological Economics*, **57**, 595–607.

Xu, W., Ouyang, Z., Viña, A., et al. (2006a) Designing a conservation plan for protecting the habitat for giant pandas in the Qionglai mountain range, China. *Diversity and Distributions*, **12**, 610–19.

Yang, W., Liu, W., Viña, A., et al. (2013a) Performance and prospects of payments for ecosystem services programs: evidence from China. *Journal of Environmental Management*, **127**, 86–95.

Yang, W., Liu, W., Viña, A., et al. (2013b) Nonlinear effects of group size on collective action and resource outcomes. *Proceedings of the National Academy of Sciences of the United States of America*, **110**, 10916–21.

CHAPTER 14

Toward a Sustainable Future

Vanessa Hull, William McConnell, Marc Linderman, and Jianguo Liu

14.1 Introduction

One of the main purposes of studying past and present dynamics of coupled human and natural systems is to help explore how they might operate in the future under different plausible scenarios (Figure 2.1). The analysis of possible future events has gained much attention in recent decades through the use of well-developed sets of conditions, such as differing levels of greenhouse gas emissions (Tuanmu et al., 2013). These global "scenarios" employ a number of "story lines" of socioeconomic activities (IPCC, 2007, Millennium Ecosystem Assessment, 2005), including production and consumption, to estimate possible socioeconomic and environmental impacts. Similar practices are employed at regional and local scales, especially in the projection of land-use changes likely to result from a range of social, cultural, and policy changes (Carlson et al., 2011).

Scenario analysis is often adopted in the context of land-use planning at local, regional, and global scales. This approach integrates a variety of relevant information to explore and compare future environmental and socioeconomic implications of land-use options (Lu et al., 2004, Van Ittersum et al., 1998). The goal is to identify key drivers for more detailed research and policy analysis for sustainability (World Commission on Environment and Development, 1987), rather than to accurately predict the future. But unlike a global scenario analysis, regional and local scenario analyses can delve deeper into the workings of coupled systems. Regional and local analyses can go beyond aggregate statistics to explore reciprocal human–nature interactions shaped by social, political, economic, and environmental contexts (An et al., 2005, 2006, An and Liu, 2010, Chen et al., 2012a, 2014).

There are a number of approaches to scenario analysis. For example, statistical models can estimate the cumulative impacts of actors' decisions. Simulation models are capable of accounting for individual endowments, opportunities, preferences, and even learning capacity (An, 2012, Chen et al., 2012a, Matthews et al., 2007). Sharing strong roots with the process of scenario analysis in the financial industry and many other fields, the land change science community has increasingly turned to agent-based modeling (ABM). This "bottom-up" approach aims to reproduce higher-level phenomena (emergent properties) by simulating the actions of multiple low-level "agents." Agents ideally are imbued with self-awareness, intelligence, autonomous behavior, and knowledge of the environment and other agents. Importantly, they may adjust their actions in response to environmental changes and one another's actions (Lim et al., 2002, Parker et al., 2003).

Despite all of the advances in the field of scenario analysis, most studies focused on one or just a few components of coupled human and natural systems. In our model coupled system of Wolong Nature Reserve, our research team has conducted scenario analyses of many components over time. The purpose of this chapter is to envision a sustainable future for Wolong by (1) highlighting some results of scenarios examined by our research team to date, (2) outlining the key factors that might be incorporated in future scenario

Pandas and People. Edited by Jianguo Liu, Vanessa Hull, Wu Yang, Andrés Viña, Xiaodong Chen, Zhiyun Ouyang, and Hemin Zhang. Published 2016 by Oxford University Press.

analyses; and (3) recommending how these scenarios should be constructed and evaluated to promote human well-being and ecological sustainability.

14.2 Overview of illustrative scenarios for Wolong's future

A large number of factors must be taken into account in envisioning a sustainable future—most importantly, just what should be sustained? This question is ultimately one for the stakeholders of Wolong to consider, but we as researchers can contribute by constructing and evaluating scenarios that outline possible future conditions. The scenario analyses formally undertaken by our group in Wolong have tackled diverse questions related to human–nature couplings. Examples include the impact of human population size and spatial distribution on panda habitat, interactions between biological cycles and human impacts on panda habitat, effects of household sociodemographic dynamics on fuelwood collection, and impacts of program design on participation in conservation programs. We highlight these studies briefly below.

14.2.1 Population–fuelwood scenarios

Some of our earliest work in Wolong tackled a central issue governing human–nature interactions in this system: the interaction between human population dynamics and fuelwood use (Liu et al., 1999). A major impetus for this research was our observations of marked increases in the human population and the number of households in the reserve (see Figure 3.3). These changes led to increased fuelwood consumption (to meet basic heating and cooking needs), which degraded forests and sometimes panda habitat (Chapters 6, 7, and 10). As discussed in Chapter 8, many of Wolong's residents are exempt from China's one-child policy because they are ethnic minorities, which opens the door to many future population scenarios not plausible elsewhere in the country.

In Liu et al. (1999), we generated future scenarios for population growth and fuelwood use in Wolong. We applied a demographic model developed by Song and Yu (1988) to observe fuelwood consumption patterns and their impacts on panda habitat. Scenarios were compared to a simulation of baseline conditions (continuity of observed population growth rate), with three scenarios involving reduced fuelwood use, reduced household size, and reduced birth rate, and two scenarios involving increased emigration by young people, or by whole families. Our baseline scenario projected that the human population would increase by 38% and panda habitat would decrease by 37% in 50 years (Liu et al., 1999). Emigration of young people was more effective than any other method at reducing population size and recovering panda habitat in the reserve (reduced by 82% at 22% emigration rate, saving 7% of panda habitat, compared to a population reduction of only 61% and panda habitat savings of 6% for full household emigration).

The result regarding the emigration of young people was particularly revealing in demonstrating the effect of having people leave the reserve during their early child-rearing years to establish households outside the reserve. This relocation approach is not only more ethical and socially acceptable than the forced relocation of entire households, but also more effective at improving panda habitat over the long term. This result shaped our appreciation of the importance of education for Wolong's youth as a potential "win–win" approach for the coupled system. In fact, our household interviews indicate that the desire for children to emigrate through access to higher education is shared and in fact strongly prioritized by almost all local families. This scenario analysis also helped set the stage for the work that we would later undertake to further understand the complexity of human–nature interactions.

14.2.2 Household–bamboo scenarios

As mentioned in Chapter 3, the diet of pandas in Wolong consists almost entirely of two bamboo species: arrow bamboo (*Bashania faberi*) and umbrella bamboo (*Fargesia robusta*). Like most bamboo species in China, these species are semelparous, meaning they undergo cyclical "mast" flowering and seeding. Flowering may occur in single bamboo patches or across landscapes, but also may occur in a mass synchronous event at the scale of an entire mountain range (Schaller et al., 1985). This massive reproductive flowering event is followed by a mass die-off before the bamboo regains its pre-flowering biomass and distribution. Pandas can survive a flowering event by shifting their foraging patterns to rely more heavily on the other (non-flowering) bamboos present (Reid et al., 1989). But if no other species are present and/or their range is already constricted by humans, pandas could starve.

To simulate bamboo dynamics in response to human impacts, we developed a more advanced spatially explicit simulation model of households and land use in Wolong (Linderman et al., 2005). Household activities such as logging and fuelwood collection change forest overstory cover, thus influencing flowering and regeneration of bamboo. The model used a number of algorithms derived from observed data to predict a probability of change in forest cover in each grid cell. We based predictions on patterns of household creation and fuelwood collection (with forest regrowth also occurring). We overlaid this information on a spatially explicit layer of bamboo distribution throughout the reserve. We projected bamboo flowering in each cell based on bamboo growth curves, flowering rates, and human and environmental influences that may interact with flowering, obtained from the literature.

We designed several scenarios to project household and bamboo dynamics during the 1997–2030 period. Our models suggest that if the two main bamboo species flowered simultaneously, around 49% of habitat could be lost. This estimate would be about 12% lower if the flowering of the two species were staggered (Linderman et al., 2006). These effects could occur in the context of an additional 3–14% loss of habitat by 2030 due to human activities (fuelwood collection). Bamboo flowering losses would be temporary because the bamboo would regrow over time. But human activities are projected to have a negative effect on bamboo regeneration. Humans were predicted to cause 9% less bamboo regrowth below 2600 m elevation at current fuelwood consumption rates (as compared to a scenario of no human activities; Linderman et al., 2006). If fuelwood consumption rates were to double, regrowth would be 15% below baseline. If this pattern were to extend several hundred years into the future, the combined impacts of fuelwood collection on bamboo regeneration would have cumulative effects that would severely deplete the bamboo population. Modeling these synergistic and cumulative effects under fuelwood collection and flowering scenarios gave us a better appreciation of the complexities of human–nature interactions in this coupled system and what they mean for long-term sustainability.

14.2.3 Conservation–development scenarios

Despite all that we learned in the previous studies, one drawback of our aforementioned scenarios is that we were not able to account for individual decision-making at the household level. Aggregating human activities up to the reserve level in our scenarios was appropriate for some types of models. But other approaches may provide more nuanced insight into the design of policies to regulate and incentivize particular land uses. For this reason we turned to the ABM approach. Much of our work on this topic has dealt with exploring household demographics and family planning in relationship to fuelwood collection. This work was described in Chapter 8 and in a series of other publications (An et al., 2005, 2006, 2011, 2014, An and Liu, 2010). Our ABM consisted of a number of complex algorithms to represent relationships between demographic characteristics and fuelwood consumption by individual agents, and ultimately land-use change. The agents and the landscape coevolved as parameters changed over time on a yearly time step.

Using the ABM, we compared a baseline projection to development and conservation scenarios to predict panda habitat dynamics over a 20-year period. For both the conservation and development scenarios, we varied the values of parameters in ways expected to improve or harm panda habitat, respectively (see Table 5 in An et al., 2005). The parameters considered included electricity use, migration, family planning, and fuelwood consumption patterns. Our simulations generated stark contrasts in the projections among these scenarios. The conservation scenario projected 600 km^2 of habitat at the end of a 20-year simulation, but the development scenario estimated only 554 km^2 (8% less). The development scenario also projected 729 (83%) more households and 3694 (141%) more people than the conservation scenario.

14.2.4 Payments for ecosystem services scenarios

Some of our most recent scenario analyses have focused on questions about the design and efficacy of payments for ecosystem services (PES) programs (Chen et al., 2009, 2010, 2012a, 2014). Even though the PES programs in Wolong have thus far helped both people and the environment (Liu et al., 2008, Yang et al., 2013a, b; see Chapter 13), we sought to use our data and scenario-building capacities to address how these programs might be improved in the future. Improvements could make the programs more efficient or cost-effective (Chen et al., 2009, 2010, 2012a, 2014, Yang et al., 2013a, b; see Chapter 13), given the limited funds available for conservation. Such information would be useful for managers in future program implementation.

One of the key PES programs in Wolong is the Grain to Green Program (GTGP), which provides subsidies for farmers to convert their croplands to forest plantations (see Chapters 5, 11, and 13). We used scenario analysis to evaluate a variety of potentially cost-effective program designs. Specifically, we considered targeting plots for varying payment schemes based on opportunity costs (costs to farmers of forgoing alternative uses of cropland). Such costs related to household (e.g., size of agricultural plot and labor supply) and environmental (e.g., topographic) characteristics (Viña et al., 2013). Results showed that such discriminative payments that maximized efficiency (minimizing opportunity costs) could be up to ten times more cost effective than flat payments (same payments to each household; Chen et al., 2010). If the total budget was 100 000 yuan, cost-effective targeting for reduction of soil erosion achieved 82% of the maximum possible soil erosion prevention compared to only 29% when plots were randomly selected for enrollment. Similar differences were also seen for cost-effective targeting for land benefits (maximizing the proportion of land enrolled in the program) and panda habitat benefits (minimizing the distance to the nearest habitat patch). Scenarios were valuable for highlighting a potential design scheme that reserve administrators might consider to make the best use of limited funds (Chen et al., 2010, Viña et al., 2013).

A recent scenario analysis looked at ways of better designing the Natural Forest Conservation Program (NFCP, see Chapters 5, 11, and 13). The purpose of the program is to prevent illegal fuelwood collection and timber harvesting by paying farmers to monitor assigned forest parcels (Yang et al., 2013a, b). We hypothesized that the program could be improved if, instead of paying farmers cash for their monitoring efforts, they were paid directly in electricity. This alternative would create a double impetus to decrease fuelwood collection (monitoring efforts that discourage collection and lack of need for fuelwood when electricity is already paid for). We designed scenarios to test this hypothesis using a spatially explicit ABM made up of interacting farmers making decisions about PES participation and fuelwood use (Chen et al., 2014). The model was parameterized with data obtained in our household interviews. By 2030, forest was projected to decrease by 31% (103 km^2) relative to 2007 under no payment. These decreases were reversed under either payment scheme, as forests were predicted to increase by 12% (42 km^2) under the current cash payment scheme and increase by 29% (98 km^2) under a direct electricity payment scheme (Chen et al., 2014). In other words, the electricity payments could result in 201 km^2 more forest than the baseline (Figure 14.1, Chen et al., 2014). This scenario analysis illustrated the potential value of transferring conservation investments from the government to ecosystem services beneficiaries (e.g., electricity provision), a design strategy that managers might consider for this coupled system.

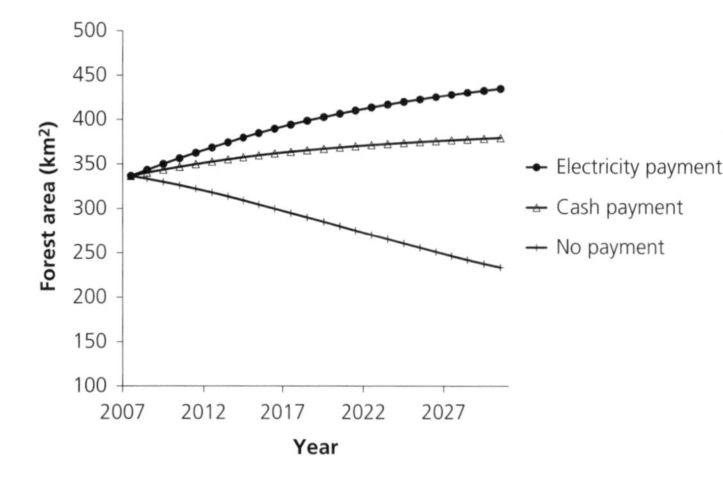

Figure 14.1 Projected amount of forest in Wolong Nature Reserve over a 23-year period. Scenarios include no NFCP cash payment, NFCP cash payment as of 2006, and hypothetical direct electricity payment. Reproduced from Chen et al. (2014).

14.3 Crafting new scenarios for a sustainable future

Our previous scenario analysis efforts taught us a great deal about the coupled system of Wolong. Our work also informed broader efforts in coupled systems research such as resource extraction patterns, population dynamics, interactions between humans and natural disturbances, and design and implementation of PES programs. But there are several new uncertainties on the horizon that need to be considered as we continue more integrative and realistic scenario analyses on this coupled system. In keeping with our conceptual framework that conceives of Wolong as a coupled human and natural system (Chapter 2), here we consider both environmental and socioeconomic dynamics that are likely to shape future outcomes. These include the effects of, and policy responses to, seismic forces, incidence of disease, and panda reintroduction, as well as a range of external and internal economic conditions and possible policy choices. Throughout, we reference the Wolong Administration's Master Plan for the Period of 2015–2025 (Sichuan Academy of Forestry, 2014) as a management instrument that can inform scenario analyses.

14.3.1 Environmental factors

Environmental factors explored in our previous scenario analyses included topographic characteristics, bamboo flowering, and forest regeneration patterns. In addition to these, a number of other environmental factors could dramatically shape the future of the coupled human and natural system in Wolong. Here we focus on three that are of prime interest: seismic activity, disease, and panda reintroduction.

Seismic activity

As Chapter 12 demonstrates, the 2008 earthquake exerted a huge effect on Wolong's landscape. The earthquake on May 12, 2008, and the tens of thousands of aftershocks that struck the reserve was a large, if not historically unprecedented, set of events. The earthquake was the result of motion on the Longmenshan thrust fault, which in geologic time was responsible for the uplift of the mountains where Wolong is situated above the Sichuan Basin to the east (see Chapters 6 and 12). The particular structure of the fault system makes it prone to infrequent, but potentially very strong, earthquakes. Even a casual inspection of the landscape of the reserve reveals ubiquitous evidence of pre-2008 landslides of varying ages and states of vegetative recovery. Thus, seismic activity needs to be considered in future scenario analyses for Wolong, both with respect to trajectories for recovery from the 2008 earthquake and the potential for further earthquakes.

Future scenario analyses should take into account the dramatic changes that have taken place in the human communities in Wolong as a result of the 2008 earthquake. For example, the government used subsidies to entice residents whose homes were damaged or destroyed to relocate to new spatially condensed, apartment-style housing located along the main road and far from their farmland (Figure 14.2). In doing so, the authorities were able to make farming and resource extraction much less practicable, further compelling residents to turn to off-farm employment. Moreover, lacking traditional hearths in their new homes, residents now mainly rely on natural gas, charcoal, and electricity for cooking and heating. Although some still use small amounts of fuelwood, this reduces the need for fuelwood from the reserve's forests. The question now is whether this experiment in social engineering via relocation will have the desired effects. It remains to be seen whether relocation will lead to the demise of resource-dependent livelihoods, through both employment substitution and out-migration. Also uncertain is whether such a change would ultimately reduce the human threat to panda extinction in Wolong.

In addition, the impact of this housing shift on human well-being needs to be addressed. We do know that some families have elected to continue farming, despite severe damage to their homes and other challenges imposed by the earthquake (Chapter 12). Those families made the choice to continue farming despite being offered government-subsidized apartments located closer to services and amenities (i.e., along the main road). Cultural attachment to and, perhaps, relative proficiency in this lifestyle likely lie behind this decision, as does skepticism concerning the potential for gainful off-farm employment. The latter is heightened by the extended delay in revitalizing the tourism sector following the earthquake, given the treacherous condition of the main road (Chapter 12). In addition, the placement of the new apartments in Wolong Township immediately adjacent to—or even within—the flood plain of the Pitiao River makes them extremely vulnerable to future seismic events. The risk of flooding is now heightened due to road construction occurring within the river channel combined with the rise in the river bed due to landslide debris. Several floods have already occurred. Future scenario analyses should test hypotheses about conditions under which people will remain in the new housing and whether the new housing arrangement benefits both the environment and human well-being considering these diverse factors.

Another earthquake-related factor that needs to be considered in future analyses is the ongoing challenge of maintaining road connections between Wolong and the cities of Dujiangyan and Chengdu, and with the rest of China and the world. The main road, which received a major upgrade beginning in 2006, continues to be affected by post-earthquake landslides despite continuous attempts to repair it, with

Figure 14.2 Proposed arrangement of new apartment housing as part of Wolong's earthquake reconstruction plan. Also depicted are four types of housing that differ by area and function (e.g., residence or small-scale hotel). Reproduced from Wolong Administration Bureau (2009). Photo taken by Wu Yang, used with permission.

injuries and fatalities still common. The effects of this on economic activity in the reserve are serious, impeding the import of agricultural inputs and the export of products, and severely hampering the recovery of the tourism industry (Chapter 12). The sharp decrease in the number of visitors to the reserve from well over a hundred thousand in 2007 to almost none after the earthquake (Liu et al., 2012) is largely attributed to safety concerns and lack of road access. The re-establishment of a reliable road connection will be crucial to the future of the human population in the reserve, and time is of the essence. Besides anticipating the continuing need for road repair, the Wolong Master Plan calls for the construction of more road tunnels to avoid the most problematic cliffside sections, but the tunnels will certainly take years to complete.

The earthquake may have ambiguous effects over the long term, and these could be sorted out using scenario analyses. While earthquakes disrupt the export of perishable produce, such as cabbages, and thereby discourage farming, this may benefit the recovery of forest on land that might otherwise have been devoted to such crops. The delay in road repair and its impact on the agricultural sector could continue to cause more residents to seek their livelihoods elsewhere. Alternatively, the delay could cause an increase in the exploitation of forest resources (e.g., raising livestock or poaching), as we have observed in some villages. Lack of employment opportunities in the tourism sector may spur further out-migration, as a workforce already seeking to leave farming pursues employment outside the reserve. On the other hand, some who purchased new apartments in hopes of wage employment may increasingly invest their labor in natural resource-based livelihood options. This investment could result in reduced rates of cropland re-enrollment in GTGP, spur more intensive extraction of forest products, and even increase illegal harvest and poaching as well as increased numbers and diversity of free-ranging livestock. Scenario analysis will help to sort through the mixture of positive and negative effects of the road conditions for sustainability.

Disease

There is a range of scenarios in which the outbreak of infectious disease could affect the coupled system in Wolong. We have not yet considered disease in our modeling efforts, but this factor could have considerable effects. Perhaps the most likely would involve human pathogens, such as Sudden Acute Respiratory Syndrome (SARS). SARS had a major negative impact on the Chinese economy generally, and in particular the tourism sector, in 2003 (*China Daily*, 2003, Clark, 2003). In fact, the tourism sector in Wolong suffered from the SARS outbreak. Tourists were not allowed to visit the reserve for several months due to concern over the spread of the disease (Liu, 2003, Liu et al., 2012). Tourist visitation sagged by nearly 20% from 82 000 in 2002 to 66 000 in 2003, later surging to 163 000 in 2004 (Figure 14.3, Liu et al., 2012).

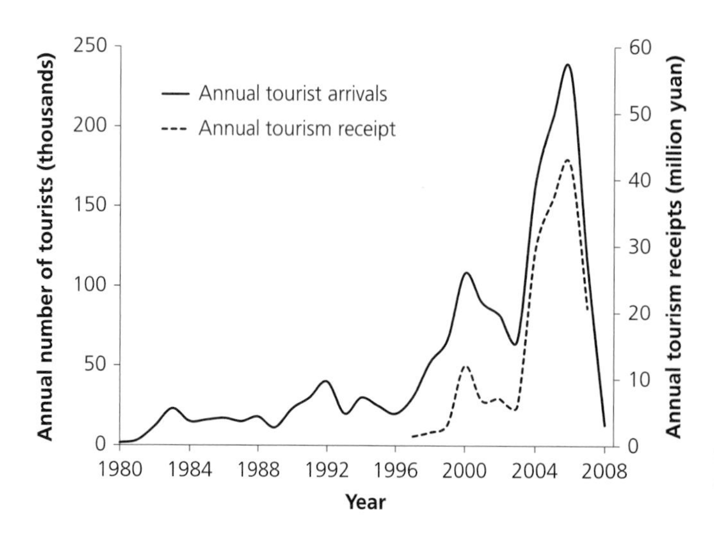

Figure 14.3 Annual tourist arrivals and revenue generated in Wolong Nature Reserve over time. Reproduced from Liu (2012).

Other possibilities involve the transmission of disease to the panda population. There has been long-standing concern about the susceptibility of pandas to zoonotic disease, prompting serologic testing of captive and wild animals. One such study tested the presence of antibodies in the blood of pandas in the breeding center in Wolong that were taken from the wild. Researchers found the pandas had been exposed to canine distemper virus (CDV), canine parvovirus-2 (CPV), canine adenovirus (CAV), and canine coronavirus (CCV; Mainka et al., 1994). Concern about the transmission of such diseases to pandas has led to a vaccination program whose effectiveness demands close scrutiny (Loeffler et al., 2007). For example, it was recently confirmed that pandas at the China Conservation and Research Center for the Giant Panda in Ya'an City, not far from Wolong, were infected with the swine flu (H1N1) virus during the outbreak in China in 2009 (Li et al., 2014). In addition, a canine distemper virus outbreak occurred at the Shaanxi Rare Wild Animal Rescue and Research Center in late 2014 and early 2015, killing three captive pandas, and sickening another three (*China Radio International*, 2015). The outbreak is believed to have been caused by transmission via visitors owning canine pets carrying the disease. In Wolong itself, we have documented cases of livestock remains and feces in water sources used by pandas. It is conceivable that diseases might cross the livestock–wildlife interface if livestock are kept in close proximity to pandas (Hull et al., 2014) or their water sources. Should panda habitat continue to be degraded, the remaining individuals will be forced into closer proximity to one another, further exacerbating the risk of disease transmission among them (Zhang et al., 2008).

While such disease outbreaks are difficult or impossible to predict, it seems certain that future pandemics will occur; their impacts will depend on the severity of the disease and the timing of the outbreak. Disease outbreaks have been a central threat to other wildlife species also facing human pressure in recent years. For example, the little brown bat (*Myotis lucifugus*) faces white nose syndrome in the USA, the koala (*Phascolarctos cinereus*) is susceptible to both a retrovirus and chlamydia, and the Tasmanian devil (*Sarcophilus harrisii*) suffers from a rare form of infectious tumor (MacPhee and Greenwood, 2013). The worst-case scenario for Wolong could involve some type of local extinction. Such an event could in turn severely impact the overall panda population. Wolong is both a stronghold for the wild population and an important linkage connecting to other pandas inhabiting other parts of the Qionglai Mountains (Xu et al., 2006). Fortunately, the Wolong Administration is sensitive to this issue, and its Master Plan calls for enhanced disease detection capabilities, including well-equipped and networked monitoring stations. Scenario analysis could inform these monitoring efforts by highlighting potential areas of concern or human and natural factors to consider in managing disease risk.

Panda reintroduction

Another factor that could shape the future of Wolong is panda reintroduction. Captive breeding of giant pandas began in the 1960s. One of the main goals of this endeavor was to establish a captive population that could one day be released to bolster or rescue declining wild populations (Mainka and Lü, 1999). After decades of unsuccessful captive breeding, there has been a surge of success in the last decade due to advances in techniques to encourage natural mating, artificial insemination, and human-assisted infant rearing. The captive population tripled from 1970 to 2000 (Zhang et al., 2006), with a current captive population of over 300 individuals (Xie and Gipps, 2011). The largest and most successful breeding center in the world is the China Conservation and Research Center for the Giant Panda (CCRCGP), based in Wolong Nature Reserve. Due to its success, and the now overflowing captive population, the center has turned its focus to reintroduction into the wild.

The Center's reintroduction program began in 2003 with the pre-release training of a captive subadult male panda named Xiang-Xiang (Hull et al., 2011a, Zhou et al., 2005). He became the first captive-born panda to be released in the wild in Wolong after a few years of training. He, unfortunately, died after several months in the wild due to conflicts with resident wild pandas, but lessons learned helped to shape the future of the program. Since then, several captive pandas have been trained at the Wolong pre-release training station for reintroduction into the wild. Three pandas have been released to the wild in another nature reserve called Liziping and are currently being monitored at the time of writing.

Reintroduction should be considered in future scenario analysis because it could shape the coupled system in a number of ways. The reintroduction program headed by the CCRCGP is poised to continue for decades, and pre-release training will likely continue in Wolong. The training project could provide opportunities for local residents to gain some small economic benefits. Mechanisms for this include growing bamboo to feed the training pandas, providing labor for maintaining and constructing training facilities, and selling food and providing hotel accommodation for researchers and journalists documenting the project. The project has garnered international attention and has been covered by several leading world media (including *Good Morning America*). This kind of press can have long-term ramifications by increasing global awareness of panda conservation issues and specifically raising the profile of Wolong as a nature reserve. This awareness could encourage greater investment (e.g., in road repair). These diverse effects need to be integrated to understand how this new program may shape the future of the model coupled system in Wolong in ways that may not otherwise be predictable.

14.3.2 Socioeconomic factors

The future of the coupled system will also depend upon a number of socioeconomic and policy conditions. While we have already examined several factors such as family planning, fuelwood collection behaviors, and PES program participation (Section 14.2), there are several new factors that need to be explored. These include both exogenous factors—such as transport and commodity prices, national migration policy, and the continued expansion of the Chinese economy—and others endogenous to the system, particularly local conservation program implementation.

External policy and economic context

Despite the considerable and lasting post-earthquake damage to the main road that has particularly isolated Wolong, this coupled system is still affected by the outside world in complex ways (Chapter 17). Many of these complex processes require scenario analyses to begin to piece them together. For instance, fuel costs affect both the farming and tourism sectors in Wolong. A drop in fuel costs could promote economic development in these sectors in the future. On the other hand, in the event that fuel costs rise faster than income in China, this could have a dampening effect on both commodity agriculture and domestic tourist visitation rates. Should this coincide with increased employment opportunities outside the reserve, the net effect is likely to be increased out-migration. Such opportunities could theoretically be spurred by the completion of high-speed rail connections between Chengdu and Europe (Gracie, 2014). On the other hand, if combined with a generalized economic slow-down that depresses industrial employment opportunities, rising real fuel costs could lead to a retrenchment of subsistence agriculture. Similar effects could result from regional commodity prices and the relative cost of alternative tourist destinations. It is conceivable that the market for agricultural products could be undercut by competition from other production areas. Such a process could reverse past gains in the local cabbage market that resulted from improvements to the main road in the early 1990s and later in 2006. These complex economic patterns and processes can best be predicted by modeling them in an integrative manner over long-term time frames.

A related topic that needs to be revisited from our earliest efforts to understand human–nature interactions in Wolong (Liu et al., 1999) is migration patterns. Even prior to the earthquake, the number of young workers venturing outside the reserve in pursuit of wage labor was already increasing (Chen et al., 2012b). This shift reflects trends in rural–urban migration throughout China, and in rural hinterlands the world over (Aide and Grau, 2004, Zhang and Song, 2003). This trend is unlikely to abate in the reserve in the foreseeable future. It might at first blush seem that these migrants express with their actions the desire *not* to sustain their agricultural livelihoods. But our survey data indicate that remittances from such labor migration are often used to replace the lost labor. Farmers also use remittances to intensify cropland cultivation (using more chemical fertilizers and pesticides), to switch from subsistence to cash crops, and to expand and diversify livestock holdings. Meanwhile, other families are investing those funds in their children's education and in small businesses, some of which cater to visitors. Ultimately, many migrants' goal is to maintain the family farm both as insurance against economic downturns that disproportionately upset the manufacturing sector, and as a locus of retirement. Even among those who have taken advantage of the new subsidized housing, many continue to cultivate their fields and maintain their livestock to some degree. Our previous work has confirmed findings elsewhere around the world concerning the importance of social networks (particularly weak ties) in a family's propensity to invest precious household labor in labor migration (Chen et al., 2012b; Chapter 11). To the degree that migrants succeed, these social networks naturally grow over time, leading to (perhaps exponential) growth in rates of participation. Of course, this also depends on continued demand for rural labor in the Chinese construction and manufacturing sectors. Such demand in turn depends on the continued profitability of Chinese exports, which further depends on maintaining the artificial undervaluation of the country's currency despite international objections.

A related factor is China's restrictive household registration (*hukou*) law. This law denies migrants access to social services (e.g., health and education) at their destinations, acting as a strong disincentive to labor migration (Cheng and Selden, 1994). Each year speculation grows that this law will undergo major revision. Depending on how such a revision is implemented, this could dramatically increase out-migration, and severely reduce the number of permanent residents in the reserve not directly engaged in tourism and conservation. China's President Xi's recent public statements about proposed *hukou* reform suggest such revision may be imminent (*The Economist*, 2014). Such a drastic change could profoundly shift population dynamics in Wolong, the effects of which could be surprising and in need of comprehensive scenario analysis.

Conservation program implementation

Perhaps the most important action the central government can take to ensure the sustainability of the coupled system in Wolong is continued support for the reserve administration. Support should come in the form of both the resources and the authority to implement conservation initiatives according to the local biophysical and socioeconomic context. Our earlier scenario analyses give some direction on ways in which the administration might consider improving upon such initiatives in the future (e.g., cost-effective targeting). Other improvements could also be explored in the future. Creative ways to build on the foundation of these programs could include the use of GTGP plots for partial harvesting of trees for fuelwood. Selective thinning could be a potential "win–win" because it would make the dense plantations potentially more amenable to understory growth (and potentially more valuable for pandas) while also fulfilling livelihood needs.

Of course, a cornerstone in this regard is local land-use policy, and this fundamentally rests on the delineation of zones of allowable uses, and the enforcement of the resulting restrictions. Our prior work has shown that the land-use zoning currently in force in Wolong does not include important parts of the panda's prime habitat in the core zone providing the greatest level of protection (Hull et al., 2011b, Chapter 4). The core zone also includes a measurable amount of land already significantly disturbed. We have put forth some preliminary recommendations for revisions to the zoning scheme that could provide benefits for all involved (Hull et al., 2011b, Figure 4.7). Scenario analysis that explores outcomes of alternative zoning schemes would build on this foundation and enrich the planning processes. Such a process would ideally examine potential scenarios constructed collaboratively by local residents directly engaged in land use.

One particular land-use topic that we have not yet explored concerns grazing lands in the reserve. Wolong's Master Plan calls for the restoration of 8000 ha of grazing land (Sichuan Academy of Forestry, 2014). This proposed idea is significant as, according to the estimates reported in Chapter 6, it would affect about half of the reserve's current grazing land. Future land-use policy in the reserve will likely grapple with issues pertaining to management of these lands. Agriculture should be managed more holistically, going beyond the traditional focus on cropping systems and pig fodder to include free-ranging livestock as integral parts of land use. Differentiating between cropping and grazing is of particular urgency. So is refining the regulation of practices such as the use of chemical fertilizers and pesticides and of controlled fire to maintain pastures. The diverse options for managing these lands require careful integrative modeling to forecast the effects of competing land-use policy alternatives.

14.3.3 Integrative scenario analyses for a sustainable future

The preceding sections have outlined a number of factors that could push the coupled system into new modes of operation. These shifts could be either toward or away from sustainability, from the standpoint of biological conservation, human well-being, or both. Integrating factors will allow us to project what kinds of combinations of changes are possible. If, for example, the next bamboo flowering event were to coincide with a panda disease outbreak prior to the implementation of habitat restoration, it could pose an existential threat to the remaining panda population in Wolong. Likewise, a significant increase in the viability of urban employment could cause rapid outmigration of the working-age population, which could lead to the demise of the traditional livelihood system. Another earthquake could threaten the lives of those left behind. The wild card factor of climate change (discussed in Chapter 19) will affect all of these scenarios by shifting the spatial distribution of bamboo and in turn pandas. At the same time, changes in climate could potentially affect the frequency and magnitude of natural disasters, agriculture in human communities, and tourism development in unexpected ways (see also Chapter 12).

In order to understand and simulate such synergistic effects, an integrative and spatially explicit ABM approach is needed. Such an approach would build upon our past ABMs for family planning (e.g., An et al., 2005, Lambin et al., 2010) and PES program design (e.g., Chen et al., 2012a, 2014, Yang et al., 2013a, b) to tackle these new and emerging sustainability challenges. Future efforts could create new components and submodels to expand upon these existing model formulations.

For example, our most recent ABM in Chen et al. (2014) included three submodels linked to one another—a demographic submodel, a policy submodel, and a landscape submodel. The demographic submodel was based largely on that in An et al. (2005), which simulates life-history characteristics of Wolong households on a yearly time-step. New parameters can be incorporated in this model to evaluate and simulate emerging processes occurring in Wolong, such as post-earthquake migration changes and shifts in income generation activities. The policy submodel used there was related to NFCP, and other configurations have involved GTGP. This submodel can be revised to account for new derivations of these policies (e.g., potential forest plantation thinning in GTGP plots). The policy submodel could also account for other policies, such as alternative zoning policies or potential changes to the *hukou* law. The landscape submodel was largely based on the distribution of forests and can be revised to account for the new post-earthquake spatial configuration of households and test scenarios about compliance with inhabiting the new housing, as well as incorporating rangelands. A new panda population submodel can also be added to directly model the dynamics of the panda population in relationship to human and natural processes occurring in other submodels. Potential scenarios can be designed using this submodel to evaluate effects of disease outbreaks and panda reintroduction. Due to the flexibility of the ABM approach, the possibilities for further simulations are vast and offer opportunities for considerable growth in this integrative modeling environment in the future (An, 2012).

14.4 Discussion and conclusion

In exploring scenarios for Wolong's future, it is important to work toward designing a coupled system that could be resilient to disturbance and threats to sustainability. It is our belief that this sort of visioning is most productively accomplished by scientists working together with stakeholders. From our experience collaborating with the reserve administration and local residents, we can sketch some preliminary outlines of such a vision. Clearly, achieving one of the prime goals of the reserve—to ensure the long-term survival of the wild panda population (Ministry of Forestry, 1998)—must be part of the vision. That means first and foremost continuing the reversal of habitat degradation (Chapter 7) and, finally, ensuring their long-term protection. Providing adequate panda habitat requires sharply reducing human impacts on existing forests and fostering the restoration of forest in areas currently or previously devoted to agricultural uses (e.g., GTGP plots).

From the standpoint of the local residents, the vision of the future is surely more ambiguous. While some, particularly today's youth, will desire opportunities to secure employment outside the reserve, others may prefer wage employment or business opportunities in the tourism sector in the reserve. Still others have expressed a strong desire to maintain their traditional agricultural lifestyle and cultural life ways (Yang et al., 2013a). The details of who might continue producing food in the reserve, perhaps limited to supplying residents and visitors, and on what land, need to be worked out among those affected. Providing information about wildlife ecology and other aspects of the functioning of the natural systems in Wolong is certainly worthwhile. Beyond this, however, we recommend that stakeholders be invited to codesign, coproduce, and coimplement future plans, as articulated in the Future Earth initiative (Future Earth, 2014). This goal can be achieved through a collaborative process, based on the revision of the zoning map (Figure 4.7). The map in turn can be expanded to encompass the specific rules governing which uses are permitted in which areas and at what times of the year. We know from our previous work (Yang et al., 2013a, b), and the growing experience in governing open-access resources like Wolong's forests and pastures, that management is most successful when the resource boundaries are well defined, the rules of appropriation and provision are adaptive to local contexts, key stakeholders are adequately engaged throughout the decision-making process, and mechanisms for their enforcement and conflict resolution are clearly understood (Ostrom, 2008, 2009, Yang et al., 2013b). The sustainability of Wolong as a coupled system will ultimately depend on choices made in attempting to balance a complex set of dynamic human and natural factors. We hope to continue to inform this process by constructing ever-evolving scenario analyses that reflect the visions and aspirations for Wolong's future.

14.5 Summary

Scenario analysis is a powerful tool for exploring alternative futures of coupled human and natural systems. In this chapter, we highlighted some scenario analyses in our model coupled system—Wolong Nature Reserve. These analyses include modeling diverse behaviors of individual households in novel ways to project forest and panda habitat change over time. Our work has also tackled key issues such as the effects of family planning, interactions between human and natural disturbances, and payments for ecosystem services (PES) programs. For example, our analyses suggest that emigration of young people was more effective than any other method at reducing population size and recovering panda habitat in the reserve. Our bamboo scenario modeling indicated that simultaneous flowering of two bamboo species in the reserve could cause a loss of 49% of panda habitat, and habitat recovery could be hampered by ongoing human activities. Recent PES program models suggest that redesigning the Grain to Green Program (GTGP) to provide households with discriminative payments based on opportunity costs could make the program up to 10 times more cost-effective. We also identified emerging environmental and socioeconomic factors that may impact future trajectories of this coupled system in unexpected ways, including seismic events, disease, panda reintroduction, market fluctuations, and conservation policy implementation. We concluded with thoughts on how to integrate these emerging factors via systems modeling such as agent-based modeling. Such an approach would set the stage for increasingly integrative and realistic scenario analyses to inform sustainability pathways in the future.

References

Aide, T.M. and Grau, H.R. (2004) Globalization, migration, and Latin American ecosystems. *Science*, **305**, 1915–16.

An, L. (2012) Modeling human decisions in coupled human and natural systems: review of agent-based models. *Ecological Modelling*, **229**, 25–36.

An, L., He, G., Liang, Z., and Liu, J. (2006) Impacts of demographic and socioeconomic factors on spatio-temporal dynamics of panda habitat. *Biodiversity and Conservation*, **15**, 2343–63.

An, L., Linderman, M., He, G., et al. (2011) Long-term ecological effects of demographic and socioeconomic factors in Wolong Nature Reserve (China). In R.P. Cincotta and L.J. Gorenflo, eds, *Human Population: Its Influences on Biological Diversity*, pp. 179–95. Springer-Verlag, Berlin, Germany.

An, L., Linderman, M., Qi, J., et al. (2005) Exploring complexity in a human-environment system: an agent-based spatial model for multidisciplinary and multiscale integration. *Annals of the Association of American Geographers*, **95**, 54–79.

An, L. and Liu, J. (2010) Long-term effects of family planning and other determinants of fertility on population and environment: agent-based modeling evidence from Wolong Nature Reserve, China. *Population and Environment*, **31**, 427–59.

An, L., Zvoleff, A., Liu, J., and Axinn, W. (2014) Agent-based modeling in coupled human and natural systems (CHANS): lessons from a comparative analysis. *Annals of the Association of American Geographers*, **104**, 723–45.

Carlson, M.J., Mitchell, R., and Rodriguez, L. (2011) Scenario analysis to identify viable conservation strategies in Paraguay's imperiled Atlantic forest. *Ecology and Society*, **16**, 8.

Chen, X., Frank, K.A., Dietz, T., and Liu, J. (2012b) Weak ties, labor migration, and environmental impacts toward a sociology of sustainability. *Organization & Environment*, **25**, 3–24.

Chen, X., Lupi, F., An, L., et al. (2012a) Agent-based modeling of the effects of social norms on enrollment in payments for ecosystem services. *Ecological Modelling*, **229**, 16–24.

Chen, X., Lupi, F., He, G., et al. (2009) Factors affecting land reconversion plans following a payment for ecosystem service program. *Biological Conservation*, **142**, 1740–47.

Chen, X., Lupi, F., Viña, A., et al. (2010) Using cost-effective targeting to enhance the efficiency of conservation investments in payments for ecosystem services. *Conservation Biology*, **24**, 1469–78.

Chen, X., Viña, A., Shortridge, A., et al. (2014) Assessing the effectiveness of payments for ecosystem services: an agent-based modeling approach. *Ecology and Society*, **19**, 7.

Cheng, T. and Selden, M. (1994) The origins and social consequences of China's hukou system. *The China Quarterly*, **139**, 644–68.

China Daily (2003) SARS Virus infects China's Economy. http://english.peopledaily.com.cn/200305/06/eng20030506_116304.shtml.

China Radio International (2015) Third Panda Dies from Virus in NW China. http://english.cri.cn/12394/2015/01/24/53s863271.htm.

Clark, E. (2003) Sars Strikes Down Asia Tourism. *BBC News Online*. http://news.bbc.co.uk/2/hi/business/3024015.stm.

Future Earth (2014) Research for Global Sustainability. http://www.futureearth.org/.

Gracie, C. (2014) All Aboard: China's Railway Dream. *BBC News—Asia*. http://www.bbc.com/news/world-asia-28289319.

Hull, V., Shortridge, A., Liu, B., et al. (2011a) The impact of giant panda foraging on bamboo dynamics in an isolated environment. *Plant Ecology*, **212**, 43–54.

Hull, V., Xu, W., Liu, W., et al. (2011b) Evaluating the efficacy of zoning designations for protected area management. *Biological Conservation*, **144**, 3028–37.

Hull, V., Zhang, J., Zhou, S., et al. (2014) Impact of livestock on giant pandas and their habitat. *Journal for Nature Conservation*, **22**, 256–64.

IPCC (2007) *Impacts, Adaptation and Vulnerability. Contribution of working group II to the Fourth Assessment Report of the Intergovernmental Panel on Climate Change*. Cambridge University Press, Cambridge, UK.

Lambin, E., Tran, A., Vanwambeke, S., et al. (2010) Pathogenic landscapes: interactions between land, people, disease vectors, and their animal hosts. *International Journal of Health Geographics*, **9**, 1–13.

Li, D., Zhu, L., Cui, H., et al. (2014) Influenza A (H1N1) pdm09 virus infection in giant pandas, China. *Emerging Infectious Diseases*, **20**, 480–83.

Lim, K., Deadman, P.J., Moran, E., et al. (2002) Agent-based simulations of household decision-making and land-use change near Altamira, Brazil. In H. R. Gimblett, ed., *Integrating Geographic Information Systems and Agent-based Modeling Techniques for Simulating Social and Ecological Processes*, pp. 137–69. Oxford University Press, Oxford, UK.

Linderman, M.A., An, L., Bearer, S., et al. (2005) Modeling the spatio-temporal dynamics and interactions of households, landscapes, and giant panda habitat. *Ecological Modelling*, **183**, 47–65.

Linderman, M.A., An, L., Bearer, S., et al. (2006) Interactive effects of natural and human disturbances on vegetation dynamics across landscapes. *Ecological Applications*, **16**, 452–63.

Liu, J. (2003) SARS, wildlife, and human health. *Science*, **302**, 53.

Liu, J., Li, S., Ouyang, Z., et al. (2008) Ecological and socioeconomic effects of China's policies for ecosystem services. *Proceedings of the National Academy of Sciences of the United States of America*, **105**, 9477–82.

Liu, J., Ouyang, Z., Taylor, W.W., et al. (1999) A framework for evaluating the effects of human factors on wildlife habitat: the case of giant pandas. *Conservation Biology*, **13**, 1360–70.

Liu, W. (2012) *Patterns and Impacts of Tourism Development in a Coupled Human and Natural System*. Doctoral Dissertation, Michigan State University, East Lansing, MI.

Liu, W., Vogt, C.A., Luo, J., et al. (2012) Drivers and socioeconomic impacts of tourism participation in protected areas. *PLoS ONE*, **7**, e35420.

Loeffler, I.K., Howard, J., Montali, R.J., et al. (2007) Serosurvey of ex situ giant pandas (*Ailuropoda melanoleuca*) and red pandas (*Ailurus fulgens*) in China with implications for species conservation. *Journal of Zoo and Wildlife Medicine*, **38**, 559–66.

Lu, C., Van Ittersum, M., and Rabbinge, R. (2004) A scenario exploration of strategic land use options for the Loess Plateau in northern China. *Agricultural Systems*, **79**, 145–70.

MacPhee, R.D., and Greenwood, A.D. (2013) Infectious disease, endangerment, and extinction. *International Journal of Evolutionary Biology*, **2013**, 571939.

Mainka, S. and Lü, Z. (1999) Introduction. In S. Mainka and Z. Lü, eds, *International Workshop on the Feasibility of Giant Panda Re-introduction*, pp. 135–37. China Forestry Publishing House, Beijing, China.

Mainka, S.A., Xianmeng, Q., Tingmei, H., and Appel, M.J. (1994) Serologic survey of giant pandas (*Ailuropoda melanoleuca*), and domestic dogs and cats in the Wolong Reserve, China. *Journal of Wildlife Disease*, **30**, 86–89.

Matthews, R.B., Gilbert, N.G., Roach, A., et al. (2007) Agent-based land-use models: a review of applications. *Landscape Ecology*, **22**, 1447–59.

Millennium Ecosystem Assessment (2005) *Ecosystems and Human Well-being: Synthesis*. Island Press, Washington, DC.

Ministry of Forestry (1998) *Wolong Nature Reserve Master Plan*. Beijing, China (in Chinese).

Ostrom, E. (2008) The challenge of common-pool resources. *Environment: Science and Policy for Sustainable Development*, **50**, 8–21.

Ostrom, E. (2009) Design principles of robust property-rights institutions: what have we learned? In K.G. Ingram and Y.-H. Hong, eds, *Property Rights and Land Policies*, pp. 25–51. Lincoln Institute of Land Policy, Cambridge, MA.

Parker, D.C., Manson, S.M., Janssen, M.A., et al. (2003) Multi-agent systems for the simulation of land-use and land-cover change: a review. *Annals of the Association of American Geographers*, **93**, 314–37.

Reid, D.G., Hu, J.C., Sai, D., et al. (1989) Giant panda *Ailuropoda melanoleuca* behavior and carrying capacity following a bamboo die-off. *Biological Conservation*, **49**, 85–104.

Schaller, G.B., Hu, J., Pan, W., and Zhu, J. (1985) *The Giant Pandas of Wolong*. University of Chicago Press, Chicago, IL.

Sichuan Academy of Forestry (2014) *The Overall Planning of Wolong National Nature Reserve*, Sichuan, China (in Chinese).

Song, J. and Yu, J. (1988) *Population System Control*. Springer-Verlag, Berlin, Germany.

The Economist (2014) The Great Transition. http://www.economist.com/news/leaders/21599360-government-right-reform-hukou-system-it-needs-be-braver-great.

Tuanmu, M.-N., Viña, A., Winkler, J.A., et al. (2013) Climate-change impacts on understorey bamboo species and giant pandas in China's Qinling Mountains. *Nature Climate Change*, **3**, 249–53.

Van Ittersum, M., Rabbinge, R., and Van Latesteijn, H. (1998) Exploratory land use studies and their role in strategic policy making. *Agricultural Systems*, **58**, 309–30.

Viña, A., Chen, X., Yang, W., et al. (2013) Improving the efficiency of conservation policies with the use of surrogates derived from remotely sensed and ancillary data. *Ecological Indicators*, **26**, 103–11.

Wolong Administration Bureau (2009) *Wolong National Nature Reserve Post-earthquake Reconstruction Master Plan* (in Chinese).

World Commission on Environment and Development (1987) *Our Common Future*. Oxford University Press, Oxford, UK.

Xie, Z. and Gipps, J. (2011) *The International Studbook for Giant Panda (Ailuropoda melanoleuca). Updated November 2011*. Beijing, China.

Xu, W., Ouyang, Z., Viña, A., et al. (2006) Designing a conservation plan for protecting the habitat for giant pandas in the Qionglai mountain range, China. *Diversity and Distributions*, **12**, 610–19.

Yang, W., Liu, W., Viña, A., et al. (2013a) Performance and prospects of payments for ecosystem services programs: evidence from China. *Journal of Environmental Management*, **127**, 86–95.

Yang, W., Liu, W., Viña, A., et al. (2013b) Non-linear effects of group size on collective action and resource outcomes. *Proceedings of the National Academy of Sciences of the United States of America*, **110**, 10916–21.

Zhang, J.-S., Daszak, P., Huang, H.-L., et al. (2008) Parasite threat to panda conservation. *Ecohealth*, **5**, 6–9.

Zhang, K.H. and Song, S. (2003) Rural–urban migration and urbanization in China: evidence from time-series and cross-section analyses. *China Economic Review*, **14**, 386–400.

Zhang, Z., Zhang, A., Hou, R., et al. (2006) Historical perspective of breeding giant pandas ex situ in China and high priorities for the future. In D.E. Wildt, A. Zhang, H. Zhang, D.L. Janssen, and S. Ellis, eds, *Giant Pandas: Biology, Veterinary Medicine, and Management*, pp. 455–68. Cambridge University Press, Cambridge, UK.

Zhou, X., Tan, Y., Song, S., et al. (2005) Comparative study on behavior and ecology between captivity and semi-nature enclosure of giant panda. *Sichuan Journal of Zoology*, **24**, 143–46.

PART III

Across Local to Global Coupled Human and Natural Systems

As demonstrated in Part II, our team's long-term efforts to understand human–nature interactions in Wolong Nature Reserve have generated many useful insights for conservation and human well-being inside the reserve. Our integrative work has also made important methodological and theoretical advances. This part of the book highlights the utility of these insights and advances in understanding and managing other coupled systems at larger scales and in different contexts. This part illustrates how some of the insights and advances developed in Wolong have been applied in different settings and how they have been scaled up to regional, national, and global levels. This part also demonstrates how Wolong is connected with the rest of the world through telecouplings (socioeconomic and environmental interactions over distances).

Viña et al. begin by applying some approaches and findings from Wolong to the entire giant panda range in China (Chapter 15). Specifically, they outline how the methods of analyzing the spatiotemporal dynamics of forest cover and giant panda habitat developed in Wolong have been upscaled to six mountain regions (Qinling, Minshan, Qionglai, Liangshan, Greater Xiangling, and Lesser Xiangling) in three provinces of China (Gansu, Shaanxi, and Sichuan). The results show that forests and panda habitat vary considerably across the geographic range. Only about 17% of the forests in this region can act as suitable habitat for the panda. Many forests lack suitable understory bamboo, a challenge that limits the success of panda conservation policies that do not explicitely consider bamboo distribution. The upscaling analysis across the giant panda's range helped to put Wolong in a larger context, and also helped identify areas in need of protection along with new strategies promoting higher connectivity between the existing nature reserves.

Next, the scope of our team's research is expanded even further to determine how generalizable the findings are to other coupled systems with different contexts. In Chapter 16, Carter et al. apply what we have learned about the coupled system in Wolong to understand the complexity of another coupled system in Chitwan National Park in Nepal. They conduct a cross-site synthesis to examine how the human–nature interactions in Wolong relate to those in Chitwan. Chitwan is similar to Wolong in many ways. Chitwan supports habitat for the tiger (another charismatic endangered species), provides key resources for local rural farmers, and faces challenges with balancing human livelihood needs with conservation goals. On the other hand, Chitwan differs from Wolong in several aspects, including an absence of residents inside the park, different types of resource conflicts, different habitat types, and different cultural underpinnings. In the cross-site synthesis, Carter et al. analyze similarities and differences across these systems with respect to several key coupled-system components, including feedbacks, time lags, non-linearities, heterogeneities, thresholds, and surprises. They also reflect on lessons learned about common properties and processes that may be driving human–nature interactions across different coupled systems. For example, in both sites, they find that within-household socioeconomic changes (e.g., fertility and marriage timing) give rise to aggregate-level dynamics, such as changes in household numbers contributing to habitat loss. In turn, they illustrate how processes at broad scales, such as regional urbanization, impact fine-scale phenomena, such as out-migration of individuals.

Liu et al. then explore various socioeconomic and environmental interactions between Wolong and other coupled systems across the globe (Chapter 17). They do so by using a novel telecoupling framework that allows for analysis of human–nature interactions over distances. Liu et al. describe how Wolong is linked to distant corners of the world via telecoupling processes such as panda loans, tourism, conservation subsidies, information dissemination, and trade of agricultural and industrial products. They outline diverse effects and feedbacks between Wolong and other systems. For instance, tourists travel from all over the world to visit Wolong and produce numerous ecological and socioeconomic impacts, encouraging future tourism development to in turn promote more tourist volume in the future. Liu et al. also integrate across the telecouplings to show how the different processes are related to one another, such as information dissemination promoting tourism, and conservation subsidies decreasing production and trade of agricultural products. This chapter illustrates that even a seemingly remote and isolated system such as Wolong is connected in numerous and often surprising ways to distant parts of the globe. This realization cements Wolong's place in global sustainability efforts and sheds light on the importance of thinking globally in long-term interdisciplinary research.

In Chapter 18, Liu et al. synthesize ten overarching lessons our team has learned over the past two decades that may be informative for others embarking on research and management of other systems worldwide. The lessons are diverse, encompassing coupled systems theories and methods, and ways to reach out to stakeholders to assist with management and education efforts for sustainability. Some examples include strategies for approaching the research such as "Put Researchers in the Local Residents' Shoes" and "Expect the Unexpected." Others are cautionary tales stemming from our research team's surprising encounters, such as "Good Intentions Can Sometimes Lead to Bad Outcomes" and "Money Isn't Everything." While many scholars have reported evidence that coupled systems are dependent on scale and context, Liu et al. find evidence for other features of coupled systems that are independent of both. This finding demonstrates the value of the model coupled system approach and the broad applicability of our team's work in Wolong to interdisciplinary efforts to tackle global sustainability challenges.

CHAPTER 15

Applying Methods from Local to Regional Scales

Andrés Viña, Weihua Xu, Zhiyun Ouyang, and Jianguo Liu

15.1 Introduction

Most detailed research on coupled human and natural systems has been conducted at the local scale (An et al., 2014, Hull et al., 2015, Liu et al., 2007a, b). Although such research has provided useful insights (e.g., Chapters 3–14), more research is needed to address important issues at broader scales. For example, long-term survival of species is dependent on abundant and well-connected suitable habitat areas across their geographic ranges (Crooks and Sanjayan, 2006, Damschen et al., 2006, Kareiva and Wennergren, 1995, Ricketts, 2001). Therefore, range-wide analyses are needed to assess habitat dynamics and human impacts at multiple scales, such as inside and across nature reserve boundaries (Loss et al., 2011, Minor and Lookingbill, 2010, Noss, 2007, Spring et al., 2010).

Range-wide habitat analyses are essential for the establishment of conservation practices that support the long-term survival of many endangered species, such as the endangered giant panda (*Ailuropoda melanoleuca*) (Xu et al., 2014). Giant pandas once were found throughout most of China, northern Vietnam, and northern Myanmar (Pan et al., 2001). But due to human activities such as agricultural expansion, logging, and infrastructure development, only 1864 wild giant panda individuals remain within six main mountain regions (Qinling, Minshan, Qionglai, Greater Xiangling, Lesser Xiangling, and Liangshan) in three provinces (Gansu, Shaanxi, and Sichuan) of China (Hu and Wei, 2004, Mackinnon and De Wulf, 1994, Reid and Gong, 1999, State Council Information Office of China, 2015, State Forestry Administration, 2006, Viña et al., 2010; Figure 1.1). These mountain regions are characterized by a high elevational range and are dissected by the valleys of perennial rivers. The complex topography, together with the substantial variability of climates and soil types, leads to diverse flora and fauna. In fact, a significant portion of the region where the pandas live has been classified as one of the world's top 25 Biodiversity Hotspots (Myers et al., 2000). The geographic range also contains a UNESCO World Heritage Site (Li et al., 2013) and several biosphere reserves for the protection of not only the giant pandas but many other endangered species.

The panda habitat across its entire geographic range includes several types of forest ecosystems (Reid and Hu, 1991, Taylor and Qin, 1993) which are home to thousands of other animal and plant species. Thus, efforts to mitigate the reduction and fragmentation of panda habitat may also assist in the conservation of other endangered species. Examples include the golden snub-nosed monkey (*Rhinopithecus roxellana*), the takin (*Budorcas taxicolor*), the red panda (*Ailurus fulgens*), the forest musk deer (*Moschus berezovskii*), and the Asiatic black bear (*Ursus thibetanus*). Therefore, assessing the dynamics of the giant panda habitat and pandas–people interactions across the entire range of the giant panda is crucial for the conservation of many other endangered species that coexist with them (Xu et al., 2014).

Pandas and People. Edited by Jianguo Liu, Vanessa Hull, Wu Yang, Andrés Viña, Xiaodong Chen, Zhiyun Ouyang, and Hemin Zhang. Published 2016 by Oxford University Press.

However, range-wide analyses and conservation efforts are not a common practice since they are inhibited mainly by the lack of resources (e.g. time, financial support, and appropriate methods). Other limitations include lack of knowledge on the overall habitat status of target species and an absence of consensus on which areas are priorities for conservation (Sanderson et al., 2002, Thorbjarnarson et al., 2006). In this chapter, we demonstrate the role of research at local scales in advancing research at broad scales by presenting the spatiotemporal dynamics of forest cover and giant panda habitat across the entire geographic range of the panda. The results were generated using methods and insights that originated from our long-term research in Wolong Nature Reserve (e.g., Chapters 3, 6, and 7). In other words, the methods developed and tested in Wolong were upscaled to the entire geographic range of the giant panda.

15.2 Forests across the current geographic range of the giant panda

Forests comprise approximately one-third of the current geographic range of the giant panda (Figure 15.1). They include three main types: broadleaf deciduous (32%), coniferous (48%), and mixed coniferous/ broadleaf deciduous (20%) (Figure 15.1; Plate 6). Broadleaf deciduous and mixed forest stands are usually distributed at lower and intermediate elevations, respectively. Coniferous forest stands usually occur at higher elevations. Some coniferous stands also occur at low and intermediate elevations, but these are generally composed of monospecific plantations (Viña et al., 2010).

Approximately 80% of the 9 million people living in the mountain regions comprising the current panda geographic range directly depend on agricultural activities for their livelihoods. The remaining inhabitants rely on non-agricultural activities such as industry and services, particularly in urban settlements (Provincial Bureau of Statistics, 2001). In these mountain regions, forest distribution is, in general terms, inversely related to human population density (Figure 15.2) and exhibits a high degree of fragmentation. Nevertheless, fragmentation is highly variable among the different mountain regions, partly due to differential human pressures exerted over the landscape. They include

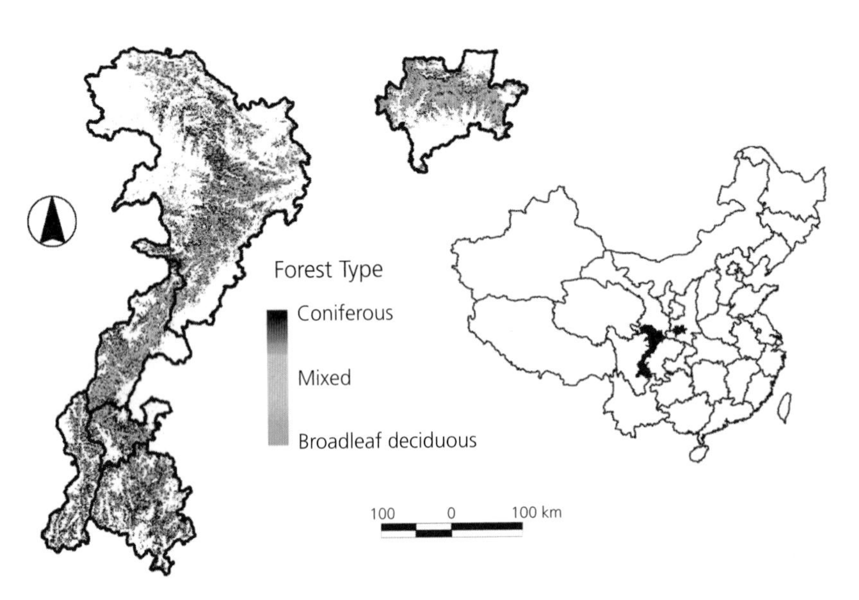

Figure 15.1 Spatial distribution in 2007 of coniferous, broadleaf deciduous, and mixed coniferous/deciduous forests in the mountain regions comprising the current geographic range of the giant panda. This map was obtained using the seasonal progression of vegetation in 2007, as measured by the Moderate Resolution Imaging Spectroradiometer (MODIS) sensor, for separating different forest types according to their different phenological signatures. Reprinted from Liu and Viña (2014) with permission from the Missouri Botanical Garden Press. For color version see color plates. [PLATE 6]

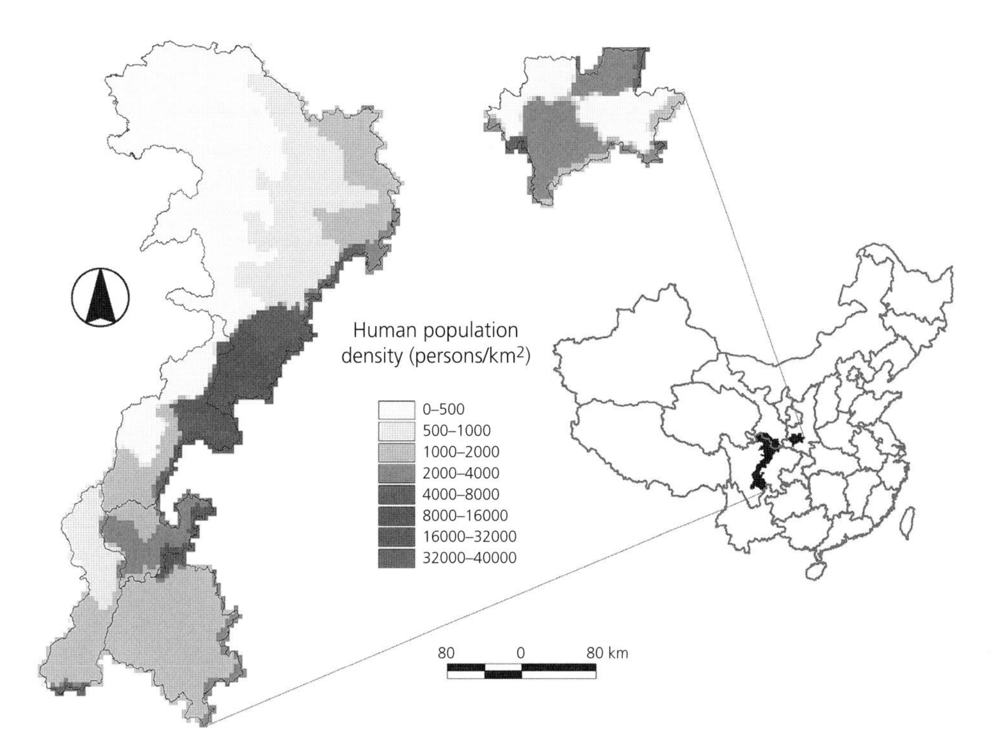

Figure 15.2 Spatial distribution of human population density during the 2000s in the mountain regions comprising the current geographic range of the giant panda. Data were obtained from the gridded population density data set (NASA Socioeconomic Data and Applications Center, 2011).

different agropastoral practices (e.g., row crops, vegetable cropping, and livestock husbandry) and different degrees of infrastructure development (e.g., roads, railroads, and buildings). Since the late 1990s, a new source of variation is differential implementation of national conservation policies such as the Natural Forest Conservation Program (NFCP) and the Grain to Green Program (GTGP; Liu et al., 2008, 2013).

During the first decade of the twenty-first century, there was a conspicuous gain in forest cover. Around 20% of the region experienced a monotonic increase in the percent of tree cover (Figure 15.3). Only ~2% of the region experienced a monotonic reduction (Figure 15.3). We performed a time-series analysis of forest cover in the region (Viña et al., 2011), albeit comprising a smaller geographic extent (i.e., Wenchuan County in Sichuan Province), and found evidence for positive effects of NFCP and GTGP. Thus, we believe that the implementation of these national conservation policies has exerted a positive effect on forest cover across the entire geographic range of the giant panda. In addition, many of the areas that exhibited a monotonic loss in the percentage of tree cover (e.g., the northeastern part of the Qionglai mountain range; Figure 15.3B) are located within the zone of influence of the May 12, 2008, Wenchuan earthquake (Viña et al., 2011). Therefore, a significant proportion of the forest-cover losses in the region could be attributed to earthquake-induced landslides rather than human activities (e.g., fuelwood collection and timber extraction) banned under NFCP (Chapter 13). Excluding these areas reinforces the notion that the increase in forest cover was the dominant land-cover dynamic observed in the region during 2001–2010. This dynamic occurred concurrently with the successful implementation of national conservation policies.

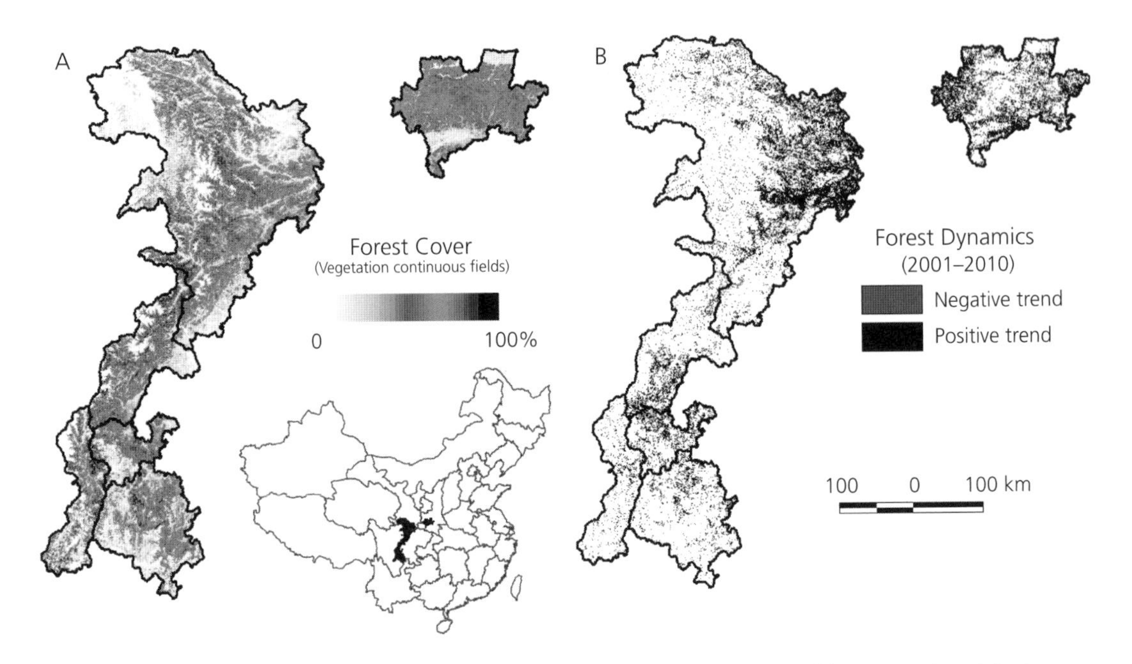

Figure 15.3 (A) Forest cover in 2001 in the mountain regions comprising the current geographic range of the giant panda, as defined by the vegetation continuous fields tree cover product derived from surface reflectance data acquired by the Moderate Resolution Imaging Spectroradiometer (MODIS; Hansen et al., 2003). (B) Spatial distribution of the areas exhibiting significant ($p < 0.05$) positive and negative trends in forest cover, as defined by the vegetation continuous fields tree cover product, between 2001 and 2010.

15.3 Habitat dynamics across the entire geographic range of the giant panda

Giant pandas are extremely difficult to encounter in the wild. Their population size is small and density is very low in an extensive geographic range with complex topography. They are also shy and tend to avoid human presence. Thus, the spatial distribution of evidence of the presence of giant pandas (e.g., tracks, feces, dens, and sleeping sites) has been used as a surrogate of species occurrence (Bearer et al., 2008, Tuanmu et al., 2011, Viña et al., 2007, 2008). Such surrogates are useful for detecting their habitat requirements and preferences, which in turn can be used to map the distribution of their habitat. For instance, giant pandas have particular topographic preferences. In Wolong Nature Reserve, their elevational range is usually between 1500 m and 3250 m, with an optimal range between 2500 m and 3000 m. They are also found on slopes of less than 45°, with optimal slopes of less than 15° (Hu and Wei, 2004, Liu et al., 1999). In addition, pandas require forest cover as shelter (Johnson et al., 1988, Reid et al., 1989, Schaller et al., 1985). Therefore, a spatially explicit characterization of these attributes is required for analyzing the spatiotemporal dynamics of their habitat.

We tested such an approach for our research in Wolong. We used topographic variables obtained from topographic maps and the digital elevation model (90 × 90 m/pixel) acquired by the Shuttle Radar Topography Mission (Berry et al., 2007). We combined these data with maps of forest cover obtained from the classification of panchromatic (e.g., Corona) and multispectral (e.g., Landsat MSS, TM, and ETM +) satellite imagery, from the 1960s to the late 2000s (Liu et al., 2001, Viña et al., 2007, 2008; see Chapter 7). These dynamics were correlated with human activities (e.g., farming and fuelwood collection) to assess their effects on forests and giant panda habitat in the reserve (An et al., 2002, 2003, 2005, An and Liu, 2010, He et al., 2009, Linderman et al., 2005, 2006). We also applied spatiotemporal forest and habitat dynamics in the reserve to assess the effects of national conservation policies, i.e., NFCP and GTGP (Chen et al., 2009, 2010, 2011, Yang et al., 2013) and division of the nature reserve into zones (Hull et al., 2011) on forests and giant pandas.

Using these same techniques beyond Wolong allowed the analysis of habitat conditions in an entire mountain region (i.e., the Qionglai mountain region), which comprises several other nature reserves administered at national and provincial levels (Xu et al., 2006). Combining these procedures with gap analysis made it possible to determine the optimal location of new nature reserves, extensions to current nature reserves, and corridors that increase the connectivity between nature reserves. Connectivity could be determined not only within the Qionglai Mountains, but also between the Qionglai Mountains and surrounding mountain regions (Xu et al., 2006). It was also possible to evaluate the effects of the implementation of NFCP and GTGP on forests. Forests of interest included those distributed across a globally important UNESCO World Heritage Site—the Sichuan Giant Panda Sanctuary located mostly within the Qionglai Mountains. The study area also included the region that exhibits the highest density of giant pandas in the wild, the Qinling Mountains in Shaanxi Province (Li et al., 2013).

These analyses have been useful for evaluating forest and giant panda habitat dynamics, as well as their relations with human activities and with conservation policies. However, they did not include suitable information on the occurrence of understory bamboo. Information on understory bamboo is very important since it is a crucial component of giant panda habitat. In fact, understory bamboo species constitute 99% of the panda's diet (Schaller et al., 1985). Thus, estimates of giant panda habitat based only on topographic characteristics combined with information on forest cover (Liu et al., 1999, 2001, Viña et al., 2007) would overestimate suitable habitat (Viña et al., 2008). Such a discrepancy occurs because not all forests have adequate understory bamboo (Chapter 7). This problem is particularly salient if the estimation of forest cover includes monospecific stands of planted forest that do not necessarily constitute habitat for the pandas. However, it is difficult to obtain information on understory bamboo due to the interference of overstory canopies (Linderman et al., 2004). The lack of information on distribution and its temporal dynamics constitutes, therefore, a major obstacle in the assessment of giant panda habitat dynamics, but particularly for assessing habitat gains.

We surmounted this major obstacle using the phenological variability of the wide dynamic range vegetation index (WDRVI). The WDRVI is a proxy of the fractional absorbed photosynthetically active radiation (Viña and Gitelson, 2005). We obtained this measure from surface reflectance data acquired by the Moderate Resolution Imaging Spectroradiometer (MODIS). The WDRVI is a non-linear transformation of the widely used Normalized Difference Vegetation Index (NDVI). The WDRVI is preferred over the NDVI because it maintains high sensitivity to changes in vegetation characterized by a high green biomass (Viña et al., 2004). The non-linear transformation is performed using the following equation (Viña and Gitelson, 2005):

$$\mathrm{WDRVI} = \frac{[(\alpha+1)\mathrm{NDVI}+(\alpha-1)]}{[(\alpha-1)\mathrm{NDVI}+(\alpha+1)]} \quad \text{(Equation 15.1)}$$

The weighting coefficient α is introduced to attenuate the contribution of the near-infrared band at moderate-to-high green biomass conditions, and to make it comparable to that of the red band (Gitelson, 2004). The specific magnitude of this coefficient depends primarily on sensor characteristics and observational conditions (Gitelson, 2004), although it can be obtained using a heuristic procedure (Henebry et al., 2004). The higher sensitivity to changes in vegetation under high green biomass conditions (Viña and Gitelson, 2005) makes the WDRVI the index of choice for isolating forests with understory bamboo (Figure 15.4), and thus for mapping giant panda habitat (Tuanmu et al., 2010, 2011, Viña et al., 2008, 2010).

This method is suitable for assessing the presence of understory bamboo but is less suitable for identifying different bamboo species. Giant pandas can feed on more than 60 bamboo species found across their geographic range, with approximately 35 of them being their preferred food (Hu and Wei, 2004, Li, 1997). But it was not essential to distinguish among bamboo species for two main reasons. First, pandas require

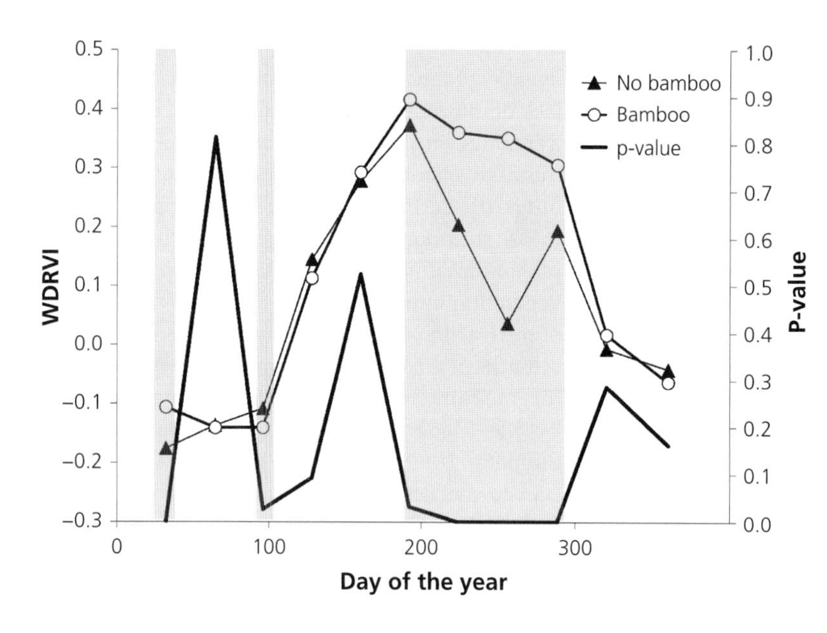

Figure 15.4 Average phenological dynamics of forests not considered panda habitat ($n = 118$ field sampling plots in forests without understory bamboo) and forests considered habitat ($n = 262$ field sampling plots in forests with understory bamboo), as measured by the Wide Dynamic Range Vegetation Index (WDRVI). Shaded areas correspond to time periods when there is a significant difference between forests with and without understory bamboo (i.e., p-values < 0.05 in a two-sample t-test, after testing for variance homogeneity).

a large amount of bamboo and feed on less preferable bamboo when the favorite bamboo species is not available (Schaller et al., 1985). Second, at a local scale, there are usually only one or two dominant bamboo species. For example, in Wolong Nature Reserve, pandas eat two major species—*Bashania faberi* and *Fargesia robusta*. They are distributed at different elevations with *B. faberi* occurring at elevations of up to 3600 m and *F. robusta* occurring at elevations of up to 2600 m (Schaller et al., 1985). Thus, topographic information can be used to separate them if needed (Linderman et al., 2006, Tuanmu et al., 2010).

To map the spatial distribution of suitable giant panda habitat across its entire geographic range, we used procedures developed for Wolong. This procedure involved a fuzzy classification algorithm supported by ecological niche theory: Maximum Entropy, MaxENT (Phillips et al., 2006). The algorithm estimates the probability of an area to be suitable habitat by finding a probability distribution of maximum entropy (i.e., maximum uniformity). The expected value of a set of environmental predictor variable matches must match its empirical average, defined by known habitat locations. This algorithm was applied to 11 phenology metrics (Tuanmu et al., 2011) derived from WDRVI imagery acquired with a high temporal resolution by MODIS. This information was combined with panda occurrence locations defined using surrogates such as feces, tracks, and dens obtained during field surveys (Viña et al., 2010). These panda habitat occurrences were used to establish the phenological characteristics (as defined by the phenology metrics) of currently suitable habitat areas. The final output was a continuous habitat suitability index (HSI) map with values ranging from 0 to 1 (Figure 15.5; Plate 7). We validated the algorithm by means of the area under the receiver operating characteristic curve (AUC; Hanley and McNeil, 1982). The AUC ranges from 0 to 1. A score of 1 indicates perfect classification, a score of 0.5 implies a classification that is not better than random, and values lower than 0.5 imply a worse than random classification. The AUC value obtained was 0.915 which denotes a high prediction success. This map shows that not all forest areas constitute suitable giant panda habitat, due mainly to a lack of understory bamboo (Viña et al., 2010). While habitat quality, defined by HSI scores, varies in different locations, suitable giant panda habitat covers ~17% of the geographic range. Suitable habitat is located particularly in the central (Qionglai mountain region) and northern (Minshan and Qinling mountain regions) portions of the geographic range. The southernmost mountain regions (i.e., Liangshan, Greater Xiangling, and Lesser Xiangling) exhibit the least amount of suitable habitat (Figure 15.5).

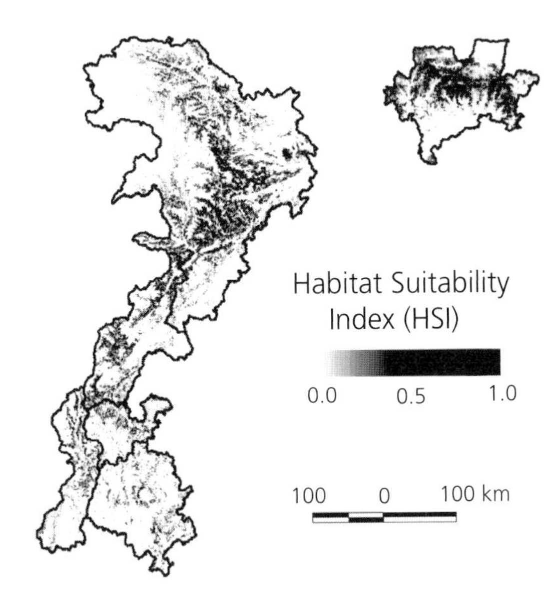

Figure 15.5 Spatial distribution in 2001 of giant panda habitat suitability index (HSI) values in the mountain regions comprising the current geographic range of the giant panda, obtained using 11 phenology metrics derived from an image time series of the Wide Dynamic Range Vegetation Index (WDRVI) calculated from data collected by the Moderate Resolution Imaging Spectroradiometer (MODIS) sensor. A higher HSI value indicates more suitable habitat. For color version see color plates. [PLATE 7]

To examine the relative importance of each phenology metric for mapping giant panda habitat suitability, we conducted a jackknife analysis on model performance. To this effect, we obtained the AUC values of models containing only one of the 11 metrics. A higher AUC value in a model containing only one metric indicates that the specific metric is more informative for mapping habitat suitability than other metrics. We also obtained the AUC values of models containing all but one of the metrics. A lower AUC value for a model lacking one specific metric indicates that the specific metric contains information for mapping habitat suitability not provided by the other metrics. Results of this jackknife analysis are similar to those obtained in Wolong (Tuanmu et al., 2011). Models with only the maximum level, the middle of season, and the start of season exhibited the highest AUC values (Figure 15.6A). These metrics thus contained the most useful information for mapping habitat suitability. In addition, the models without the increase rate, the date of the start of the season, and the maximum level exhibited the lowest AUC values (i.e., largest loss in AUC; Figure 15.6B). Therefore, these metrics contained unique information for mapping habitat suitability.

Using the calibration coefficients obtained for 2007, we mapped the distribution of giant panda habitat annually from 2001 to 2010. We then assessed on a per-pixel basis if there were significant trends in the HSI values. Results of this analysis show that ~1.6% of the region exhibited a significant ($p < 0.05$) positive trend in HSI values, while ~2.5% exhibited a significant negative trend (Figure 15.7; Plate 8). The frequency distribution of these areas among the HSI values shows that the pixels exhibiting a positive trend were characterized by having mostly low HSI values (< 0.4). Those exhibiting a negative trend had HSI values ranging from 0.01 to 0.92 (Figure 15.8). This finding suggests that while negative trends are widespread across the entire HSI variability, positive trends tend to occur more often in areas with low habitat suitability, thus with a lower probability of supporting panda individuals. Yet, these positive trends bring hope that if they continue over the next 10+ years, they will allow these areas to become suitable for pandas. Indeed, Wolong pandas have been observed to use secondary forests if adequate understory bamboo is present (Bearer et al., 2008, Viña et al., 2007).

15.4 Effects of conservation activities on panda habitat across their entire geographic range

We found that ~40% of the suitable habitat is located inside the 63 nature reserves established specifically for the conservation of giant pandas and their habitat (Viña et al., 2010). The Qinling mountain region had the highest proportion of habitat inside nature reserves. A lower proportion of habitat inside nature reserves occurred in the Qionglai, Greater Xiangling, and Lesser Xiangling mountain ranges (Viña et al., 2010). Panda populations in the last two are on the brink of extinction (Hu and Wei, 2004). Nature reserves, although not immune to human threat (Liu et al., 2001, Viña et al., 2007), may help protect the remaining habitat. Gap analysis allowed

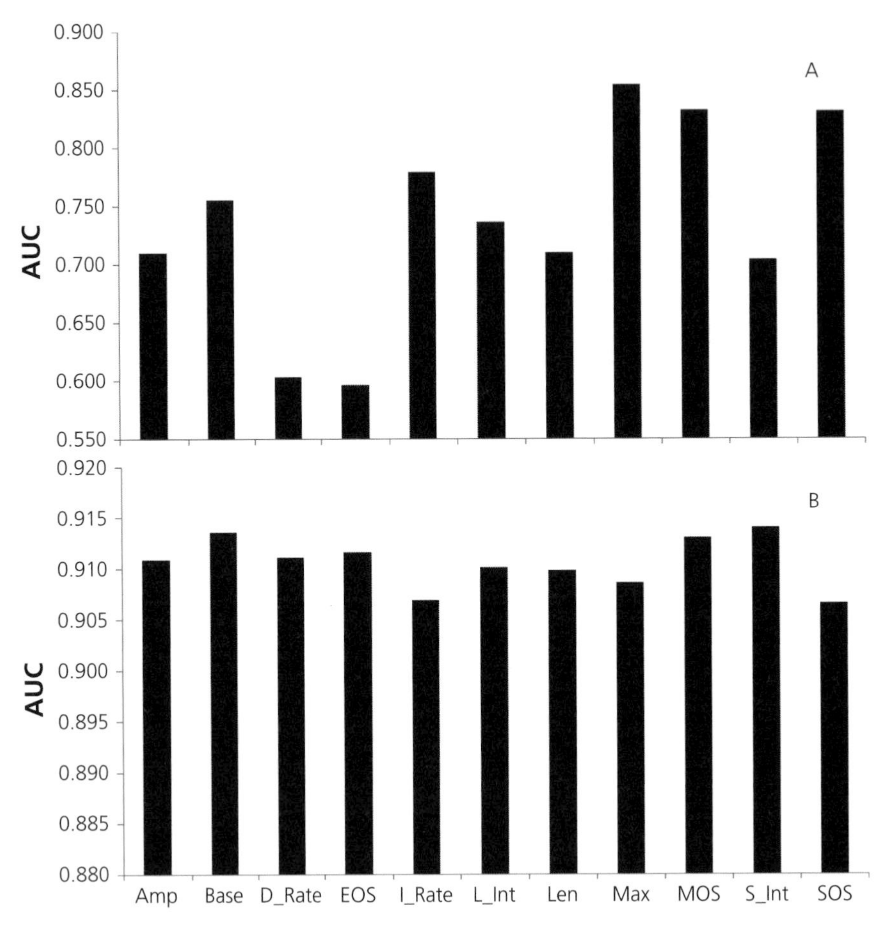

Figure 15.6 Mean values in area under the receiver operating characteristic curve (AUC) of panda habitat models (A) with only one of the 11 phenology metrics and (B) with all but one metric. The phenology metrics used were the following: Amp, amplitude; Base, Base level; D_Rate, decrease rate; EOS, date of the end of the season; I_Rate, increase rate; L_Int, large integral; Len, length of the season; Max, maximum level; MOS, date of the middle of the season; S_Int, small integral; and SOS, date of the start of the season (Tuanmu et al., 2010, 2011).

assessment of the optimal locations for new nature reserves or additions to current nature reserves to increase habitat connectivity across the entire geographic range of the species (Viña et al., 2010). Such additions could potentially benefit the entire population of wild giant pandas. However, the final selection of areas for conservation also needs to explicitly consider human factors (e.g. land use, land tenure, and infrastructure) in order to reduce potential socioeconomic impacts. Therefore, fine-scale socioeconomic studies, such as those performed in Wolong (An et al., 2002, 2003, 2005, An and Liu, 2010, He et al., 2009, Linderman et al., 2005, 2006), need to be applied across the entire geographic range of the pandas. Only then can feasible and efficient conservation measures be successfully implemented. Results from broad-scale evaluations such as the ones developed here provide the biophysical foundation to help delineate areas for further socioeconomic feasibility assessments.

The proportion of habitat inside nature reserves increased with the value of the HSI in all mountain regions except Greater Xiangling (Figure 15.9). This result implies that the current nature reserves have been preferentially located in areas with high HSI scores, which makes them isolated and thus prone to local extinctions. In fact, the third national giant panda survey (field data collected in 2000–2001)

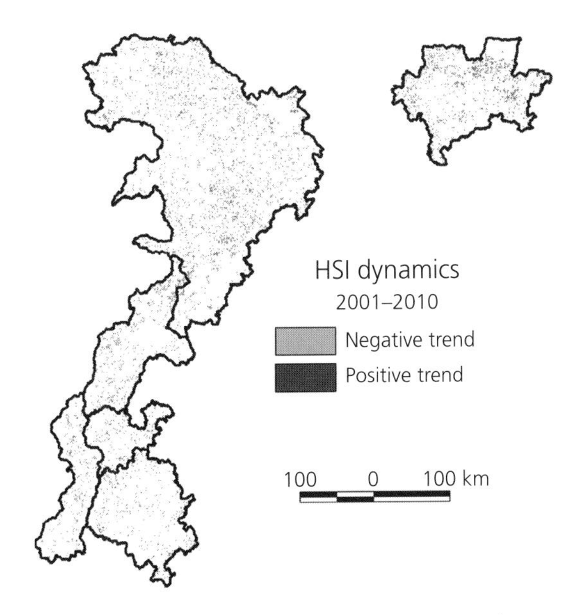

Figure 15.7 Spatial distribution of the areas exhibiting significant ($p < 0.05$) positive and negative trends in habitat suitability index (HSI) values between 2001 and 2010, in the mountain regions comprising the current geographic range of the giant panda. For color version see color plates. [PLATE 8]

used the actual presence of giant pandas as one of the major criteria to define an area as habitat and to establish new nature reserves (State Forestry Administration, 2006). This practice overlooked the availability of suitable habitat elsewhere. Such a procedure also undermined the conservation of corridors that allow for genetic exchange among nature reserves, although some studies have suggested their inclusion (Loucks et al., 2001, Viña et al., 2010, Xu et al., 2006, Yin et al., 2006). It also ignored the potential for conserving the habitat of many other endangered vertebrates that share their habitat with the pandas (Xu et al., 2014). One example is the red panda (*Ailurus fulgens*), which also depends on understory bamboo as a staple food. Therefore, efforts to mitigate the reduction and fragmentation of giant panda habitat can also promote the conservation of other endangered species (Xu et al., 2014).

To analyze the degree of habitat connectivity within each mountain region, we calculated clumpiness, a landscape metric related to habitat aggregation and independent of habitat area (Neel et al.,

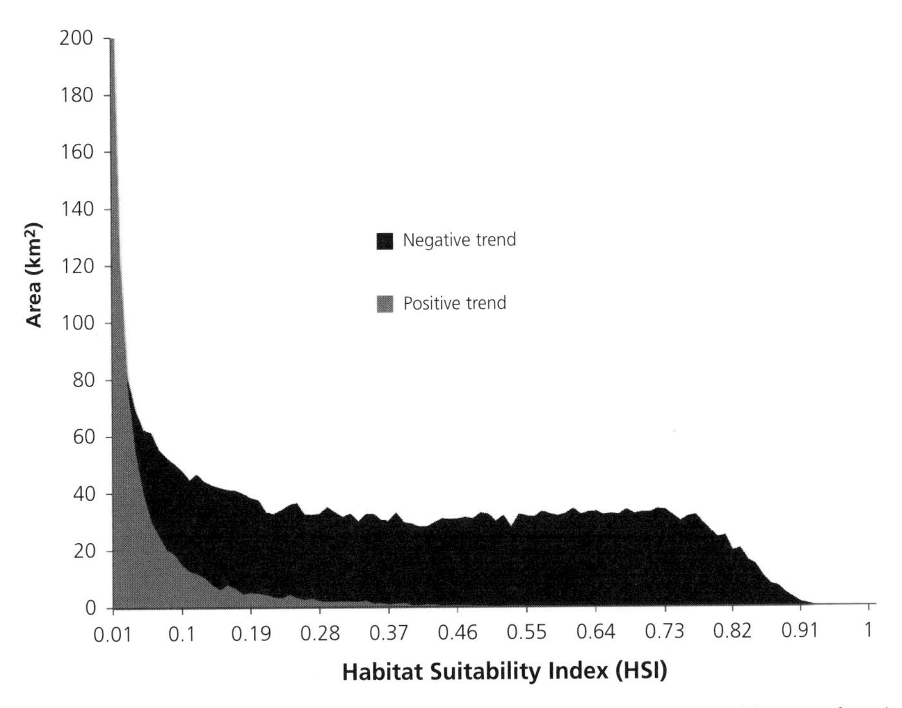

Figure 15.8 Frequency distribution (expressed as area) of habitat suitability index (HSI) values among areas exhibiting significant ($p < 0.05$) positive and negative trends in HSI values between 2001 and 2010.

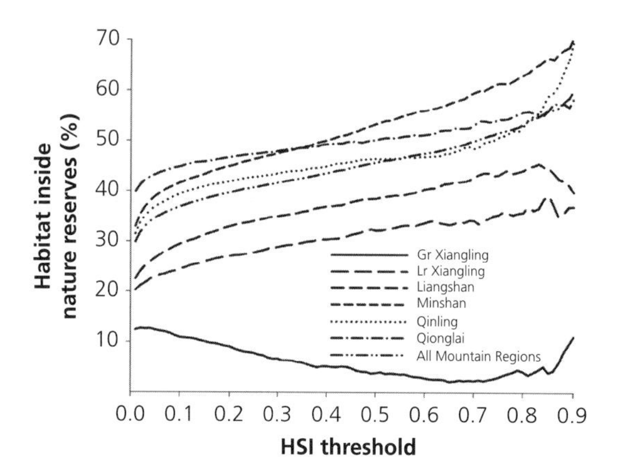

Figure 15.9 Proportion (expressed in percent) of giant panda habitat inside current nature reserves under different cumulative habitat suitability index (HSI) thresholds.

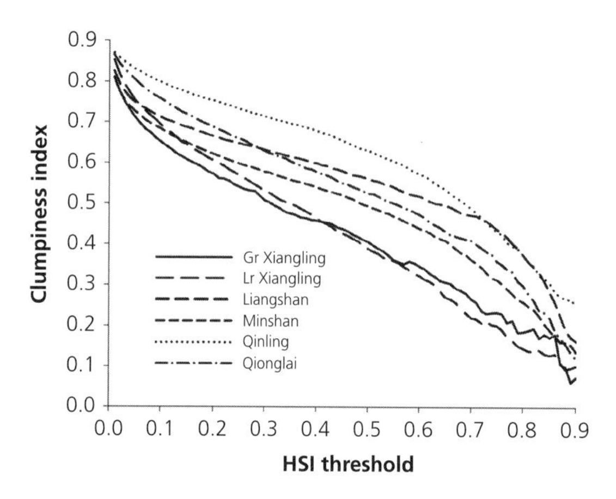

Figure 15.10 Degree of habitat connectivity under different cumulative habitat suitability index (HSI) thresholds, as measured by the clumpiness index, a landscape metric related to habitat aggregation and independent of habitat area (Neel et al., 2004).

2004). The calculations were conducted at different cumulative habitat suitability index thresholds using FRAGSTATS (McGarigal et al., 2002). Through this analysis, we found a monotonic reduction in the amount of habitat connectivity with a cumulative increase in the HSI threshold (Figure 15.10). This finding indicates that highly suitable habitat areas tend to be isolated and suggests that current panda populations may also be subjected to isolation and thus are more prone to local extinctions. Approximately 180 000 km^2 of land (~2% of China's territory) will be added to China's nature reserve system within the next 15 years (Liu and Raven, 2010). Thus, opportunities exist to increase the size of current nature reserves so they include more suitable habitat areas, as well as to create additional nature reserves and corridors that connect presently isolated habitat areas. Such approaches will benefit not only the pandas, but also the numerous other species (some endangered) that coexist with them (Xu et al., 2014).

15.5 Lessons and rewards of scaling up from local to regional scales

This chapter has shown the successful implementation of procedures developed and tested in the Wolong Nature Reserve for analyzing the spatiotemporal dynamics of giant panda habitat across the entire geographic range of the species. Such analyses can be used for assessing the effects of human activities (including conservation actions) on the entire population of a target species. A major focus of such a perspective has been on the ecological connectivity—the degree to which the movement of species is either facilitated or impeded depending on the spatial configuration of suitable habitat (Taylor et al., 1993). Connectivity is an increasingly important issue in ecology and wildlife conservation (Crooks and Sanjayan, 2006). This is important since the long-term survival of a species depends on its habitat being not only suitable and abundant, but also easily reachable, i.e., well connected through dispersal (Crooks and Sanjayan, 2006, Damschen et al., 2006, Kareiva and Wennergren, 1995, Ricketts, 2001, Taylor et al., 1993). It is also important for conservation practice because areas suitable for enhancing connectivity (e.g., corridors) can be defined, identified, and mapped as conservation targets for enhancing ecological connectivity (McRae et al., 2012). Therefore, the procedures developed in Wolong are significant from both theoretical and practical perspectives.

The pace of current environmental degradation worldwide is exacerbating the need for conservation actions across broad geographic regions. Analyses performed in Wolong suggest that the implementation of NFCP has had significant positive effects on giant panda habitat suitability, particularly when local people are actively involved in monitoring (Chapter 7). However, across the entire geographic range this is less clear, since most of the region exhibited no significant changes during the first decade of the twenty-first century. Therefore, despite NFCP and GTGP and their associated gains in forest cover across the entire geographic range of the species (Figure 15.3B), the observed forest-cover gains may not have translated into a widespread increase in suitable giant panda habitat. Nevertheless, as some areas are experiencing positive trends in habitat suitability, there is hope that the widespread implementation of conservation policies will eventually translate into giant panda habitat recovery. Thus, the continuation of habitat monitoring across the entire geographic range of the species using the procedures described here will be crucial in assessing the effects of conservation policies over the long run.

Given proper modifications, the procedures developed for panda habitat in Wolong may also be useful for assessing the habitat conditions of many other endangered species around the world. Results from these analyses would be helpful in proposing conservation strategies for endangered species using a more holistic conservation practice targeting entire geographic ranges. Example strategies include land-use zones, establishment of new nature reserves and corridors, and extensions to current reserves.

15.6 Summary

To demonstrate the utility of methods developed at local scales in the analyses at regional scales, we applied novel panda habitat models, developed for our model system of Wolong Nature Reserve, to the entire geographic range of giant pandas that spans six mountain regions in three provinces of China (Gansu, Shaanxi, and Sichuan). Our analyses indicated that the quantity and quality of panda habitat vary greatly across the geographic range. While forests occupy about one-third of the range, only ~17% constitutes suitable habitat for the pandas because many forests lack suitable understory bamboo. A conspicuous increase in forest cover was observed during the first decade of the twenty-first century, in part due to national conservation policy implementation. But this forest recovery has not fully translated into a widespread panda habitat recovery. The upscaling procedures allowed for assessment of the connectivity of suitable habitat and identification of areas that require further conservation actions. We found that only ~40% of the suitable habitat was located inside the 63 existing nature reserves, as of 2010. Our approach showed that current nature reserves need to be expanded and to be better connected through creating dispersal corridors. Upscaling can also help promote the conservation of other species, because panda habitat comprises several types of forest ecosystems that are home to thousands of other animal and plant species. Finally, with proper modifications, the procedures developed in Wolong will also be useful for assessing the spatiotemporal habitat dynamics of numerous species in many other parts of the world.

References

An, L., Linderman, M., Qi, J., et al. (2005) Exploring complexity in a human-environment system: an agent-based spatial model for multidisciplinary and multiscale integration. *Annals of the Association of American Geographers*, **95**, 54–79.

An, L. and Liu, J. (2010) Long-term effects of family planning and other determinants of fertility on population and environment: agent-based modeling evidence from Wolong Nature Reserve, China. *Population and Environment*, **31**, 427–59.

An, L., Lupi, F., Liu, J., et al. (2002) Modeling the choice to switch from fuelwood to electricity: implications for giant panda habitat conservation. *Ecological Economics*, **42**, 445–57.

An, L., Mertig, A.G., and Liu, J. (2003) Adolescents leaving parental home: psychosocial correlates and implications for conservation. *Population and Environment*, **24**, 415–44.

An, L., Zvoleff, A., Liu, J., and Axinn, W. (2014) Agent-based modeling in coupled human and natural systems (CHANS): lessons from a comparative analysis. *Annals of the Association of American Geographers*, **104**, 723–45.

Bearer, S., Linderman, M., Huang, J., et al. (2008) Effects of fuelwood collection and timber harvesting on giant panda habitat use. *Biological Conservation*, **141**, 385–93.

Berry, P.A.M., Garlick, J.D., and Smith, R.G. (2007) Near-global validation of the SRTM DEM using satellite radar altimetry. *Remote Sensing of Environment*, **106**, 17–27.

Chen, X., Lupi, F., An, L., et al. (2011) Agent-based modeling of the effects of social norms on enrollment in payments for ecosystem services. *Ecological Modelling*, **229**, 16–24.

Chen, X., Lupi, F., He, G., et al. (2009) Factors affecting land reconversion plans following a payment for ecosystem service program. *Biological Conservation*, **142**, 1740–47.

Chen, X., Lupi, F., Viña, A., et al. (2010) Using cost-effective targeting to enhance the efficiency of conservation investments in payments for ecosystem services. *Conservation Biology*, **24**, 1469–78.

Crooks, K.R. and Sanjayan, M.A. (2006) *Connectivity Conservation*. Cambridge University Press, Cambridge, UK.

Damschen, E.I., Haddad, N.M., Orrock, J.L., et al. (2006) Corridors increase plant species richness at large scales. *Science*, **313**, 1284–86.

Gitelson, A.A. (2004) Wide dynamic range vegetation index for remote quantification of biophysical characteristics of vegetation. *Journal of Plant Physiology*, **161**, 165–73.

Hanley, J.A. and McNeil, B.J. (1982) The meaning and use of the area under a receiver operating characteristic (ROC) curve. *Radiology*, **143**, 29–36.

Hansen, M.C., DeFries, R.S., Townshend, J.R.G., et al. (2003) Global percent tree cover at a spatial resolution of 500 meters: first results of the MODIS vegetation continuous fields algorithm. *Earth Interactions*, **7**, 1–15.

He, G., Chen, X., Bearer, S., et al. (2009) Spatial and temporal patterns of fuel collection in Wolong Nature Reserve: implications for panda conservation. *Landscape and Urban Planning*, **92**, 1–9.

Henebry, G.M., Viña, A., and Gitelson, A.A. (2004) The Wide Dynamic Range Vegetation Index and its potential utility for Gap Analysis. *GAP Analysis Program Bulletin*, **12**, 50–56.

Hu, J. and Wei, F. (2004) Comparative ecology of giant pandas in five mountain ranges of their distribution in China. In D. Lindburg and K. Baragona, eds, *Giant Pandas: Biology and Conservation*, pp. 137–47. University of California Press, Berkeley, CA.

Hull, V., Tuanmu, M.N., and Liu, J. (2015) Synthesis of human-nature feedbacks. *Ecology and Society* **20**(3), 17.

Hull, V., Xu, W., Liu, W., et al. (2011) Evaluating the efficacy of zoning designations for protected area management. *Biological Conservation*, **144**, 3028–37.

Johnson, K.G., Schaller, G.B., and Hu, J. (1988) Responses of giant pandas to a bamboo die-off. *National Geographic Research*, **4**, 161–77.

Kareiva, P. and Wennergren, U. (1995) Connecting landscape patterns to ecosystem and population processes. *Nature*, **373**, 299–302.

Li, C. (1997) *A Study of Staple Food Bamboo for the Giant Panda*. Guizhou Scientific Publishing House, Guiyang, China (in Chinese).

Li, Y., Viña, A., Yang, W., et al. (2013) Effects of conservation policies on forest cover change in giant panda habitat regions, China. *Land Use Policy*, **33**, 42–53.

Linderman, M., Bearer, S., An, L., et al. (2006) Interactive effects of natural and human disturbances on vegetation dynamics across landscapes. *Ecological Applications*, **16**, 452–63.

Linderman, M.A., An, L., Bearer, S., et al. (2005) Modeling the spatio-temporal dynamics and interactions of households, landscape, and giant panda habitat. *Ecological Modelling*, **183**, 47–65.

Linderman, M.A., Liu, J., Qi, J., et al. (2004) Using artificial neural networks to map the spatial distribution of understorey bamboo from remote sensing data. *International Journal of Remote Sensing*, **25**, 1685–700.

Liu, J., Dietz, T., Carpenter, S.R., et al. (2007a) Complexity of coupled human and natural systems. *Science*, **317**, 1513–16.

Liu, J., Dietz, T., Carpenter, S.R., et al. (2007b) Coupled human and natural systems. *Ambio*, **36**, 639–49.

Liu, J., Li, S., Ouyang, Z., et al. (2008) Ecological and socioeconomic effects of China's policies for ecosystem services. *Proceedings of the National Academy of Sciences of the United States of America*, **105**, 9477–82.

Liu, J., Linderman, M., Ouyang, Z., et al. (2001) Ecological degradation in protected areas: the case of Wolong Nature Reserve for giant pandas. *Science*, **292**, 98–101.

Liu, J., Ouyang, Z., Taylor, W.W., et al. (1999) A framework for evaluating effects of human factors on wildlife habitats: the case of giant pandas. *Conservation Biology*, **13**, 1360–70.

Liu, J., Ouyang, Z., Yang, W., et al. (2013) Evaluation of ecosystem service policies from biophysical and social perspectives: the case of China. In S.A. Levin, ed., *Encyclopedia of Biodiversity* (second edition), **vol. 3**, pp. 372–84. Academic Press, Waltham, MA.

Liu, J. and Raven, P.H. (2010) China's environmental challenges and implications for the world. *Critical Reviews in Environmental Science and Technology*, **40**, 823–51.

Liu, J. and Viña, A. (2014) Pandas, plants, and people. *Annals of the Missouri Botanical Garden*, **100**, 108–25.

Loss, S.R., Terwilliger, L.A., and Peterson, A.C. (2011) Assisted colonization: integrating conservation strategies in the face of climate change. *Biological Conservation*, **144**, 92–100.

Loucks, C.J., Lü, Z., Dinerstein, E., et al. (2001) Giant pandas in a changing landscape. *Science*, **294**, 1465–65.

Mackinnon, J. and De Wulf, R. (1994) Designing protected areas for giant pandas in China. In R.I. Miller, ed., *Mapping the Diversity of Nature*, pp. 127–42. Chapman & Hall, London, UK.

McGarigal, K., Cushman, S., Neel, M., and Ene, E. (2002) *FRAGSTATS Version 3: spatial pattern analysis program for categorical maps*. www.umass.edu/landeco/research/fragstats/fragstats.html, Amherst, Massachusetts.

McRae, B.H., Hall, S.A., Beier, P., and Theobald, D.M. (2012) Where to restore ecological connectivity? Detecting barriers and quantifying restoration benefits. *PLoS ONE*, **7**, e52604.

Minor, E.S. and Lookingbill, T.R. (2010) A multiscale network analysis of protected-area connectivity for mammals in the United States. *Conservation Biology*, **24**, 1549–58.

Myers, N., Mittermeier, R.A., Mittermeier, C.G., et al. (2000) Biodiversity hotspots for conservation priorities. *Nature*, **403**, 853–58.

NASA Socioeconomic Data and Applications Center (2011) *Global Rural-Urban Mapping Project, Version 1 (GRUMPv1): population count grid*. http://sedac.ciesin.columbia.edu/theme/population.

Neel, M.C., McGarigal, K., and Cushman, S.A. (2004) Behavior of class-level landscape metrics across gradients of class aggregation and area. *Landscape Ecology*, **19**, 435–55.

Noss, R.F. (2007) Focal species for determining connectivity requirements in conservation planning. In D.B. Lindenmayer and R.J. Hobbs, eds, *Managing and Designing Landscapes for Conservation: moving from perspectives to principles*, pp. 263–79. Blackwell Publishing, Ltd., Oxford, UK.

Pan, W., Lü, Z., Wang, D., and Wang, H. (2001) *A Lasting Chance for Survival*. Peking University Press, Beijing, China (in Chinese).

Phillips, S.J., Anderson, R.P., and Schapire, R.E. (2006) Maximum entropy modeling of species geographic distributions. *Ecological Modelling*, **190**, 231–59.

Provincial Bureau of Statistics (2001) *Sichuan, Shaanxi, and Gansu Statistical Yearbooks* (in Chinese).

Reid, D.G. and Gong, J. (1999) Giant panda conservation action plan. In C. Servheen, S. Herrero and B. Peyton, eds, *Bears Status Survey and Conservation Action Plan*, pp. 241–45. IUCN/SSC Bear and Polar Bear Specialist Groups, Gland, Switzerland.

Reid, D.G. and Hu, J. (1991) Giant panda selection between *Bashania fangiana* bamboo habitats in Wolong Reserve, Sichuan, China. *Journal of Applied Ecology*, **28**, 228–43.

Reid, D.G., Hu, J., Sai, D., et al. (1989) Giant panda *Ailuropoda melanoleuca* behavior and carrying-capacity following a bamboo die-off. *Biological Conservation*, **49**, 85–104.

Ricketts, T.H. (2001) The matrix matters: effective isolation in fragmented landscapes. *American Naturalist*, **158**, 87–99.

Sanderson, E.W., Redford, K.H., Chetkiewicz, C.L.B., et al. (2002) Planning to save a species: the jaguar as a model. *Conservation Biology*, **16**, 58–72.

Schaller, G.B., Hu, J., Pan, W., and Zhu, J. (1985) *The Giant Pandas of Wolong*. University of Chicago Press, Chicago, IL.

Spring, D., Baum, J., MacNally, R., et al. (2010) Building a regionally connected reserve network in a changing and uncertain world. *Conservation Biology*, **24**, 691–700.

State Council Information Office of China (2015) *Press Conference on the Fourth National Panda Survey Results*. http://www.scio.gov.cn/xwfbh/gbwxwfbh/fbh/Document/1395514/1395514.htm (in Chinese).

State Forestry Administration (2006) *Report of the Third National Giant Panda Census*. Science Publishing House, Beijing, China (in Chinese).

Taylor, A.H. and Qin, Z. (1993) Bamboo regeneration after flowering in the Wolong Giant Panda Reserve, China. *Biological Conservation*, **63**, 231–34.

Taylor, P.D., Fahrig, L., Henein, K., and Merriam, G. (1993) Connectivity is a vital element of landscape structure. *Oikos*, **68**, 571–73.

Thorbjarnarson, J., Mazzotti, F., Sanderson, E., et al. (2006) Regional habitat conservation priorities for the American crocodile. *Biological Conservation*, **128**, 25–36.

Tuanmu, M.N., Viña, A., Bearer, S., et al. (2010) Mapping understory vegetation using phenological characteristics derived from remotely sensed data. *Remote Sensing of Environment*, **114**, 1833–44.

Tuanmu, M.N., Viña, A., Roloff, G.J., et al. (2011) Temporal transferability of wildlife habitat models: implications for habitat monitoring. *Journal of Biogeography*, **38**, 1510–23.

Viña, A., Bearer, S., Chen, X., et al. (2007) Temporal changes in giant panda habitat connectivity across boundaries of Wolong Nature Reserve, China. *Ecological Applications*, **17**, 1019–30.

Viña, A., Bearer, S., Zhang, H., et al. (2008) Evaluating MODIS data for mapping wildlife habitat distribution. *Remote Sensing of Environment*, **112**, 2160–69.

Viña, A., Chen, X.D., McConnell, W.J., et al. (2011) Effects of natural disasters on conservation policies: the case of the 2008 Wenchuan Earthquake, China. *Ambio*, **40**, 274–84.

Viña, A. and Gitelson, A.A. (2005) New developments in the remote estimation of the fraction of absorbed photosynthetically active radiation in crops. *Geophysical Research Letters*, **32**, L17403.

Viña, A., Henebry, G.M., and Gitelson, A.A. (2004) Satellite monitoring of vegetation dynamics: sensitivity enhancement by the wide dynamic range vegetation index. *Geophysical Research Letters*, **31**, L04503.

Viña, A., Tuanmu, M.-N., Xu, W., et al. (2010) Range-wide analysis of wildlife habitat: implications for conservation. *Biological Conservation*, **143**, 1960–63.

Xu, W., Ouyang, Z., Viña, A., et al. (2006) Designing a conservation plan for protecting the habitat for giant pandas in the Qionglai mountain range, China. *Diversity and Distributions*, **12**, 610–19.

Xu, W., Viña, A., Qi, Z., et al. (2014) Evaluating conservation effectiveness of nature reserves established for surrogate species: case of a giant panda nature reserve in the Qinling Mountains, China. *Chinese Geographical Science*, **24**, 60–70.

Yang, W., Liu, W., Viña, A., et al. (2013) Performance and prospects of payments for ecosystem services programs: evidence from China. *Journal of Environmental Management*, **127**, 86–95.

Yin, K., Xie, Y., and Wu, N. (2006) Corridor connecting giant panda habitats from north to south in the Min Mountains, Sichuan, China. *Integrative Zoology*, **1**, 170–78.

CHAPTER 16

Cross-Site Synthesis of Complexity in Coupled Human and Natural Systems

Neil Carter, Li An, and Jianguo Liu

16.1 Introduction

Coupled human and natural systems (CHANS) are integrated systems in which human and natural components, including wildlife, interact with each other (Liu et al., 2007a; Chapter 2). Previous chapters (Chapters 3–14) have extensively explored complex human–nature interactions in a single coupled system—Wolong Nature Reserve—from numerous angles and by integrating information from multiple disciplines. Nonetheless, to reach even broader and more generalizable insights about the dynamics of coupled systems, findings from site-specific coupled system studies in different ecological, socioeconomic, political, demographic, and cultural settings should be synthesized (Acevedo et al., 2008, Carter et al., 2014a, Liu et al., 2007a, Parker et al., 2003, Rindfuss et al., 2008, Turner et al., 2003). Such cross-site syntheses can facilitate knowledge exchange among researchers, managers, policy makers, and local residents, and enhance their capacity to address conservation and sustainability challenges in coupled systems around the world.

As such, in this chapter we apply what we have learned about the coupled system in Wolong Nature Reserve (hereafter Wolong) to understand another complex coupled system in Chitwan National Park (hereafter Chitwan) in Nepal. This application is facilitated by the fact that studies on human–wildlife interactions in Chitwan were inspired by those in Wolong (Carter et al., 2014a). We chose to investigate Chitwan because it has several commonalities with Wolong but also provides a different local context to help illustrate the diversity of coupled systems. Like Wolong, Chitwan is a "flagship" protected area within a global biodiversity hotspot (Myers et al., 2000). Similar to the way in which Wolong supports the giant panda (*Ailuropoda melanoleuca*), Chitwan supports another important wildlife population—the tiger (*Panthera tigris*). Both species are globally endangered conservation icons. Like Wolong, long-term empirical, interdisciplinary data exist for Chitwan, giving us a more holistic perspective of the various interconnections between components of coupled systems. The interactions between people and nature, institutional arrangements, and socioeconomic and demographic changes at both sites are also very similar to those in many other coupled systems around the world (Frost and Bond, 2008). Here, we take a comparative approach to discussing several key features of coupled systems occurring across both sites. Many important patterns and processes observed in Chitwan would have been missed had an integrated approach not been used. Further, we highlight several lessons learned that may be useful for fostering human–wildlife coexistence not only in China and Nepal but also in many other places (Chapron et al., 2014).

16.2 The homes of two wildlife conservation icons

Information about the home of giant pandas, Wolong Nature Reserve, is provided in Chapter 3 and many other previous chapters. Below we briefly describe the home of the tiger.

Pandas and People. Edited by Jianguo Liu, Vanessa Hull, Wu Yang, Andrés Viña, Xiaodong Chen, Zhiyun Ouyang, and Hemin Zhang. © Oxford University Press 2016. Published 2016 by Oxford University Press.

Chitwan National Park (Figure 16.1), established in 1973, comprises an area of 1,000 km^2 and is situated in Chitwan District at the base of the Himalayas in Nepal. It was initially established to protect a rapidly diminishing population of the one-horned rhino, but it is now also a globally important region for the conservation of the tiger (Sanderson et al., 2006). The park has one of the largest wild populations of tigers (~125 adults) in South Asia (Karki et al., 2013). The park also affords protection for many other endangered species such as the gharial crocodile (*Gavialis gangeticus*), gaur (*Bos gaurus*), and Indian rock python (*Python molurus*). Chitwan's climate is subtropical. A summer monsoon occurs from mid-June to late September, followed by a cool, dry winter. Average annual rainfall is 240 cm, 90% of which falls during the summer monsoon. Temperatures peak (maximum 38°C) during the monsoon and drop to a low of 6°C afterward (October to January; Laurie, 1982). Chitwan ranges in elevation from 150 m to 815 m. Natural forests include moist deciduous forests dominated by Sal (*Shorea robusta*), with some mixed deciduous/evergreen forests mainly along river banks (i.e., riverine). Other natural land-cover types include grasslands (e.g., wooded grasslands, phantas, and floodplain grasslands; Carter et al., 2013).

Like Wolong, local livelihoods in Chitwan are primarily based on subsistence agriculture with dependence on forest resources (Table 16.1). Unlike Wolong, however, no one lives inside Chitwan National Park. In 2011, the human population living adjacent to the park was approximately 550 000 local residents in over 130 000 households (Nepal Central Bureau of Statistics, 2012). Many of those residents adjacent to the park use resources inside the park. From the perspective of resource use, there is little difference from residents inside Wolong. As in Wolong, household activities such as forest conversion to cropland and livestock grazing

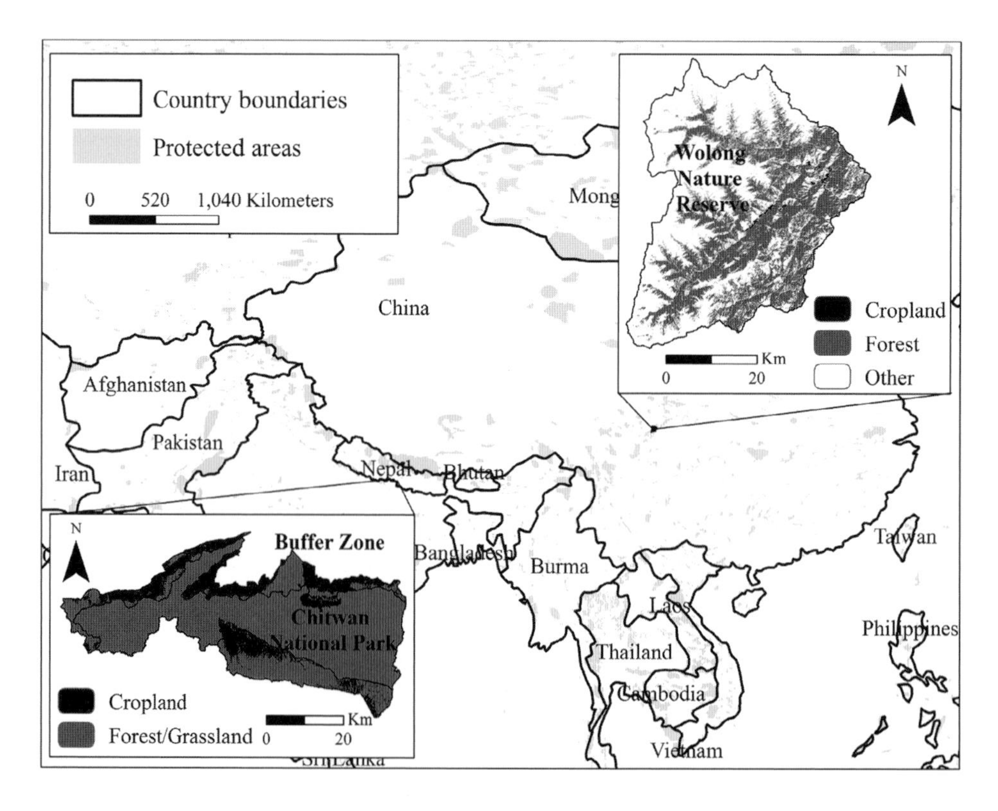

Figure 16.1 Locations and major land-cover types of the two focal systems: Wolong Nature Reserve in China and Chitwan National Park in Nepal.

Table 16.1 Major features of key components in coupled human and natural systems in Wolong, China, and Chitwan, Nepal.

Components	Major features	Wolong (China)	Chitwan (Nepal)
Local residents	Main crops	Cabbage, maize, potatoes, turnips	Rice, maize, wheat, mustard, lentils
	Main livestock	Cattle, goats, horses, yaks, pigs, chickens	Buffalo, goats, chickens
	Timber and non-timber forest products	Timber, fuelwood, fodder, medicinal herbs	Timber, fuelwood, fodder, thatch, medicinal herbs
	Sources of off-farm income	Tourism, wage labor, labor migration, commercial businesses	Tourism, wage labor, labor migration, commercial businesses
Forests	Major land-cover classes	Mainly coniferous forest, deciduous broad leaf forest, mixed deciduous-coniferous forest, grassland above the tree line	Deciduous forest (e.g., Sal forest), mixed deciduous and evergreen forest (e.g., riverine forest), grassland (mostly in river banks)
Wildlife	Endangered charismatic megafauna	Giant panda, golden monkey, takin, snow leopard	Bengal tiger, one-horned rhino, wild elephant, gharial crocodile, gaur, leopard
Policies	Conservation policies	Resource extraction bans in nature reserve, Natural Forest Conservation Program (collective forest monitoring), Grain to Green Program	Resource extraction bans in national park, grass-cutting program, community forest comanagement in buffer zone
Contextual factors	Macrolevel socioeconomics	Opportunities for tourism, off-farm jobs, access to markets, infrastructure	Opportunities for tourism, off-farm jobs, access to markets, infrastructure

in forests negatively affect tiger habitat and behavior (Carter et al., 2013, 2012a). Hunting pandas and tigers in both sites, now illegal, was more common in the past (Hu, 1989, Nowell, 2012). Although diminished, tiger hunting in Chitwan continues to be a constant threat given their small population size (Chapron et al., 2008). Tigers, like pandas, increasingly provide economic benefits to local residents through tourism rather than through hunting (Liu et al., 2012, Spiteri and Nepal, 2008). Tourism benefits thus provide a rationale for panda and tiger conservation. However, residents in both Wolong and Chitwan can incur the indirect costs of conservation. Examples include constraints on resource use and increased crop predation by growing numbers of other wild animals such as wild boars. The latter occurs because conservation and restoration of habitats for pandas and tigers are also good for many other wild animals (Liu et al., 1999a). In addition, residents in Chitwan can incur significant direct costs associated with tiger conservation, such as tiger attacks on livestock and people (Gurung et al., 2008). It is clear that fostering long-term coexistence in both sites necessitates a holistic understanding of how people and wildlife are interconnected.

16.3 The complexity of coupled systems

The components of coupled systems (see examples in Table 16.1) form complex webs of interactions. As a result, coupled systems are characterized by features of complex systems. Examples include reciprocal interactions and feedback loops, non-linear relations and thresholds, surprises, heterogeneity, telecoupling (Chapters 2 and 17), vulnerability, and time lags and legacy effects (Liu et al., 2007a, b, 2013a; Chapters 2, 13, and 17). In this section, we integrate findings across the two coupled systems with respect to each of these features (examples shown in Table 16.2), with an emphasis on impacts on pandas and tigers.

16.3.1 Reciprocal interactions and feedback loops

In coupled systems, people and nature interact reciprocally. As such, the effects of human activities on forests and wildlife often generate feedback loops that affect humans and their activities (Liu et al., 2007b). For example, the growth and expansion of natural resource-dependent human

Table 16.2 Examples of complexity features for coupled human and natural systems in Wolong Nature Reserve in China and Chitwan National Park in Nepal. For definitions of each feature, see Table 2.1.

Complexity features	Examples in Wolong (China)	Examples in Chitwan (Nepal)
Reciprocal interactions and feedback loops	People collect fuelwood → degrade panda habitat → people go farther to collect fuelwood → increases area of panda habitat loss.	Forest conservation policies → more tigers → more tiger–human conflicts → possibly lose local support of conservation policies.
Non-linearity and thresholds	Collection of fuelwood up to 1,800 m from household decreases area of panda habitat, though impact is negligible beyond 1,800 m.	Tolerance to impacts from tigers (e.g., livestock depredation and attacks on people) has thresholds, beyond which people may kill tigers.
Surprises	Loss of panda habitat increased after the reserve was established due to synergistic effects of factors such as human population growth, household proliferation, and increased tourism.	Grazing restrictions increased tiger prey numbers in park's buffer zone but also likely increased negative human–tiger interactions because people more frequently enter forests to collect fodder for stall-fed livestock.
Heterogeneity	Household locations and resource consumption activities vary in different parts of the reserve.	Household locations, being outside park, and resource consumption activities differ across space.
Embedment and telecoupling	Local residents migrate out of Wolong to find employment in other areas, and often send remittances back home.	Chitwan is located on a transit route and is a point of origin for poached tigers, whose parts are sold on the international black market.
Vulnerability	Earthquake in 2008 caused severe landslides, disrupted agricultural trade and tourism, and reduced panda habitat.	Nepal civil war (1996–2006) displaced local residents and increased poaching of tigers.
Time lags and legacy effects	Past logging locations affect current forest type and panda habitat quality.	Past migration policies affect spatial patterns of human activities (e.g., land use) with respect to tiger habitat.

communities in both sites are strongly linked to declines in panda and tiger habitats and population sizes (Axinn et al., 2010, Axinn and Ghimire, 2011, Chen et al., 2010, Matthews et al., 2000, Tuanmu et al., 2011). However, as forests and grasslands shrink, they become more distant from households. This spatial shift makes the extraction of timber and non-timber forest products more difficult and more time-consuming (Axinn and Ghimire, 2011, He et al., 2009). In Chitwan, such changes reciprocally influence human population parameters, including childbearing and migration, which in turn, exert different effects on wildlife habitat. For example, increasing costs and time in collecting forest products are linked to larger households. Couples facing such challenges have more children to help collect forest resources to support the household (Biddlecom et al., 2005, Liu et al., 1999b, c). Each additional birth places more pressure on vegetation and thus wildlife habitat (Axinn and Ghimire, 2011, Linderman et al., 2006). This phenomenon is an example of a positive feedback loop. In contrast, less access to fuelwood and fodder increases the difficulties of an agricultural lifestyle. Such challenges may cause people to find a different means of living such as ecotourism or move somewhere else for jobs through rural–urban migration (Chen et al., 2012, Massey et al., 2010). These trends have occurred in both sites. This phenomenon is an example of a negative feedback loop.

Policies are key feedback mechanisms (Chapter 13). In Wolong and Chitwan, degradation of forests and wildlife habitat prompted policy makers to develop and implement new policies (Adhikari, 2002, Liu et al., 2001, Nagendra et al., 2008, Viña et al., 2007). Policies can change human activities, both directly and indirectly. Direct changes may include preventing timber extraction and fuelwood collection or spurring tree planting. Indirect changes may include incentives to use alternatives to fuelwood, such as electricity and natural gas (Entwisle et al., 1996, Homewood et al., 2001, Li et al., 2013, Liu et al., 2005). For instance, to counter the loss of panda habitat, and to restore forests for other benefits (e.g., reduction in soil erosion), two major national conservation programs began in Wolong in

2000 and 2001 (see also Chapter 13). The Grain to Green Program (GTGP) provides cash, grain, and tree seedlings to farmers if they return cropland to forest (Chen et al., 2009, Liu et al., 2008). The Natural Forest Conservation Program (NFCP) bans logging and provides cash for households and communities to monitor forests to prevent illegal harvesting (Chen et al., 2014; see also Chapter 13). The implementation of these conservation policies has reversed a more than 30-year trend of panda habitat degradation in the reserve (Viña et al., 2007, 2011; see also Chapter 7).

Similarly, to reduce local resentment toward the exclusion policies of Chitwan National Park, a "grass-cutting" program was initiated in 1976. This program allows local residents to legally enter the park for a limited number of days each year to collect thatch grass, reeds, rope bark, and rope grass (Stræde and Helles, 2000). Furthermore, to mitigate human pressure on Chitwan's forests, a buffer zone (~750 km^2) surrounding the park was created in 1996 with the goal of restoring ecosystem integrity while also improving human livelihoods. Approximately 30–50% of the park's annual revenue from tourism must be invested in the buffer zone for community development programs (Government of Nepal, 1993). Examples include alternative income opportunities and infrastructure improvement. Furthermore, livestock grazing was prohibited in the buffer zone forests. In addition, resource management responsibility for several forest tracts was devolved to local community user groups (Gurung et al., 2008, Nagendra et al., 2005). These forest conservation policies likely enabled forests outside Chitwan National Park to support greater densities of wild prey animals and provide better coverage for tigers (Carter et al., 2013). As a result, tiger habitat quality improved from 1999 to 2009 after the implementation of those policies (Figure 16.2).

However, as the positive effects of conservation policies in both coupled systems become manifest, unanticipated feedbacks such as human–wildlife conflicts are also emerging. For example, the increase in forest cover in Wolong has caused an increase in native wildlife that raid cropland (Yang et al., 2013). In Chitwan, forest recovery is supporting greater numbers of tigers (Barlow et al., 2009). This increase in tigers has resulted in an increase of

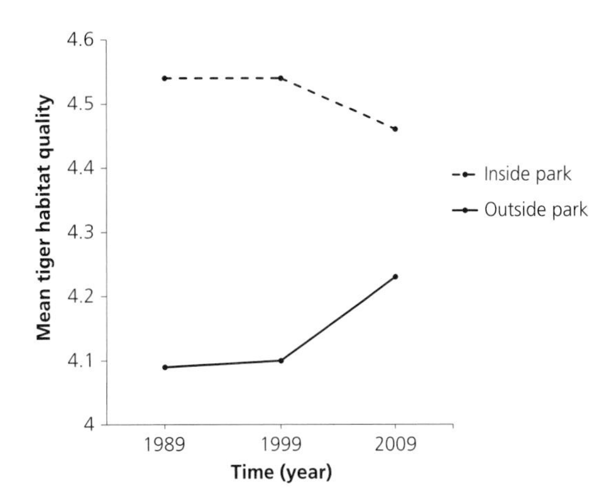

Figure 16.2 Mean tiger habitat quality inside and outside the northern portion of Chitwan National Park in 1989, 1999, and 2009. Tiger habitat quality decreased inside the park from 1999 to 2009, while it increased outside the park over the same time period. The increase in habitat quality outside the park was likely due to forest conservation policies, such as the prohibition of livestock grazing and community forestry, which were implemented in the late 1990s. Data from Carter et al. (2013).

attacks on people (Carter et al., 2012b). For example, 65 local residents were killed from 1998 to 2006 compared to 6 from 1989 to 1997 (Gurung et al., 2008). A feedback occurs when the crop/livestock losses or human attacks become too great. Local residents may be driven to further participate in off-farm economic activities, such as wage labor opportunities in the cities. Residents may also decide to abandon farming altogether, a phenomenon we saw occurring in both sites.

16.3.2 Non-linearity and thresholds

Multiple and reciprocal interactions within a coupled system typically result in non-linear relationships between and among its components. For instance, changes in household-level fertility patterns occur, such as the interval between successive births and the age at which a woman has her first child (Chapter 8). These changes have non-linear effects on the number of households in a given area. Simulation results indicate that fuelwood consumption resulting from such changes in household numbers, in turn, has a non-linear effect on panda habitat over time (Dussault et al., 2005).

A specific yet common type of non-linear relationship is a threshold or tipping point, beyond which one state or regime abruptly changes to another (Liu et al., 2007a). For instance, human effects on wildlife may exceed a threshold, after which wildlife habitat or behavior alters drastically. In Wolong, the distance between household and fuelwood collection site has a threshold effect on panda habitat (Dussault et al., 2005). The area of panda habitat is negatively related to the distance between household and fuelwood collection site until that distance reaches approximately 1800 m (Figure 16.3A). Beyond that distance, the area of panda habitat stabilizes because the impacts of fuelwood collection on forest and bamboo (the main food source for the panda) become sparsely distributed. In Chitwan, a similar threshold appears at approximately 600 m from the human-settled area (Figure 16.3B).

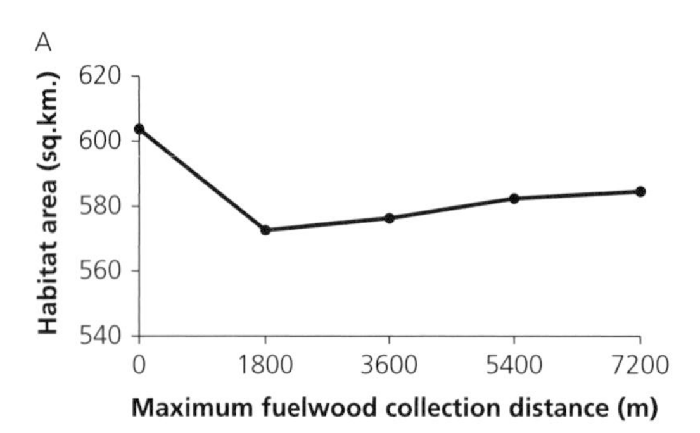

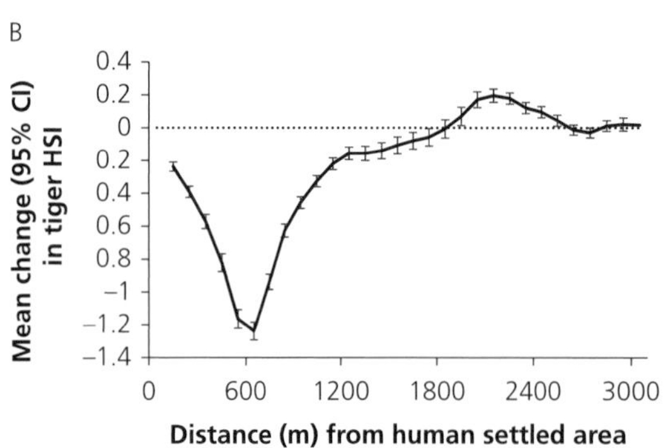

Figure 16.3 (A) Example of threshold (1,800 m) where effect of distance between household and fuelwood collection location on panda habitat in Wolong changes. Source: adapted from An et al. (2005). Reprinted by permission of the Association of American Geographers. (B) Change in Bengal tiger estimated habitat suitability index (HSI) in the 1990s with distance from human-settled areas inside the northern portion of Chitwan National Park, Nepal. Source: adapted from Carter et al. (2013).

The effect of wildlife on people also exhibits thresholds. For example, in Chitwan, local people may tolerate tiger-related risks to a certain degree. But when that tolerance threshold is exceeded, they will likely take action to reduce the threat to human livelihood or safety (Slovic, 1987). As a result, poisoning or poaching of tigers may suddenly occur when it had not occurred in the past, although this has not yet been empirically detected.

16.3.3 Surprises

When coupled systems are not understood, surprising dynamics may occur, with negative consequences for wildlife and their habitat. For instance, panda habitat degraded faster inside Wolong after the reserve was established even though the reserve was expressly designed to protect panda habitat from degradation (Liu et al., 2001). In part, this was due to the human population inside the reserve, mostly of minority ethnicities and thus not encumbered by the one-child policy, continuing to increase after the reserve was established. In addition, between 1975 and 2012, the number of people inside Wolong Nature Reserve increased by 92% but the number of households increased by 241% (Chapter 8). The tourism industry also grew substantially, leading local people to use more fuelwood to produce marketable goods. The synergistic effects of population growth, household proliferation, and tourism catalyzed the surprising decline in panda habitat detailed in Chapter 7 (Liu et al., 2001). As with Wolong, household proliferation in Chitwan appears to be strongly linked to environmental degradation (Carter et al., 2013). From 1991 to 2011, the number of people in Chitwan District increased 64% but the number of households increased by 103% (Nepal Central Bureau of Statistics, 2012). These changes perhaps contributed to a decreasing trend in tiger habitat suitability inside the park over the same period (Figure 16.2; Carter et al., 2013).

Exclusion policies in Wolong and Chitwan also have led to surprising changes in human livelihoods, particularly with respect to livestock numbers and husbandry, which have impacts on panda and tiger habitats. As access to fodder has become increasingly difficult in Wolong, local residents purchased more horses and let them range freely inside the reserve. As Chapter 4 showed, pandas avoided free-ranging herds of horses inside the reserve. Horses may disturb important panda behaviors such as feeding (grazing for food), mating, and raising young (Hull et al., 2011, 2014). In Chitwan, restrictions on livestock grazing have had different consequences from those in Wolong. To adjust to changes in grazing policies, households in Chitwan have reduced their holdings of large livestock, like buffalo, in favor of goats (Gurung et al., 2009). Households also now stall feed their livestock comparatively more than they did in the past when they could let the livestock range freely in the forests. As such, the direct impact of livestock on forest cover and structure is less than in the past. Nonetheless, fodder collection is now more common, which increases the likelihood of tiger attacks on people (Gurung et al., 2009).

The exact reasons for the difference in response to grazing restrictions between Wolong and Chitwan have yet to be determined. However, it is likely related to the respective balance of costs and benefits associated with holding free-ranging livestock in both sites. For instance, maintaining free-ranging livestock may be easier in Wolong. There, people are already living inside the reserve, natural livestock predators are not common, and the cost of collecting fodder by hand is quite high due to the mountainous topography. In contrast, goat milk and meat are highly valuable in Chitwan while the traction benefits of water buffalo are growing less important as mechanized farm equipment becomes more common and agricultural landholding per household decreases over time. Surprising patterns and divergence in livestock husbandry practices stemming from changes in policies and socioeconomic conditions were evident in both Wolong and Chitwan. These patterns and their potential effects on pandas and tigers highlight the need to take an integrated approach to studying coupled systems.

16.3.4 Heterogeneity

The number and strength of couplings between human and natural systems vary across spatial, temporal, and organizational scales. Such heterogeneity has implications for wildlife and their habitat. For instance, at a broad spatial scale, dense populations of people and wildlife in both sites generally do not inhabit the exact same areas. Human settlers usually clear forests (and thus destroy tiger and panda habitats) for cultivation. However, at a fine spatial scale along the interface between people and wildlife (e.g., forest–agriculture edge), the story is different. Tigers and pandas are frequently using the same space as people who are entering the forest on foot and vehicles. Tigers and pandas are both naturally shy and elusive. In Chitwan tigers offset their activity patterns to be much less active during the day when human activity (e.g., local residents collecting forest products and tourists on vehicle safaris) peaks (Carter et al., 2012a). Temporal rather than spatial displacement at the fine scale allows tigers to continue using the prey-rich habitats in Chitwan despite the ubiquitous presence of people.

The strength of couplings between human and natural systems also varies over time. For instance, in recent years, local people in both Wolong and Chitwan have been progressively moving away from agriculture toward off-farm employment. They take jobs in the booming tourism industry and seek wage employment in construction of infrastructure (e.g., buildings and roads; Axinn and Ghimire, 2011, He et al., 2008). Shifts away from agriculture in Wolong and Chitwan likely mean that local residents are less directly dependent on nearby natural resources (e.g. forest products). Less dependence on natural resources lessens the strength of the direct coupling between local residents and natural ecosystems. Over the short term, this decoupling will likely reduce human pressure on panda and tiger habitats; however, it is unclear what the impacts of the decoupling will be in the future. Uncertainty about the long-term effects of dynamic couplings on people and wildlife underscores the urgent need to continue research on coupled systems for long time frames.

The creation of institutions or institutional change also can modify the strength of couplings. Institutional arrangement (e.g., top-down versus bottom-up) with regard to land management policies and practices can have a particularly large effect on wildlife and their habitat. In Wolong and Chitwan, forests are primarily managed by the central government. The Chinese government exerts strict control over the resources extracted from Wolong (including both timber and non-timber forest products). However, forest harvesting still occurs (Liu et al., 2001). Likewise, the Nepalese government controls access to forests within Chitwan National Park, and strictly prohibits natural resource extraction except during a very limited period. Yet illegal collection of forest products in Chitwan occurs throughout the year (Stræde and Treue, 2006). Notably, in state-controlled forests, panda habitat was lost in Wolong and tiger habitat degraded in Chitwan. However, forest management regimes have changed recently in Wolong and Chitwan, with direct implications for wildlife habitat.

In Wolong, NFCP, initiated in 2001, departed from the traditional top-down model. NFCP in Wolong does not rely on state agencies to monitor certain forest parcels for infractions (e.g., illegal logging) but instead devolved those responsibilities to local households. Specifically, monitoring activities of large forest parcels were assigned to groups ranging in size from one to 16 households (Chapter 13). Households received payment from the government for effective protection of the forest parcels. Residents suffer payment reduction if illegal activities (e.g., logging, hunting, mining, or grazing in restricted areas) are detected during the government's biannual field assessments (Yang et al., 2013). Land management regimes have also changed in Chitwan. Management of forested areas outside the park in Chitwan District, which previously were part of the state-controlled national forest system, was handed over to local user-group committees. These committees have responsibility and control over resource use. For instance, committees dictate the amount and times of year that local people can collect or purchase fuelwood, timber, and fodder from community forests (Nagendra et al., 2005). Decentralizing some of the monitoring and land management responsibilities to local institutions in both sites seems to be aiding the recovery of panda and tiger habitats (Carter et al., 2013). This pattern occurs because decentralization can encourage greater participation by those who depend on forests, greater accountability of decision-makers, and stronger enforcement of property rights and governance arrangements (Agrawal et al., 2008).

16.3.5 Telecoupling

The degree to which coupled systems are embedded within other systems (e.g., embedment) or connected with distant systems (e.g., telecoupling, see Chapter 17) also varies. For example, Chitwan is part of a broader international effort to link protected areas along the base of the Himalayas (*Terai*) in India and Nepal through forest corridors (Dinerstein et al., 2007). As a result, conservation interventions that strengthen institutional support for community-managed forestry in "stepping stone" forests between protected areas have been implemented. Additionally, a number of conservation measures have been conducted in priority forest corridors and tracts outside protected areas. These efforts include reducing local reliance on forest products, providing alternative income opportunities for local residents, and raising awareness about the benefits of intact forests and tigers for local communities. This endeavor has modified the way people use forests and interact with tigers, and has helped expand the available land base for tigers, particularly outside protected areas. Similarly, national-level policies in China connect the human and natural systems in Wolong to larger and distant coupled systems. A severe drought in 1997 and catastrophic floods in 1998 affected much of China. These events triggered the development of NFCP and GTGP, which were designed to reduce soil erosion and the likelihood of drought due to poor water retention in soils devoid of vegetation (Liu et al., 2013b). Both programs have modified the connections between people and nature in Wolong, and as described above, have helped restore panda habitat in Wolong.

The movement of people inside and outside of Wolong and Chitwan links those coupled systems to other systems via telecoupling (e.g., socioeconomic and environmental interactions over distances; Liu et al., 2013a). Out-migration physically removes people from the coupled systems, and remittances from migrants connect the coupled systems of Wolong and Chitwan to the broader national and global economies. In general, out-migration and household goods purchased with remittances reduce human pressure on habitat (Chen et al., 2012). Tourism, on the other hand, brings people from around the world into the coupled systems of Wolong and Chitwan. Although tourism does support local livelihoods to an extent, there is concern that tourist activity in the protected areas may disturb pandas and tigers (Curry et al., 2001, Liu and Viña, 2014). Uncontrolled resource collection and development (e.g., building lodges and tea houses) for the sake of the tourist industry depletes forest resources and negatively impacts panda and tiger habitats (Liu et al., 2001, UNEP/WCMC, 2011).

The movement and transport of tigers and pandas to areas outside Chitwan and Wolong, respectively, also connect these coupled systems to other systems. For example, tigers are sometimes poached from Chitwan and trafficked to and sold in black markets in places all over the world, but mainly China. In addition, Chitwan sits on a main transit route for traffickers moving tiger parts over the Himalayas from India to China or elsewhere in the Himalayan region (Nowell, 2012). The trafficking of poached tigers within and across international boundaries is highly illegal and is also enormously lucrative. The trade of wild animals for profit ranks as the world's third-largest (US$20 billion) illicit activity behind drug and weapon smuggling (Wyler and Sheikh, 2008). Pandas, like tigers, are highly sought by zoos all over the world. As such, the panda breeding center in Wolong loans their captive-bred pandas, typically at very high costs, to zoos in other countries (Chapter 17). The selection of zoos for panda visits is entwined in national and international politics. Changes in these two coupled systems due to telecoupling create spillover effects on other coupled systems (Liu et al., 2013a). Examples of these spillover systems include Chengdu in China and Kathmandu in Nepal, both stopovers of wildlife transported from and people traveling to Wolong and Chitwan. Systematic assessments of these spillover systems and explicit consideration of them are important when developing and implementing policies.

16.3.6 Vulnerability

Vulnerability is the likelihood that coupled systems experience harm due to changes in their dynamics from internal or external forces (Liu et al., 2007a). For instance, in Chitwan, the introduction of invasive species from outside the system has altered vegetation composition and structure. Specifically, *Mikania micrantha*, or "mile-a-minute weed," is increasingly common in Chitwan and rapidly grows over and kills vegetation by restricting its access to sunlight. As such, *Mikania* is considered a major and imminent threat to natural ecosystems in Chitwan. In addition, *Mikania* is generally unpalatable to wild ungulates. By killing other palatable plant species for tiger prey species, *Mikania* can have a cascading negative impact on tiger habitat. Invasive species have not yet had this magnitude of impact on panda habitat in Wolong, but managers should be vigilant for future invasions given the impacts seen in Chitwan.

Coupled systems can be vulnerable to natural and anthropogenic events. A strong earthquake in Wolong in 2008 significantly disrupted the coupled systems there (Chapter 12). With respect to the natural system in Wolong, the earthquake caused many severe landslides that have exacerbated flooding in the region and reduced panda habitat (Viña et al., 2010). With respect to the human system, the earthquake destroyed the main road leading into the reserve. This event brought tourism to a halt and severely crippled access to outside markets. The earthquake also forced many people whose homes were destroyed to relocate to other areas inside the reserve (Chapter 12). A different type of disturbance had similar overarching effects in Chitwan. A ten-year-long civil war (1996–2006) took place between the military of the monarchy and an insurgent group with a Maoist political philosophy. This war had direct implications for both the human and natural systems as well as their interactions. Many people were displaced from Chitwan, and people changed their daily activity patterns to avoid potentially dangerous situations. Importantly, enforcement of the National Park rules broke down. Park guards (who belong to the Nepalese army) were removed from the park in order to fight insurgents in other parts of the country. As a result, poaching rates of tigers and rhinos spiked during the civil war (Baral and Heinen, 2005). Additionally, tiger conservation actions organized and led by international conservation agencies were put on hold or halted indefinitely because it was too dangerous in Nepal. These two very different disturbances highlight the vulnerability of coupled systems and demonstrate the profound effects of disturbances on coupled systems. Both resulted in changes that reverberated throughout the entire systems (not just one sector or subsystem), and both caused the systems to enter into a new system state. Although one disturbance was acute and the other sustained, both resulted in long-term vulnerabilities.

Among protected areas, Wolong and Chitwan support large numbers of pandas and tigers relative to other areas; however, the populations of both species in their respective coupled systems are very small in demographic terms. This fact is largely due to habitat fragmentation that occurred in both landscapes (Smith et al., 1998, Viña et al., 2010). Smaller local animal populations are inherently more vulnerable than larger populations to demographic and environmental stochasticity. For instance, genetic drift, inbreeding, and catastrophic losses to disease place smaller populations at a greater risk of extinction than larger populations (Lande, 1993). Thus, human impacts, such as poaching, loss of habitat, and exposure to disease from livestock increase the vulnerability of the small populations of pandas and tigers (Kenney et al., 2014). Life-history characteristics of the animal can also affect their vulnerability. For example, tigers may be more vulnerable to habitat fragmentation than pandas because their home ranges of 20–240 km^2 (Bengal tigers, Goodrich et al., 2010) are several orders of magnitude larger than the 3–10 km^2 occupied by pandas (Hull et al., 2015, Pan et al., 2001, Schaller et al., 1985, Yong et al., 2004, Zhang et al., 2014).

16.3.7 Time lags and legacy effects

The mechanisms linking human and natural systems are also temporal (Chapter 2). Thus, the effects of one component on another may not become apparent until after a certain amount of time, or time lag. In Wolong, simulation results indicate that increasing fertility (number of children) increases the number of households about 20 years later as it takes time for children to mature and establish their own households. Increasing household numbers increases fuelwood consumption and reduces panda habitat in about 30 years (An and Liu, 2010; Chapter 8). A similar relationship between human population processes, household number, resource demand, and tiger habitat degradation likely also occurs in Chitwan, but it has not been empirically demonstrated. A much shorter time lag exists between changes in the prices of electricity or non-wood fuels (e.g., kerosene) and impacts on habitat. If prices for electricity or non-wood fuels rise, people immediately collect more fuelwood to cook with and heat their homes, and thus destroy more habitats (Liu et al., 2007a).

The impacts of forest conservation policies on wildlife habitat also take time to manifest because the process of forest growth is relatively slow. In Chitwan, the effects of different forest management regimes on tiger habitat are only recently becoming apparent. It has been 10–15 years since forests outside the park were handed over to local communities to manage in the late 1990s. In this time, tiger and tiger prey numbers have increased because forest conditions have improved (Carter et al., 2013). In Wolong, time delays in forest recovery after implementation of policies involving local communities (NFCP, GTGP) have been shorter than in Chitwan. Perhaps this difference may be because they involved replanting fast-growing exotic tree plantations. However, the shorter time lag may be offset in the long term due to the inadequacy of the exotic plantations to provide suitable habitat for pandas.

Another temporal feature of coupled systems is legacy effects. Legacy effects are the impacts of past interactions in coupled systems on later conditions (Liu et al., 2007b). For instance, prior to the 1950s, the indigenous Tharu people of Chitwan were sparsely distributed throughout the forests and were subsistence hunters and gatherers. However, hunting was prohibited inside the park once it was established and the large-scale conversion of forest to agriculture starting in the 1950s forced indigenous people to rapidly modify their lifestyles. Thus, the present spatial patterns of land cover and land use, human population distribution, and human activities, all of which affect tigers and their habitat, are the legacy of past policies. Similarly, in Wolong, past logging locations affect current forest type and panda habitat quality. The frequency of pandas using an area is reduced for several decades after harvest of timber (Bearer et al., 2008).

16.4 Some lessons learned

Knowledge about complex human–nature interactions in Wolong helped us explain similarly complex dynamics between people and wildlife in Chitwan. For example, multilevel or multiscalar patterns and processes observed in Wolong helped us understand how similar processes may be at play in Chitwan. In both sites, we found that dynamics at the household level, such as fertility and marriage timing, underlie important aggregate-level patterns, such as the association between habitat loss and household number at the reserve level. Likewise, processes at broad scales, such as regional or global industrialization and urbanization, appear to influence fine-scale behaviors in both sites, such as out-migration of individuals or households. These findings highlight that some processes occurring in coupled human and natural systems are independent of context, a key conclusion that could not be made without cross-site application and synthesis. Such knowledge not only improves our understanding of complex systems more generally, but also informs policy makers on which specific socioeconomic and demographic factors drive changes in wildlife population and habitat dynamics.

Work in Wolong also highlighted the importance of policies as feedback mechanisms, which we observed in Chitwan as well. Considering policies as feedback mechanisms can help managers and policy makers anticipate the potential impacts that may emerge from various policies. Otherwise, conservation policies can have unintended or undesirable consequences. This was the case with Wolong. The synergistic effects of population growth, household proliferation, and tourism led to faster declines in panda habitat after the reserve was established than before (Liu et al., 2001). Likewise, the increase in tiger attacks on people in Chitwan over the last two decades could have been anticipated (and addressed more effectively) if it had been interpreted as a feedback. Such a feedback emerged due to land management policies that occurred outside the park in the mid-1990s.

Similarly, our experience in Wolong indicates that spatial and temporal thresholds are common features, and as such, should be anticipated and accounted for in policy making. The exact point at which the threshold may occur may not always be precisely identified. But simply being aware of an imminent threshold provides rationale for implementing or strengthening policies that better protect panda or tiger habitat or mitigate human–wildlife conflict before the threshold point is crossed. For example, using various policies to proactively increase tolerance among local communities toward tigers may reduce retaliatory killing of tigers likely to occur once tiger attacks on people and livestock exceed a certain threshold (Carter et al., 2014b). An example of how this could be done is by creating conflict response teams comprising local people and government authorities.

Our work also suggests the importance of collaborative management and protection of natural ecosystems. Local people should be partners in the design, implementation, and enforcement of resource management to aid with the recovery of imperiled wildlife. In Wolong (Chapter 7) and in Chitwan's buffer zone, a habitat "transition" from degradation to recovery was observed after institutions implemented policies involving local people in conservation. It is important to note, however, that policies alone may not be enough. Whether conservation policies, especially those developed from outside the focal system, will overlay on preexisting community institutions and networks likely determines the success of those policies (Ostrom et al., 1999). For example, grazing restrictions in Wolong, implemented at the state level, were ignored by many local people in part because grazing livestock was still considered socially acceptable among local communities. In contrast, grazing restrictions were linked with community forestry in the buffer zone outside Chitwan, thus grazing livestock in community forests was very uncommon because it violated community-held norms.

Over the last few decades, the telecoupling processes between Wolong and outside coupled systems have grown in strength and number (Chapter 17). As in Wolong, we found that telecoupling processes, such as tourism and migration, are also growing in strength in Chitwan. Such telecouplings have cascading and complex effects on tigers and people (Carter et al., 2014a). The increasing influences of telecoupling processes on human–wildlife interactions in both sites suggest that a similar trend may be occurring in many coupled systems around the world, with yet unknown and potentially negative consequences on wildlife. Thus, more research on telecoupling processes and their spillover effects on other systems is urgently needed. This gap needs to be addressed to explain telecoupling impacts on wildlife and potential for human–wildlife coexistence not only in Wolong and Chitwan but also many other sites around the world.

Finally, complex features and processes do not occur independently from one another, but rather interact. Moreover, the outcomes of these interactions on human–wildlife coexistence are mediated by local site differences, underscoring the need to understand local context and history in addition to more general features of complex systems. For example, both sites are characterized by vulnerability, legacy effects, and heterogeneity. However, differences in how these features interacted with each other and shaped human–wildlife dynamics in both sites required an understanding of each site's unique land-use, anthropological, and institutional histories. Such information helps reveal why certain policies, such as grazing restrictions, may be more or less successful in seemingly similar sites.

16.5 Summary

Research on coupled human and natural systems has enriched knowledge on how humans and nature interact and how such interactions in turn affect global sustainability. But to move this emerging field forward, more generalizable theories, approaches, and conclusions about human–nature interactions are needed. To this end, intensive research conducted across multiple different model coupled systems needs to be synthesized. In this chapter, we applied what we have learned about human–wildlife interactions in Wolong Nature Reserve in China to understanding complex processes and dynamics in the coupled system in Chitwan National Park in Nepal. Like Wolong, Chitwan is also a "flagship" protected area within a global biodiversity hotspot that supports a globally endangered conservation icon—the tiger (*Panthera tigris*). The interactions among wildlife, institutional arrangements, and socioeconomic and demographic changes at both sites are also similar to those in many other coupled systems around the world. We conducted an in-depth analysis of key properties across the two sites, including reciprocal interactions and feedback loops, non-linear relations and thresholds, surprises, heterogeneity, telecoupling, vulnerability, time lags, and legacy effects. We found similarities and differences across the two sites and discussed what the findings mean for understanding complex human–wildlife systems. For example, in both systems collaborative policies that involved local people were more effective than exclusionary policies that forcibly limited people's activities. Explicating the effects and interactions of complex features on human–wildlife coexistence will only become more pertinent in the future as the world is expected to grow ever more crowded and interconnected.

References

Acevedo, M.F., Baird-Callicott, J., Monticino, M., et al. (2008) Models of natural and human dynamics in forest landscapes: cross-site and cross-cultural synthesis. *Geoforum*, **39**, 846–66.

Adhikari, T.R. (2002) The curse of success. *Habitat Himalaya*, **9**, 1–4.

Agrawal, A., Chhatre, A., and Hardin, R. (2008) Changing governance of the world's forests. *Science*, **320**, 1460–62.

An, L. and Liu, J. (2010) Long-term effects of family planning and other determinants of fertility on population and environment: agent-based modeling evidence from Wolong Nature Reserve, China. *Population and Environment*, **31**, 427–59.

Axinn, W.G., Barber, J.S. and Biddlecom, A.E. (2010) Social organization and the transition from direct to indirect consumption. *Social Science Research*, **39**, 357–68.

Axinn, W.G. and Ghimire, D.J. (2011) Social organization, population, and land use. *American Journal of Sociology*, **117**, 209–58.

Baral, N. and Heinen, J.T. (2005) The Maoist people's war and conservation in Nepal. *Politics and the Life Sciences*, **24**, 2–11.

Barlow, A.C.D., McDougal, C., Smith, J.L.D., et al. (2009) Temporal variation in tiger (*Panthera tigris*) populations and its implications for monitoring. *Journal of Mammalogy*, **90**, 472–78.

Bearer, S., Linderman, M., Huang, J., et al. (2008) Effects of fuelwood collection and timber harvesting on giant panda habitat use. *Biological Conservation*, **141**, 385–93.

Biddlecom, A.E., Axinn, W.G., and Barber, J.S. (2005) Environmental effects on family size preferences and subsequent reproductive behavior in Nepal. *Population and Environment*, **26**, 583–621.

Carter, N.H., Gurung, B., Viña, A., et al. (2013) Assessing spatiotemporal changes in tiger habitat across different land management regimes. *Ecosphere*, **4**: art124.

Carter, N.H., Riley, S.J., and Liu, J. (2012b) Utility of a psychological framework for carnivore conservation. *Oryx*, **46**, 525–35.

Carter, N.H., Riley, S.J., Shortridge, A., et al. (2014b) Spatial assessment of attitudes toward tigers in Nepal. *Ambio*, **43**, 125–37.

Carter, N.H., Shrestha, B.K., Karki, J.B., et al. (2012a) Coexistence between wildlife and humans at fine spatial scales. *Proceedings of the National Academy of Sciences of the United States of America*, **109**, 15360–65.

Carter, N.H., Viña, A., Hull, V., et al. (2014a) Coupled human and natural systems approach to wildlife research and conservation. *Ecology and Society*, **19**, 43.

Chapron, G., Kaczensky, P., Linnell, J.D.C., et al. (2014) Recovery of large carnivores in Europe's modern human-dominated landscapes. *Science*, **346**, 1517–19.

Chapron, G., Miquelle, D.G., Lambert, A., et al. (2008) The impact on tigers of poaching versus prey depletion. *Journal of Applied Ecology*, **45**, 1667–74.

Chen, X., Frank, K.A., Dietz, T., and Liu, J. (2012) Weak ties, labor migration, and environmental impacts toward a sociology of sustainability. *Organization & Environment*, **25**, 3–24.

Chen, X., Lupi, F., He, G., et al. (2009) Factors affecting land reconversion plans following a payment for ecosystem service program. *Biological Conservation*, **142**, 1740–47.

Chen, X., Lupi, F., Viña, A., et al. (2010) Using cost-effective targeting to enhance the efficiency of conservation investments in payments for ecosystem services. *Conservation Biology*, **24**, 1469–78.

Chen, X., Viña, A., Shortridge, A., et al. (2014) Assessing the effectiveness of payments for ecosystem services: an agent-based modeling approach. *Ecology and Society*, **19**, 7.

Curry, B., Moore, W., Bauer, J., et al. (2001) Modelling impacts of wildlife tourism on animal communities: a case study from Royal Chitwan National Park, Nepal. *Journal of Sustainable Tourism*, **9**, 514–29.

Dinerstein, E., Loucks, C., Wikramanayake, E., et al. (2007) The fate of wild tigers. *BioScience*, **57**, 508–14.

Dussault, C., Ouellet, J.-P., Courtois, R., et al. (2005) Linking moose habitat selection to limiting factors. *Ecography*, **28**, 619–28.

Entwisle, B., Rindfuss, R.R., Guilkey, D.K., et al. (1996) Community and contraceptive choice in rural Thailand: a case study of Nang Rong. *Demography*, **33**, 1–11.

Frost, P.G.H. and Bond, I. (2008) The CAMPFIRE programme in Zimbabwe: payments for wildlife services. *Ecological Economics*, **65**, 776–87.

Government of Nepal (1993) Fourth amendment to the national parks and wildlife conservation act (2029). *Nepal Gazette*, **43**(Suppl.).

Goodrich, J.M., Miquelle, D.G., Smirnov, E.N., et al. (2010) Spatial structure of Amur (Siberian) tigers (*Panthera tigris altaica*) on Sikhote-Alin Biosphere Zapovednik, Russia. *Journal of Mammalogy*, **91**, 737–48.

Gurung, B., Nelson, K.C., and Smith, J.L.D. (2009) Impact of grazing restrictions on livestock composition and husbandry practices in Madi Valley, Chitwan National Park, Nepal. *Environmental Conservation*, **36**, 338–47.

Gurung, B., Smith, J.L.D., McDougal, C., et al. (2008) Factors associated with human-killing tigers in Chitwan National Park, Nepal. *Biological Conservation*, **141**, 3069–78.

He, G., Chen, X., Bearer, S., et al. (2009) Spatial and temporal patterns of fuelwood collection in Wolong Nature Reserve: implications for panda conservation. *Landscape and Urban Planning*, **92**, 1–9.

He, G., Chen, X., Liu, W., et al. (2008) Distribution of economic benefits from ecotourism: a case study of Wolong Nature Reserve for Giant Pandas in China. *Environmental Management*, **42**, 1017–25.

Homewood, K., Lambin, E.F., Coast, E., et al. (2001) Long-term changes in Serengeti-Mara wildebeest and land cover: pastoralism, population, or policies? *Proceedings of the National Academy of Sciences of the United States of America*, **98**, 12544–49.

Hu, J.C. (1989) *Life of the Giant Panda*. Chongqing University Press, Chongqing, Sichuan, China (in Chinese).

Hull, V., Shortridge, A., Liu, B., et al. (2011) The impact of giant panda foraging on bamboo dynamics in an isolated environment. *Plant Ecology*, **212**, 43–54.

Hull, V., Zhang, J., Zhou, S., et al. (2014) Impact of livestock on giant pandas and their habitat. *Journal for Nature Conservation*, **22**, 256–64.

Hull, V., Zhang, J., Zhou, S., et al. (2015) Space use by endangered giant pandas. *Journal of Mammalogy*, **96**, 230–36.

Karki, J.B., Pandav, B., Jnawali, S.R., et al. (2013) Estimating the abundance of Nepal's largest population of tigers *Panthera tigris*. *Oryx*, **49**, 150–56.

Kenney, J., Allendorf, F.W., McDougal, C., and Smith, J.L.D. (2014) How much gene flow is needed to avoid inbreeding depression in wild tiger populations? *Proceedings of the Royal Society B: Biological Sciences*, **281**, 20133337.

Lande, R. (1993) Risks of population extinction from demographic and environmental stochasticity and random catastrophes. *The American Naturalist*, **142**, 911.

Laurie, A. (1982) Behavioural ecology of the greater one horned rhinoceros (*Rhinoceros unicornis*). *Journal of Zoology*, **196**, 307–41.

Li, Y., Viña, A., Yang, W., et al. (2013) Effects of conservation policies on forest cover change in giant panda habitat regions, China. *Land Use Policy*, **33**, 42–53.

Linderman, M.A., An, L., Bearer, S., et al. (2006) Interactive effects of natural and human disturbances on vegetation dynamics across landscapes. *Ecological Applications*, **16**, 452–63.

Liu, J., An, L., Batie, S.S., et al. (2005) Beyond population size: examining intricate interactions among population structure, land use, and environment in Wolong Nature Reserve (China). In B. Entwisle and P.C. Stern, eds., *Population, Land Use, and Environment: Research Directions*, pp. 217–37. The National Academies Press, Washington, DC.

Liu, J., Dietz, T., Carpenter, S.R., et al. (2007a) Complexity of coupled human and natural systems. *Science*, **317**, 1513–16.

Liu, J., Dietz, T., Carpenter, S.R., et al. (2007b) Coupled human and natural systems. *Ambio*, **36**, 639–49.

Liu, J., Hull, V., Batistella, M., et al. (2013a) Framing sustainability in a telecoupled world. *Ecology and Society*, **18**, 26.

Liu, J., Ickes, K., Ashton, P.S., et al. (1999a) Spatial and temporal impacts of adjacent areas on the dynamics of species diversity in a primary forest. In D. Mladenoff and W. Baker, eds., *Spatial Modeling of Forest Landscape Change: Approaches and Applications*, pp. 42–69. Cambridge University Press, Cambridge, UK.

Liu, J., Li, S., Ouyang, Z., et al. (2008) Ecological and socioeconomic effects of China's policies for ecosystem services. *Proceedings of the National Academy of Sciences of the United States of America*, **105**, 9477–82.

Liu, J., Linderman, M., Ouyang, Z., et al. (2001) Ecological degradation in protected areas: the case of Wolong Nature Reserve for giant pandas. *Science*, **292**, 98–101.

Liu, J., Ouyang, Z., Tan, Y., et al. (1999b) Changes in human population structure: implications for biodiversity conservation. *Population and Environment*, **21**, 45–58.

Liu, J., Ouyang, Z., Taylor, W.W., et al. (1999c) A framework for evaluating the effects of human factors on wildlife habitat: the case of giant pandas. *Conservation Biology*, **13**, 1360–70.

Liu, J., Ouyang, Z., Yang, W., et al. (2013b) Evaluation of ecosystem service policies from biophysical and social perspectives: the case of China. In S.A. Levin, ed., *Encyclopedia of Biodiversity* (second edition), vol. **3**, pp. 372–84. Academic Press, Waltham, MA.

Liu, J. and Viña, A. (2014) Panda, plants, and people. *Annals of the Missouri Botanical Garden*, **100**, 108–25.

Liu, W., Vogt, C.A., Luo, J., et al. (2012) Drivers and socioeconomic impacts of tourism participation in protected areas. *PLoS ONE*, **7**, e35420.

Massey, D.S., Axinn, W.G., and Ghimire, D.J. (2010) Environmental change and out-migration: evidence from Nepal. *Population and Environment*, **32**, 109–36.

Matthews, S.A., Shivakoti, G.P., and Chhetri, N. (2000) Population forces and environmental change: observations from western Chitwan, Nepal. *Society and Natural Resources*, **13**, 763–75.

Myers, N., Mittermeier, R.A., Mittermeier, C.G., et al. (2000) Biodiversity hotspots for conservation priorities. *Nature*, **403**, 853–58.

Nagendra, H., Karmacharya, M., and Karna, B. (2005) Evaluating forest management in Nepal: views across space and time. *Ecology and Society*, **10**, 24.

Nagendra, H., Pareeth, S., Sharma, B., et al. (2008) Forest fragmentation and regrowth in an institutional mosaic of community, government and private ownership in Nepal. *Landscape Ecology*, **23**, 41–54.

Nepal Central Bureau of Statistics (2012) *National Planning Comission Secretariat, Government of Nepal*. https://sites.google.com/site/nepalcensus.

Nowell, K. (2012) Wildlife Crime Scorecard: Assessing compliance with and enforcement of CITES commitments for tigers, rhinos and elephants. WWF Report. http://awsassets.panda.org/downloads/wwf_wildlife_crime_scorecard_report.pdf.

Ostrom, E., Burger, J., Field, C.B., et al. (1999) Revisiting the commons: local lessons, global challenges. *Science*, **284**, 278–82.

Pan, W., Lü, Z., Zhu, X.J., et al. (2001) *A Chance for Lasting Survival*. Beijing University Press, Beijing, China (in Chinese).

Parker, D.C., Manson, S.M., Janssen, M.A., et al. (2003) Multi-agent systems for the simulation of land-use and land-cover change: a review. *Annals of the Association of American Geographers*, **93**, 314–37.

Rindfuss, R.R., Entwisle, B., Walsh, S.J., et al. (2008) Land use change: complexity and comparisons. *Journal of Land Use Science*, **3**, 1–10.

Sanderson, E., Forrest, J., Loucks, C., et al. (2006) *Setting Priorities for the Conservation and Recovery of Wild Tigers: 2005–2015*. The technical assessment. WCS, WWF, Smithsonian, and NFWF-STF, New York–Washington, DC.

Schaller, G.B., Hu, J., Pan, W., and Zhu, J. (1985) *The Giant Pandas of Wolong*. University of Chicago Press, Chicago, IL.

Slovic, P. (1987) Perception of risk. *Science*, **236**, 280–85.

Smith, J.L.D., Ahearn, S.C., and McDougal, C. (1998) Landscape analysis of tiger distribution and habitat quality in Nepal. *Conservation Biology*, **12**, 1338–46.

Spiteri, A. and Nepal, S.K. (2008) Distributing conservation incentives in the buffer zone of Chitwan National Park, Nepal. *Environmental Conservation*, **35**, 76–86.

Stræde, S. and Helles, F. (2000) Park-people conflict resolution in Royal Chitwan National Park, Nepal: buying time at high cost? *Environmental Conservation*, **27**, 368–81.
Stræde, S. and Treue, T. (2006) Beyond buffer zone protection: a comparative study of park and buffer zone products' importance to villagers living inside Royal Chitwan National Park and to villagers living in its buffer zone. *Journal of Environmental Management*, **78**, 251–67.
Tuanmu, M.-N., Viña, A., Roloff, G.J., et al. (2011) Temporal transferability of wildlife habitat models: implications for habitat monitoring. *Journal of Biogeography*, **38**, 1510–23.
Turner, B.L., Kasperson, R.E., Matson, P.A., et al. (2003) A framework for vulnerability analysis in sustainability science. *Proceedings of the National Academy of Sciences of the United States of America*, **100**, 8074–79.
UNEP/WCMC (2011) *Royal Chitwan National Park, Nepal.* United Nations Environment Programme/World Conservation Monitoring Centre. http://www.unep-wcmc.org/.
Viña, A., Bearer, S., Chen, X., et al. (2007) Temporal changes in giant panda habitat connectivity across boundaries of Wolong Nature Reserve, China. *Ecological Applications*, **17**, 1019–30.
Viña, A., Chen, X., McConnell, W.J., et al. (2011) Effects of natural disasters on conservation policies: the case of the 2008 Wenchuan Earthquake, China. *Ambio*, **40**, 274–84.
Viña, A., Tuanmu, M.N., Xu, W., et al. (2010) Range-wide analysis of wildlife habitat: implications for conservation. *Biological Conservation*, **143**, 1960–69.
Wyler, L.S. and Sheikh, P.A. (2008) *International Illegal Trade in Wildlife: threats and US policy*. Congressional Research Service, Library of Congress, Washington, DC.
Yang, W., Liu, W., Viña, A., et al. (2013) Performance and prospects on payments for ecosystem services programs: evidence from China. *Journal of Environmental Management*, **127**, 86–95.
Yong, Y., Liu, X., Wang, T., et al. (2004) Giant panda migration and habitat utilization in Foping Nature Reserve, China. In D. Lindburg and K. Baragona, eds., *Giant Pandas: Biology and Conservation*, pp. 159–69. University of California Press, Berkeley, CA.
Zhang, Z., Sheppard, J.K., Swaisgood, R.R., et al. (2014) Ecological scale and seasonal heterogeneity in the spatial behaviors of giant pandas. *Integrative Zoology*, **9**, 46–60.

CHAPTER 17

Human-Nature Interactions over Distances

Jianguo Liu, Vanessa Hull, Junyan Luo, Wu Yang, Wei Liu, Andrés Viña, Christine Vogt, Zhenci Xu, Hongbo Yang, Jindong Zhang, Li An, Xiaodong Chen, Shuxin Li, Zhiyun Ouyang, Weihua Xu, and Hemin Zhang

17.1 Introduction

Human–nature interactions are key factors shaping global sustainability and human well-being. They have been widely studied within a particular area, but the world has become increasingly connected over distances, both socioeconomically and environmentally. Distant interactions such as trade, migration, and spread of invasive species are now more widespread than they ever were before (Liu et al., 2013a). For instance, human societies obtained much of their food, water, and fuel locally in the past, but now increasingly rely on sources at opposite ends of the Earth via global trade (Kastner et al., 2011, Konar et al., 2011). Global food exports have increased tenfold over the past several decades (United Nations Statistics Division, 2012). Water scarcity has also led to large-scale water transfer schemes, such as China's South-North Water Transfer Scheme, which aims to transfer 45 billion m^3 of water across the nation each year (Liu and Yang, 2012). These distant interactions have profound implications for sustainability and human well-being, often exacerbating social and environmental problems such as climate change, famine, land degradation, species extinctions, and social unrest (Liu et al., 2013a).

These and other distant connections have often been separately studied. For example, studies on climate teleconnections concentrate on linkages between climate systems that are hundreds and even thousands of kilometers apart (Avissar and Werth, 2005) but largely ignore relevant socioeconomic linkages. On the other hand, studies on economic globalization (e.g., Levitt, 1982) focus on distant socioeconomic relationships. They pay relatively little attention to environmental interactions, although there have been some separate recent studies on greenhouse gas (GHG) emissions associated with trade (Peters et al., 2011). Furthermore, previous studies often treat distant factors as drivers of changes in a particular area (e.g., DeFries, 2010, Lambin and Meyfroidt, 2011, Stevens et al., 2014). But little research has been done on the feedback and impacts on other areas (Folke et al., 2011). While previous studies provide useful information, the results may be incomplete or partial. This shortcoming occurs due to the lack of simultaneous consideration of distant socioeconomic and environmental interactions, feedbacks, impacts beyond the systems being focused on, and relationships among various distant linkages.

To address these crucial issues, umbrella concepts need to be developed to bring together multiple different interactions and disciplines. A prominent example of the umbrella concept is ecosystem services (Daily, 1997), which include diverse benefits that nature provides to humans (e.g.,

Pandas and People. Edited by Jianguo Liu, Vanessa Hull, Wu Yang, Andrés Viña, Xiaodong Chen, Zhiyun Ouyang, and Hemin Zhang. © Oxford University Press 2016. Published 2016 by Oxford University Press.

water and air quality, nutrient cycling, and spiritual and recreational benefits). Another good example is the environmental footprints concept, which integrates different types of human impacts on nature (Hoekstra and Wiedmann, 2014).While these and other umbrella concepts are valuable, none of them is able to systematically integrate human–nature interactions that occur over distances.

A new framework of telecoupling (socioeconomic and environmental interactions over distances) has been proposed to facilitate such integrated research (Liu et al., 2013a). The framework (Liu et al., 2013a) treats each place as a coupled human and natural system in which humans and natural components interact (Liu et al., 2007). It provides an explicit approach to account for and internalize socioeconomic and environmental externalities across space. The framework consists of five major interrelated components (coupled human and natural systems; flows of material, information, and energy among systems; agents that facilitate the flows; causes that drive the flows; and effects that result from the flows; Liu et al., 2013a). Depending on the direction of flows, systems can be classified as three different types. They include sending systems (e.g., exporting countries), receiving systems (e.g., importing countries), and spillover systems (e.g., countries other than the trade partners). Spillover systems are those that affect and are affected by the interactions between sending and receiving systems. A single system can be classified as one type of system for one telecoupling and another type of system for another telecoupling. For instance, a city can be a receiving system for food and be a sending system for industrial products. Within each coupled system, there are internal or local couplings. The framework also explicitly considers feedbacks (e.g., human–nature feedbacks within a coupled system and across telecoupled systems). There has been increasing interest in the telecoupling framework. For example, the framework has been conceptually applied to land change science (Eakin et al., 2014, Gasparri et al., 2015, Liu et al., 2014); species invasion (Liu et al., 2014); payments for ecosystem services programs (Liu and Yang, 2013); conservation (Carter et al., 2014, Gasparri and de Waroux, 2014); and the trade of food (Garrett et al., 2013), forest products (Liu, 2014), energy (Liu et al., 2015b), and virtual water (Liu et al., 2015b). However, there is a lack of quantification of the telecoupling framework.

To fill such an important knowledge gap, we apply the telecoupling framework to examine telecouplings between Wolong Nature Reserve (hereafter "Wolong") and the rest of the world. As stated in Chapter 3, Wolong is a 2000-km^2 protected area located within a global biodiversity hotspot (Liu et al., 2003, Myers et al., 2000) in southwestern China (Ministry of Forestry, 1998; Figure 17.1). It is a home of the endangered giant panda (*Ailuropoda melanoleuca*) and more than 6000 other animal and plant species. Wolong also encompasses approximately 5000 local residents, mostly farmers (State Forestry Administration, 2006) who grow crops, raise livestock, and collect timber and non-timber forest products (Li et al., 1992). In this chapter, which is largely based on Liu et al. (2015a), we analyze major components of several telecoupling processes. These include panda loans, tourism, information dissemination, conservation subsidies, and trade of agricultural and industrial products (Figure 17.1). We discuss their similarities, differences, and interrelationships. We chose these example processes due to their prominence in the coupled system of Wolong and their data availability, although there are also other telecouplings (e.g., labor migration from Wolong to cities; Chen et al., 2012; Chapter 11) which may also interact with these analyzed in this chapter.

17.2 Telecouplings

17.2.1 Panda loans

Wolong is a stronghold for the wild giant panda population. It is also home to the China Conservation and Research Center for the Giant Panda (CCRCGP), a research and captive breeding base that houses over 200 pandas, the largest captive population in the world. The panda loan program enables zoos inside and outside China to borrow pandas from the center over extended periods of time (from one to many years) and generally involves the payment of a fee. The total number of panda loans from Wolong to other places in China and other countries increased from fewer than 20 in 1998 to 85 in 2010. There was a total of 63 new panda loans between 2004 and 2010 (Figure 17.2A).

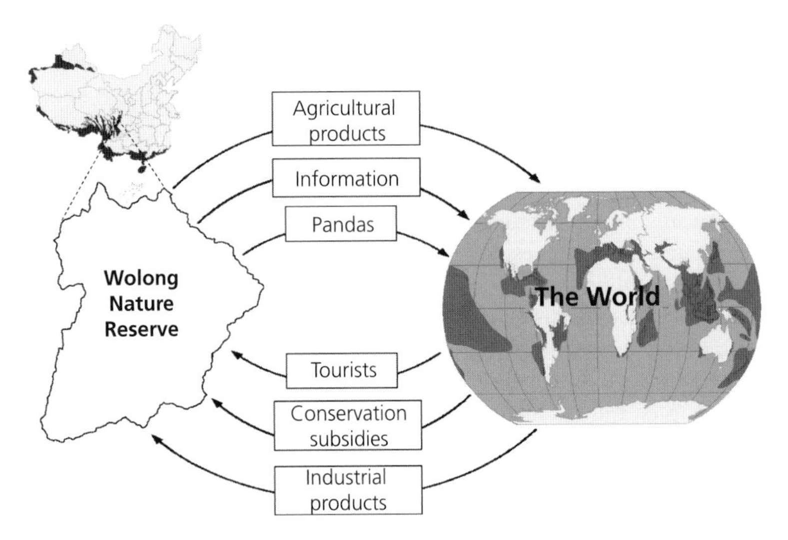

Figure 17.1 Schematic illustration of telecoupling processes between Wolong Nature Reserve in southwest China and the rest of the world. Shaded areas represent global biodiversity hotspots (Conservation International, 2011). Adapted from Liu et al. (2015a).

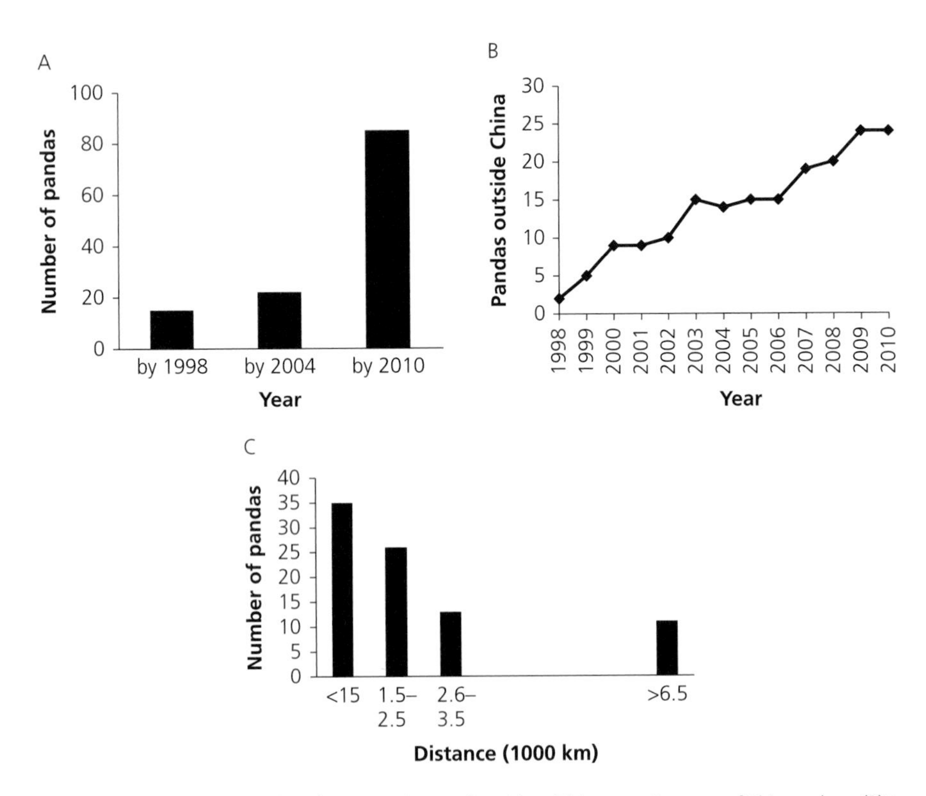

Figure 17.2 (A) Temporal changes in the number of giant pandas transferred from Wolong to other areas of China on loan; (B) temporal changes in the number of giant pandas transferred from Wolong to outside China on loan; (C) distances between Wolong and zoos hosting pandas from Wolong. Adapted from Liu et al. (2015a).

Telecoupled systems

In this case, the sending system for pandas is Wolong. The receiving systems are zoos inside and outside China, such as the Beijing Zoo, the San Diego Zoo, the National Zoo in Washington DC, and zoos in Europe (e.g., London) and in Asia (e.g., Kobe). A total of 22 cities in mainland China, Hong Kong, and Taiwan received at least one panda from Wolong, with Guangzhou hosting the largest number (i.e., ten; Figure 17.3). Furthermore, Wolong's pandas were sent to seven cities in five other countries (Japan, Australia, USA, Thailand, and Austria). A total of 12 went to Japan and seven to the United States in 2010 (Figure 17.3). The spillover systems are many, including areas from which people travel to see the pandas in those receiving systems. Spillover systems also include areas that provide funding for the loans, grow bamboo to feed the pandas, and are affected by other activities related to panda loans.

Agents

The agents include people and organizations that make panda loans possible. In the sending system, agents include the China Society for Wildlife Conservation and the State Forestry Administration that develop policies and agreements. Another agent is the Wolong Nature Reserve Administration Bureau that implements the policies (e.g., it selects which pandas are to be loaned). In the receiving systems, agents include people and organizations that lobby and find resources for panda loans. Zoos often seek

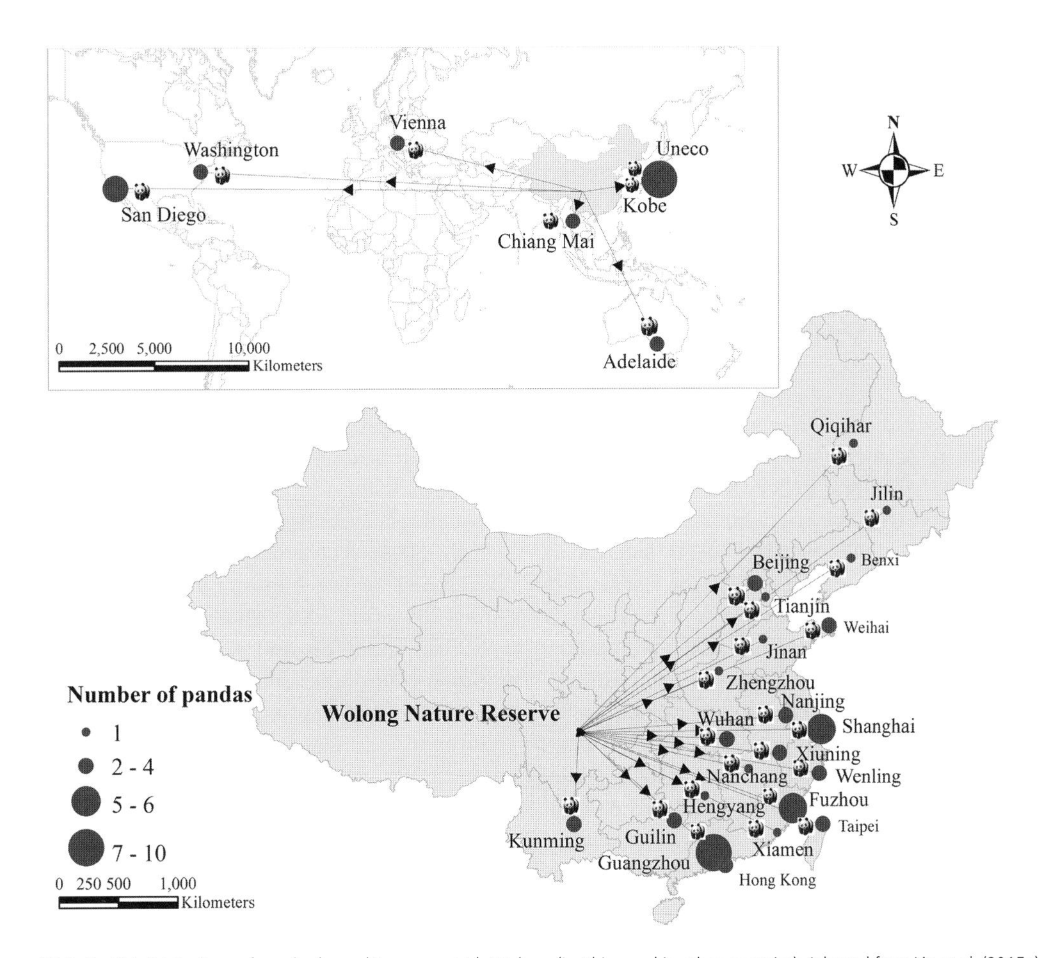

Figure 17.3 Spatial distributions of pandas loaned to zoos outside Wolong (in China and in other countries). Adapted from Liu et al. (2015a).

corporate sponsors to help fund the loans (*National Post*, 2010). There also may be agents from the spillover systems outside Wolong and the receiving systems that help negotiate panda loans. Negotiations often involve individuals with high-level positions in receiving countries' governments. For example, the Prime Minister in the UK visited China, while China's Vice Premier traveled to the UK as part of the negotiations on the panda loans to the Edinburgh Zoo (*The Guardian*, 2011).

Flows

The flows include the movement of pandas and people involved in the panda loans. The number of panda loans to other countries increased from two in 1998 to 24 in 2010 (Figure 17.2B). The numbers of pandas and the distances between Wolong and the receiving systems show non-linear relationships. Within 3500 km, as distances increase, the numbers of pandas decrease (Figure 17.2C). There were no pandas in places between 3500 km and 6500 km from Wolong, but beyond 6500 km the numbers of pandas increased (Figure 17.2C). Regarding flows of people, panda experts from Wolong provide training to the staff in the receiving systems. The information flows include exchanges of agreements and, in many cases, money transactions (fees amount to US$1 million per panda pair per year).

Causes

The causes behind panda loans include a variety of factors. The receiving systems have strong interests in pandas due to a long history of cultural affinity and fascination with the charismatic panda worldwide (Ellis et al., 2006, Schaller, 1994). Interest in scientific research has also grown in the last several decades due to many unique aspects of the panda's biology. Examples include the panda's adaptation to bamboo, narrow reproductive estrus window, and scent communication (Swaisgood et al., 2009). As threats to the panda's wild population increase, the impetus also increases to establish a sustainable population of captive pandas. These pandas could theoretically be used for population rescue via reintroduction (Wildt et al., 2006). A technological cause is rooted in recent improvements in captive breeding and infant care that have allowed for a tripling of the captive panda population from 1970 to 2000 (Zhang et al., 2006). Such success has allowed for more individuals to be available for loans. There are also economic causes in both sending and receiving systems, as zoos can substantially increase their visitation rates, which translate into gains of millions of dollars. The sending system can also be spurred to participate due to expected economic benefits through the loan deals (Buckingham et al., 2013). In addition, there is political will for panda loans by relevant leaders (Buckingham et al., 2013).

Effects

There are both environmental and socioeconomic effects in the sending, receiving, and spillover systems. The socioeconomic effects include publicity about sending and receiving systems, and economic benefits from panda loans. Costs of keeping pandas in zoos are considerable, as the construction of new facilities for one pair of pandas alone (e.g., indoor/outdoor enclosures, specialized heating/cooling systems, and educational exhibits) cost US$10.3 million at the Adelaide Zoo and US$14.5 million at the Toronto Zoo (Buckingham et al., 2013), in addition to operational costs. For the spillover systems, those visiting pandas in the receiving systems pay for entrance fees and travel costs. Visitors to the National Zoo in Washington DC to see pandas come from not only the USA, but also many other countries (Smithsonian National Zoological Park, 2012).

In addition, new panda loans improve social networks for scientific collaboration across countries via participation in international networks such as the Conservation Breeding Specialist Group (Wildt et al., 2006). The environmental impacts include awareness of the importance of panda conservation (Ellis et al., 2006, Schaller, 1994) by residents in sending, receiving, and spillover systems. A lesser known environmental impact on the receiving system is the large amount of bamboo required to sustain captive pandas (up to 32 kg per day per panda). Zoos often establish bamboo plantations to meet this large and specialized food requirement. Edinburgh Zoo imported bamboo from Holland (a spillover system) at over US$100 000 per year (Brown, 2011, Buckingham et al., 2013). Other zoos seek to collaborate with local citizens to grow bamboo on their properties (Buckingham et al., 2013).

Another environmental impact is the CO_2 emissions of transporting pandas from sending to receiving systems as well as of tourists traveling to see the pandas in the receiving systems. For example, a Boeing 777 jet flight emits roughly 29 kg of CO_2 per km (BlueSkyModel, 2014). Given the 8000-km distance between Chengdu and Edinburgh, Scotland (a recent receiving system), transporting a pair of pandas in a Boeing 777 could emit 232 000 kg of CO_2 one way. However, Edinburgh did not have enough bamboo to support the pandas for the trip. Therefore, they financed a US-operated plane originating in Memphis, Tennessee, to fully load up with bamboo and then travel to Chengdu to pick up the pandas prior to the Edinburgh leg (BBC, 2011), an additional 12 550 km and over 360 000 kg of CO_2. The amount of CO_2 emissions caused by tourists may vary depending on the distance and mode of traveling. But emissions from tourism would be less per capita than a panda because several hundred tourists can travel on one plane, while one or two pandas take up an entire plane on their own. For example, a passenger traveling in economy class from Detroit, USA, to Beijing, China, and then Chengdu, China (the closest airport to Wolong) would generate roughly 1705 kg of CO_2 (International Civil Aviation Organization, 2014).

Feedbacks from panda loans are many. For example, the first pair of pandas sent to the National Zoo in Washington DC garnered such widespread appeal that a second pair of pandas was later sent after the first died. Some of the revenues from panda loans have been targeted for panda conservation in the sending system and spillover systems in other areas of panda habitat. Receiving systems have also undertaken capacity-building endeavors in Wolong, sending experts to both train and learn from Wolong's scientists. On the environmental side, the continued interests in panda loans have raised concerns about the well-being of captive pandas, and there were appeals and discussions to limit the number of panda loans (Schaller, 1994).

17.2.2 Tourism

Tourism is one of the largest industries in the world (World Travel and Tourism Council, 2014). With nearly 266 million direct jobs, tourism and its related economic activities produce 9.5% of the World Gross Domestic Product with almost one billion international travelers in 2013 (World Travel and Tourism Council, 2014). Nature-based tourism (focused on observing and appreciating nature) has been the fastest growing sector of tourism since the 1980s (Newsome et al., 2002). To meet the increasing demands for nature-based tourism, many nature reserves around the world have hosted tourists. For example, by the late 1990s ~80% of nature reserves in China had developed ecotourism (a type of nature-based tourism). Almost 16% of the nature reserves each hosted more than 100 000 tourists annually (Chinese National Committee for Man and the Biosphere, 1998, Li and Han, 2001). Like many other nature reserves, Wolong has attracted a large number of tourists since the early 1980s (Liu, 2012).

Telecoupled systems

For tourists, Wolong is the receiving system. The sending systems are the places in the rest of the country (for domestic tourists, Figure 17.4) and the rest of the world (Figure 17.5) where tourists visiting Wolong originate. During the summers of 2006 and 2007 we surveyed tourists visiting captive pandas held at the CCRCGP in Wolong. Our tourist surveys indicated that more than half (651) of the 1063 sampled tourists in 2006 and 2007 were from at least 30 provinces and cities in mainland China, Taiwan, and Hong Kong (Figure 17.4). While a large number of Chinese tourists lived in the province of Sichuan where Wolong is located (28.6%) and the neighboring municipality of Chongqing (15.8%), more distant provinces and cities were also represented. Guangdong (6.4%), Beijing (2.7%), and Shanghai (2.0%) also sent large numbers of tourists to Wolong. Internationally, we recorded 26 different countries of origin. The majority of international tourists were from Japan (13.4%), the USA (7.9%), the UK (5.0%), France (2.8%), and The Netherlands (2.6%; Figure 17.5). The spillover systems are areas in the rest of the world that support the supply chain industry of tourism. Spillover systems might also include the stopover cities along the travel route to Wolong, such as Beijing, Shanghai, and Chengdu, which provide services to tourists.

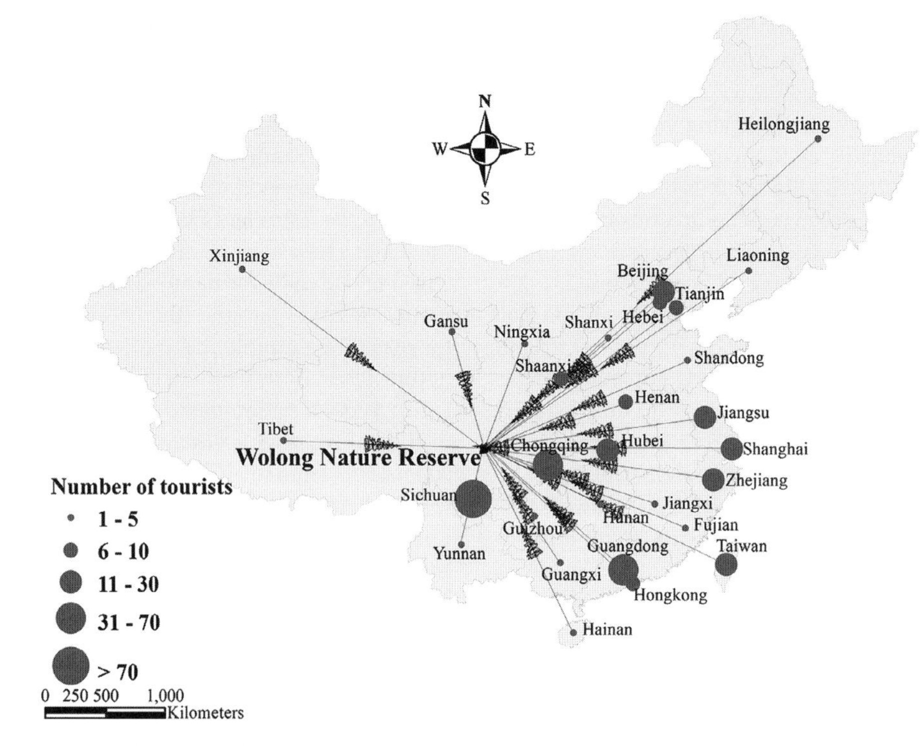

Figure 17.4 Spatial distribution of sampled tourists to Wolong from other parts of China, 2006–2007. Adapted from Liu et al. (2015a).

Agents

Diverse agents are involved in tourism in Wolong. They include government agencies and officials (e.g., Sichuan Tourism Bureau, Sichuan Forestry Department, and Wolong Administration Bureau) who develop and implement tourism policies. Other agents include tourism agencies that facilitate and attract tourists (e.g., Jiuzhaigou Scenic Area Administration, and investment companies such as Luneng Xinyi Ltd. Co., a subsidiary

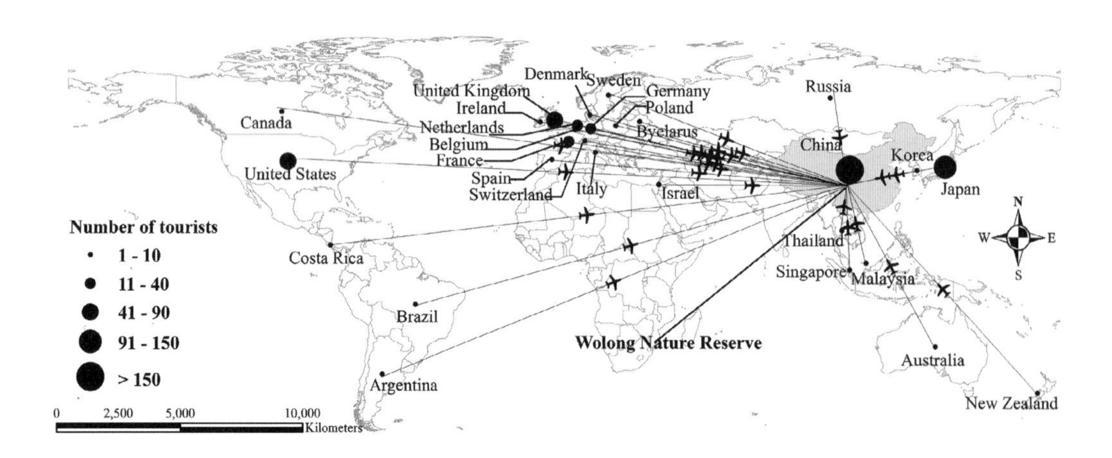

Figure 17.5 Spatial distribution of sampled tourists traveling to Wolong from other countries, 2006–2007. Adapted from Liu et al. (2015a).

of a large state-owned enterprise in Shandong Province). Wolong Tourism Development, Inc., a government-owned company, was established in 1991 to organize and promote visitation to Wolong. In 1997, the company was converted into the Department of Tourism, an official governmental agency under the Wolong Administration Bureau, to be responsible for all tourism planning and management issues (Liu et al., 2015b). Agents in the receiving system also include local residents engaged in supporting the industry. Other agents in sending and spillover systems include those who arrange trips for tourists (e.g., tourist agents who make travel arrangements, foreign affairs officers who issue passports, and embassies and consulates general that issue visas). People who provide services to tourists in stopovers such as Beijing (for international tourists) and Chengdu (for international tourists and domestic tourists coming from outside Sichuan Province) are also agents.

Flows

The number of tourists to Wolong increased dramatically over time (Figure 17.6A). Tourism started in the early 1980s and reached a peak in 2006, with an annual visitation of 220 000 visitors (Figure 17.6A). There was a drop in 2003 and a complete stop in tourism after the 2008 Wenchuan earthquake. For domestic

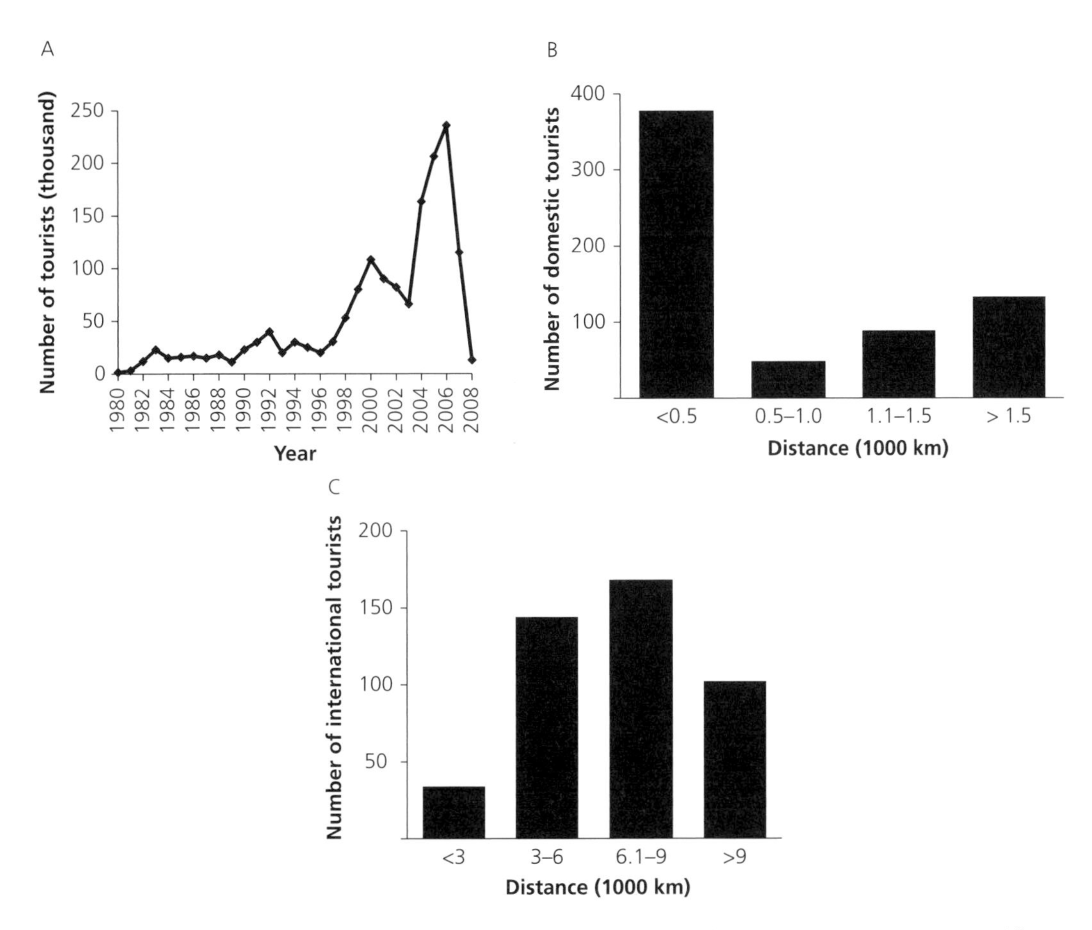

Figure 17.6 (A) Changes in the total numbers of tourists to Wolong over time; (B) number of sampled Chinese domestic tourists at different distances between their origins and Wolong, 2006–2007; (C) number of sampled international tourists at different distances between their origins and Wolong, 2006–2007. Adapted from Liu et al. (2015a).

tourists, the largest number of visitors came from nearby areas (< 500-kilometer range, e.g., from Sichuan and Chongqing, Figure 17.6B). Beyond this distance, the number of tourists dropped markedly by 87% and then increased slightly with distance. For foreign tourists (Figure 17.6C), the largest numbers occurred at intermediate distances. Aside from tourists, money also flows into Wolong for tourism-related infrastructure. In 2002, for example, the Luneng Company in Shandong Province invested 42 million yuan to upgrade the Wolong Hotel to four-star level with 668 beds (Wolong Nature Reserve, 2005).

Causes

Tourism is affected by economic, political, technological, cultural, and ecological factors. On one hand, many local people in Wolong are enthusiastic about participating in tourism as they have a strong desire for income from tourism. Government agencies have been actively promoting tourism, and technological advances, especially communication technologies and transportation, have played important roles in promoting Wolong tourism. On the other hand, there are demands for nature-based tourism by people in other parts of China and the world. Wild pandas, natural forests and wildlife, and clean air and water were the top three reasons that motivated domestic tourists to come to the reserve. For international tourists, the top three were natural forests and wildlife, wild pandas, and pandas in captivity (Liu, 2012). Both external and internal disturbances shaped tourism over time. For example, the Severe Acute Respiratory Syndrome (SARS) epidemic occurring across China led to a drastic drop in the number of tourists in 2003 (Figure 17.6A). In 2008, the 8.0 M_s Wenchuan earthquake completely stopped tourism (Figure 17.6A), as Wolong was near the epicenter of the earthquake. Most of the infrastructure was destroyed by this disaster and associated landslides (see Chapter 12). Destroyed infrastructure included tourism facilities and the main road connecting Wolong to outside markets. Furthermore, landslides have become much more frequent after the earthquake. The road has been repeatedly destroyed even after many rounds of repairs. As a result, so far tourism has not yet recovered.

Effects

There are a variety of socioeconomic and environmental effects of tourism. Here we use the data from 220 households in Wolong that we have sampled since 1998 to illustrate the socioeconomic effects of tourism in Wolong. The number of sampled households that directly participated in tourism-related activities increased from nine (4%) in 1998 to 60 (28%) in 2007. Approximately 77% of local rural households received income associated with tourism directly or indirectly (Liu, 2012). For example, a total of 116 households (including 87 households that were not directly involved in tourism) reported having received some income from temporary labor jobs on infrastructure construction. A number of households also reported having earned income from selling locally produced products such as local medicinal herbs (14 tourism households and 25 non-tourism households). Other households sold honey (6 tourism households and 19 non-tourism households), and smoked pork (10 tourism households and 12 non-tourism households). Some of these products were sold to restaurants, shops, and street vendors in Wolong, much of which was later sold to tourists. The locals used the remaining products themselves or sold them directly to tourists and markets outside Wolong. The composition of income between tourism households and non-tourism households differed substantially. For tourism households, direct and indirect tourism income was most important, and their non-farm income percentage increased from 40% to 66% between the late 1990s and mid-2000s (Liu, 2012). In contrast, non-tourism households generally earned more farm income, while their non-farm income percentage remained basically the same (~36% to 38%) from the late 1990s to mid-2000s. Furthermore, the development of tourism has motivated the community to upgrade local infrastructure. Tourism contributed to the transformation of Wolong's traditional subsistence agriculture economy into a diverse, modern, and more service-oriented one (He et al., 2008, Liu, 2012).

Environmental effects of tourism on the receiving system are direct or indirect. For direct effects, tourists influence vegetation along trails. To estimate the impacts of tourists using trails on vegetation, we conducted field sampling along four main hiking trails in Wolong between June and August 2007. Sites (n = 64)were sampled at regular intervals of 200–300 m. We found that there were more plant species occurring at trailsides than in the forest interior in the shrub, sapling, and seedling layers. Herbaceous species richness at trailsides was also higher than in the forest interior (Liu, 2012). This effect may be attributed to new niches opening up due to the disturbance created by people and livestock trampling the soil. Fecal deposition by livestock along trails also added nutrients to the soil (Liu, 2012).

Some tourists from all over the world also donate to Wolong after visiting to provide support for captive breeding and research, helping it to become the largest captive giant panda breeding center in the world. Besides regular donations in the form of cash and goods, donations are also delivered through an adoption program that allows donors to "adopt" a captive panda with the donated funds.

Tourists also have many indirect effects. The construction of roads, facilities, and other infrastructure has a negative impact on the local ecosystem (e.g., fragmentation of giant panda habitat; Hull et al., 2011). On the other hand, tourists affect the environment positively through altering the livelihoods of local residents. Some tourists purchase local products, create job opportunities (e.g., having locals as guides or employees at tourism facilities), and bring in new information (e.g., about markets, jobs, and technology). By helping to increase local residents' income, tourism has helped increase the affordability of electricity and thus reduce fuelwood collection in panda habitat. For instance, households engaged in restaurant or hotel operation were more likely to decrease their fuelwood consumption compared to those who were not (Liu, 2012). Engagement in nature-based tourism activities has enhanced local residents' awareness of conservation issues (Liu et al., 2012).

In terms of effects on sending systems, visits to Wolong may improve tourists' quality of life, e.g., enriching experience (Neal et al., 2007). Visits may also reduce household spending and environmental impacts in sending systems when tourists are away. Although travel by tourists leaves an environmental footprint (Gössling et al., 2002), it may also have positive environmental effects. For example, many residents from the city of Chengdu in Sichuan Province spend an extended period of the summer in Wolong due to its cool weather (Liu, 2012). If they switch off the air conditioners at their homes when they are away, they can reduce the consumption of energy and GHG emissions. However, the net impact is unknown due to unavailable data on GHG emissions from transport and other activities.

For the spillover systems, tourism creates economic benefits that ripple through the chain of tourism-related industries (Balmford et al., 2009). For instance, tourism-related industries outside receiving and sending systems benefit from selling goods (e.g., outdoor clothes and hiking boots) and providing services to tourists. On the other hand, GHG emissions as a result of both national and international travel contribute to global climate change and thus affect all systems, including the spillover systems.

Tourism also generates feedback effects. After tourists visit Wolong, they disseminate information to friends and colleagues, thus potentially affecting the probability of visits by others. The past success of tourism within Wolong also provides an incentive that has attracted a large amount of outside investment for further tourism infrastructure development. One example is the recent construction of a new captive panda breeding center (Hong Kong SAR Government, 2012). According to Wolong's Master Plan for 2015–2025, proposals have also been put forth for the construction of a large hotel and mushroom plantation to support future tourism growth (Sichuan Academy of Forestry, 2014).

17.2.3 Information dissemination

Wolong has become increasingly known both nationally and internationally through the news media and publication of books and articles, as well as through visitors. Of course, information is also constantly entering Wolong from the outside, but due to data constraints here we focus only on the information emanating from the system.

Telecoupled systems

Wolong is the sending system. The receiving systems are places that publish articles and books on, create television programming about, and send visitors to Wolong. Spillover systems are other places that know about Wolong through reading relevant articles and books, watching the news or other programs, and receiving information from visitors to Wolong, as well as the Internet and social media.

Agents

In the receiving system, the main agents are news media outlets and book publishers as well as authors in disseminating information about Wolong. For news media outlets, there are usually major national and international media organizations, such as *The New York Times* (Simons, 2003). Book publishers also include major international publishing houses, such as University of Chicago Press (e.g., *The Last Panda* by Schaller (1994)). A variety of scholars and reporters, as well as writers, have written articles about Wolong. Many news agencies have also created specials on television programs, such as short stories on *Good Morning America* and documentaries on nature channels such as *Animal Planet* (Feldon, 2008). In the sending system, local scientists, conservation organizations, reserve managers, and residents are the main agents in providing information to the news media organizations and publishers. In the spillover systems, the main agents are readers and audiences who read news articles and books or watch videos related to Wolong.

Flows

Flows include the movement of information about Wolong and people who visited Wolong to collect such information. We examined the frequency of the phrase "Wolong Nature Reserve" in English books published since 1980 based on the n-gram corpus data set provided by Google™ Books (Brants and Franz, 2006). We also performed a search for "Wolong Nature Reserve" in all international news articles published in the English language since 1980 using the LexisNexis® Academic search engine. The frequency of the phrase "Wolong Nature Reserve" in English books has increased since 1980, especially after 2000 (Figure 17.7A). We found a total of 806 articles using the term "Wolong Nature Reserve" published since 1980 in international news media published in the English language. The number of articles published each year remained under 11 between 1980 and 1998 and then increased to 51 during a three-year period from 1999 to 2001 (Figure 17.7B). The numbers of articles fluctuated between 20 and 60 in each of the remaining years, except for a sharp peak in 2008 (213 articles; Figure 17.7B) due to the devastating earthquake.

Causes

A number of factors have led to the flows of information from Wolong to distant systems. Because of greater emphasis on sustainability worldwide, many global citizens are concerned about endangered and charismatic species like the giant panda, which results in a large amount of research about pandas in Wolong. What is more, as an emblem of China, the panda's important role in diplomacy has also attracted the media and scientists' attention. Being home to the panda as a global conservation icon, Wolong is an important place for

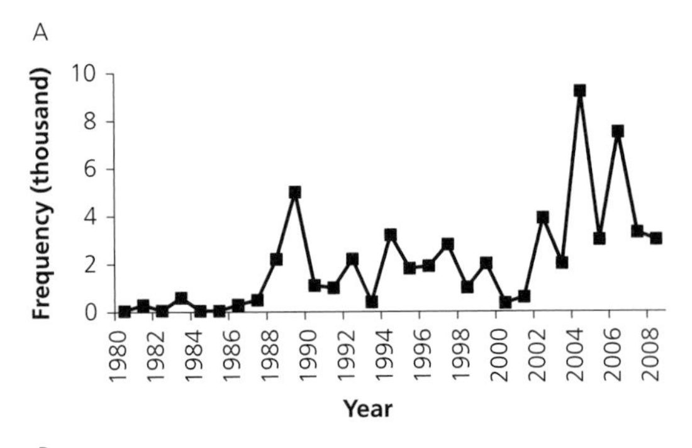

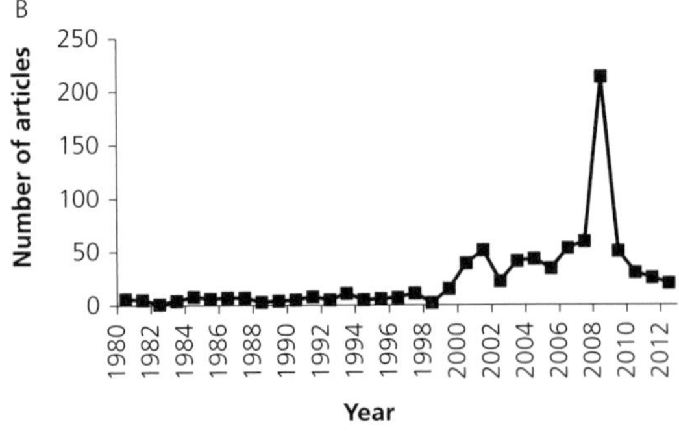

Figure 17.7 (A) Temporal changes in the frequency of "Wolong Nature Reserve" in published English books (unsmoothed) between 1980 and 2008; (B) number of international news articles containing the words "Wolong Nature Reserve" published in the English language from 1980 to 2012, as documented using the LexisNexis® Academic search engine. Adapted from Liu et al. (2015a).

panda conservation. As a flagship national nature reserve, it has received exceptional national and international financial and technical assistance (Liu et al., 2001). Although there are 67 nature reserves for panda conservation (State Council Information Office of China, 2015, Liu, 2015), Wolong is one of the first and one of the largest. Tourism is another mechanism that has helped spread information about Wolong. In addition, Wolong hosts a number of captive pandas in the breeding center and sends many of these pandas to places around the world (through the panda loan program described in Section 17.2.1). A number of events collectively led to the initial increase in media attention in the 1999–2001 period. These included a boom in captive panda breeding (Mouland, 2001), the arrival of the second pair of Wolong's pandas at the National Zoo in Washington DC (Pan, 2000), and the publication of a paper documenting the rapid decline of panda habitat in Wolong despite its protected status (Liu et al., 2001). The reason for the rapid increase in publications in 2008 stems from the Wenchuan earthquake (see example research articles on the earthquake impacts by Viña et al. (2011), Yang et al. (2013a), and Yang et al. (2015)).

Effects

The effects of information flows out of Wolong are numerous. Information shared about Wolong helps raise awareness among the general public regarding the plight of the endangered giant panda, as well as broader conservation issues in China and those facing other wildlife species around the world. Furthermore, scientific research related to giant panda conservation can inform decision-making surrounding conservation of other endangered species, not only in Wolong but also across the globe (Liu et al., 2003, Schaller, 1994). In Wolong, visits from journalists promote the local economy. Also, some researchers like us pay local field workers to assist with data collection and analysis.

Feedbacks also can occur because the information disseminated from Wolong helps attract tourists to visit Wolong from receiving systems, who engage in various activities that affect the sending system. Another feedback occurred when people around the world who had heard about or visited Wolong sent donations to support disaster relief after the Wenchuan earthquake (Liu, 2012). Also, the Chinese government referred to results from scientific research in Wolong (Liu et al., 2001) for more effective panda conservation.

17.2.4 Conservation subsidies

Wolong has received substantial external financial support from the Chinese government and the international community since its establishment as a nature reserve in 1963. Wolong is a national-level nature reserve and overseen by both the central government's State Forestry Administration in Beijing and the Forestry Department of Sichuan Province in Chengdu. Therefore, financial support from the governments is supplied regularly. For instance, a major road construction project was initiated in 1992 financed by a 35 million yuan investment from the central government (Liu, 2012). The goal of this project was to link Wolong to neighboring rural communities and outside markets. International organizations such as the World Wildlife Fund (WWF) have also invested in Wolong since the 1980s. Since the earthquake, the Hong Kong special administrative region government has committed a total of HK$1.58 billion (US$204 million as of May 2012) investment for earthquake reconstruction efforts in Wolong (*News.Gov.Hk*, 2014).

Conservation subsidies have also been provided to local residents in Wolong to compensate them for participation in conservation efforts. These include the Grain to Green Program (GTGP, since 2000) and the Natural Forest Conservation Program (NFCP, since 2001) (Liu et al., 2008). GTGP provides subsidies to farmers to convert their cropland on steep slopes to forests (see Chapter 13). In Wolong, NFCP offers subsidies for local households to monitor the forests to prevent illegal harvesting. In this section, we focus on the subsidies from NFCP and GTGP.

Telecoupled systems

Wolong is the receiving system for conservation subsidies while the rest of China (represented by the Chinese government) is the sending system. The rest of the world is the spillover system, as ecosystem services (e.g., carbon sequestration by forests) provided by Wolong can abate global climate change.

Agents

Government officials in China and farmers in Wolong are the agents who provide and receive financial support, respectively. Agents in the spillover system include the general public affected by ecosystem services provided by the programs.

Flows

The main flows for external financial support are monetary funds. The cumulative amount of NFCP payments between 2001 and 2007 was almost 18 million yuan (ranging from 2.2 to 2.6 million yuan per year). The cumulative amount of GTGP payments between 2000 and 2007 was over 10 million yuan (ranging from 0.96 to 1.32 million yuan per year; Figure 17.8).

Causes

The implementation of GTGP and NFCP in Wolong was due to both national and local factors. At the national level, the huge floods in 1998 prompted the establishment of these two conservation programs in order to improve soil water retention and prevent erosion (Liu et al., 2008). At the local level, the degradation of forests and panda habitat in Wolong caused the local government to seek ways to minimize the continuing destruction of forests and panda habitat. There were also economic causes. At the national level, the government recognized the need to improve the livelihoods of the millions of rural poor. At the local level, the administration recognized the need to provide alternative forms of income for residents living inside Wolong, who were economically limited by related conservation initiatives (e.g., the ban on timber harvesting).

Effects

In the receiving system, external financial support has had many positive effects. The majority of local residents identified positive effects from NFCP, such as economic gains and environmental benefits (e.g., reducing soil erosion and landslides; Yang et al., 2013c). For example, the subsidies from NFCP and GTGP respectively accounted for 5% and 8% of the average household income as of 2005 (Liu et al., 2013b). On the other hand, NFCP restricts forest use by locals. After NFCP started, electricity consumption in the reserve doubled, and the labor spent on fuelwood collection decreased by nearly half between 1998 and 2001 (Yang et al., 2013c). These programs also have helped forest and panda habitat recovery (Viña et al., 2011). Forest cover in Wolong decreased from 106 000 ha in 1965 to 70 000 ha in 2001 but recovered to 79 000 ha by 2007 through the implementation of these programs (Yang et al., 2013c). However, the negative effects of GTGP on

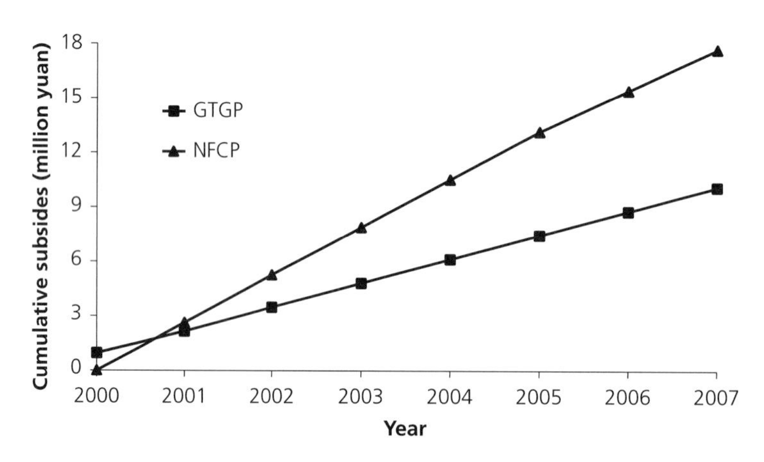

Figure 17.8 Cumulative amounts of subsidies to Wolong from the Chinese central government through the Natural Forest Conservation Program (NFCP) and Grain to Green Program (GTGP). Data are from government records on investment in NFCP and GTGP (Wolong Nature Reserve, 1998–2010, Wolong Nature Reserve, 2005). NFCP and GTGP payment rates are flat (i.e., the present values in each corresponding year are not discounted). Adapted from Liu et al. (2015a).

household income might outweigh its positive effects. The compensation level (3150 yuan per ha per year) might be too low to cover the potential income loss due to the lost cropland. Based on a household survey in 2006, 1 ha of cropland planted with off-seasonal cabbage (a cash crop) could bring as much as 15 times more income than the subsidies provided by GTGP (Liu et al., 2013b).

Feedbacks also occur. NFCP and GTGP have led to increases in forest cover, which enhances the capacity of carbon sequestration and climate change mitigation. More forest cover may also attract more tourists (Liu et al., 2012). The recovery of forests also improved habitat for many wildlife species, including crop-raiding species such as wild pigs. Increased crop raiding may motivate farmers to convert more farmland to forest land to reduce future crop losses and to find jobs in cities.

17.2.5 Trade of agricultural products

Agriculture has been the central livelihood strategy in Wolong for centuries (Ghimire, 1997). Households grow subsistence crops for direct household consumption and cash crops to sell to outside markets. Some of the main cash crops include cabbages, carrots, corn, and potatoes. Wolong residents also engage in livestock production, including raising yak, goats, sheep, cattle, pigs, and horses for both subsistence and selling meat to outside markets.

Telecoupled systems

Wolong is the sending system for trade of agricultural products as Wolong farmers sell products. The receiving systems are diverse, including the cities of Chengdu and Dujiangyan, which are 130 and 50 kilometers from Wolong, respectively. The spillover systems include other rural areas around China that are affected by the trade of agricultural products between Wolong and the receiving systems.

Agents

The agents in the trade of agricultural products include farmers in Wolong, traders, and consumers of agricultural products. Around 80% (306 of 381 laborers from 180 surveyed households) of Wolong laborers were involved in cash crop production in 2007 (Yang et al., 2013b). Agents in the spillover systems include agricultural sellers and buyers affected by changes in the market as a result of Wolong's trade activities.

Flows

Flows include food products transported out of Wolong and the money coming into Wolong. The total income earned from cabbages and livestock has been generally increasing from around 4.5 million yuan in 1998 to nearly 9.8 million yuan in 2007, or 11.7 million in 2007 after taking into account China's consumer price index increase of 19.6% during this period and standardizing to 1998 rates (Sichuan Provincial Bureau of Statistics, 2007; Figure 17.9A). However, the percentage of households with these income sources declined from over 90% to just over 60% during the same period (Figure 17.9B).

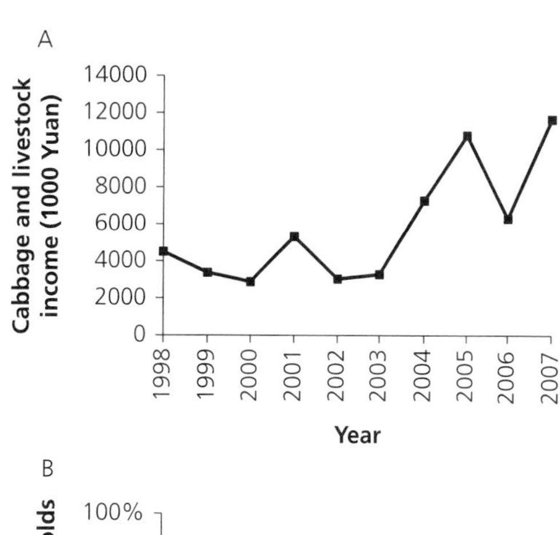

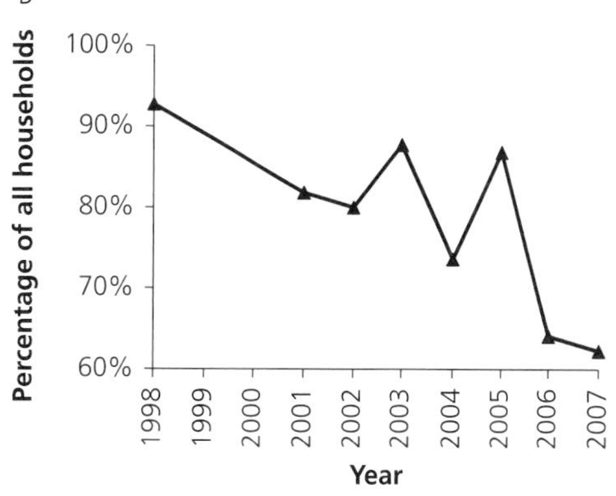

Figure 17.9 (A) Total income from cash crops and livestock in Wolong (× 1000 yuan), discounted to 1998 rates using consumer price index statistics from Sichuan Provincial Bureau of Statistics (2007); (B) percentage of sampled households with income from both cash crops and livestock. Adapted from Liu et al. (2015a).

Causes

There is a cultural cause stemming from the long history of agriculture in the reserve and the traditions passed down in family lineages. One of the other main causes of agricultural trade is economic. An important source of income for Wolong residents has come from producing agricultural products for outside markets. The increase in agricultural production in Wolong is in part attributed to the construction of the road that provides reliable access to outside markets. However, destruction of the road during and after the 2008 earthquake has severely impacted cash crop production in Wolong because farmers no longer have reliable means to transport their crops.

Effects

The trade of agricultural products has helped farmers in Wolong earn cash and improve their economic conditions. For example, in 2007, Wolong farmers in 159 surveyed households earned 1.34 million yuan from selling cabbages, potatoes, and livestock (Liu et al., 2013b). At the same time, those agricultural products helped meet the demands of people in cities. In the trade of agricultural products, money from selling the products is a strong feedback. Some of the income generated from the trade of agricultural products allows for the purchase of agricultural technology (e.g., industrial products) which can help produce more agricultural products.

17.2.6 Trade of industrial products

Demand for industrial products manufactured outside Wolong has also substantially increased over time. For example, in recent years, local people have invested more on agricultural inputs to improve crop production. This investment includes chemical fertilizers, e.g., carbamide, phosphate, and potash (Wolong Administration Bureau Department of Social and Economic Development, 2008). Also included is plastic film purchased to cover soils to decrease weeds, maintain temperature, and increase water retention.

Telecoupled systems

Wolong is the receiving system for industrial products. There are many sending systems, including the same cities that are the receiving systems for agricultural products. Other sending systems are technology hubs in cities farther away, such as Beijing, Shanghai, and even as far away as some in the USA. Spillover systems include areas under environmental impacts from technology production and application. For instance, fertilizers applied in Wolong may leach into the Pitiao River, which feeds into the 735-km Minjiang River stretching across Sichuan Province.

Agents

Agents include manufacturers and suppliers of industrial products such as fertilizers and plastic films that Wolong farmers purchase. Two of China's leading fertilizer production companies with plants in nearby cities include Sichuan Chemicals and Sichuan Meifeng Chemical, companies that are part of a greater network of China's industrial sector. Agents in spillover systems include farmers and the general public who are affected by accumulation of chemical fertilizers, as well as agricultural traders competing with Wolong farmers on the market.

Flows

Flows include money going out of Wolong and industrial products coming into Wolong. The total amount of fertilizers purchased increased from a little over 300 tons to 440 tons per year between 1998 and 2007 (Figure 17.10A). Just as the percentages of households with incomes from agricultural products declined, the percentages of households that purchased fertilizers each year also declined over time from more than 90% to just above 60% (Figure 17.10B). The total amount of plastic film purchased per year for the entire reserve also decreased over time from nearly 13 000 kg to 5000 kg, with some fluctuations (Figure 17.10C).

Causes

Many of the aforementioned causes of increased trade of agricultural products also apply to trade of industrial products. To increase agricultural production, large amounts of fertilizers and plastic films were needed. One other important cause of imported agricultural technology is that technology can improve the efficiency of land use and produce more agricultural products per unit of land. Such efficiency can help people earn more money and

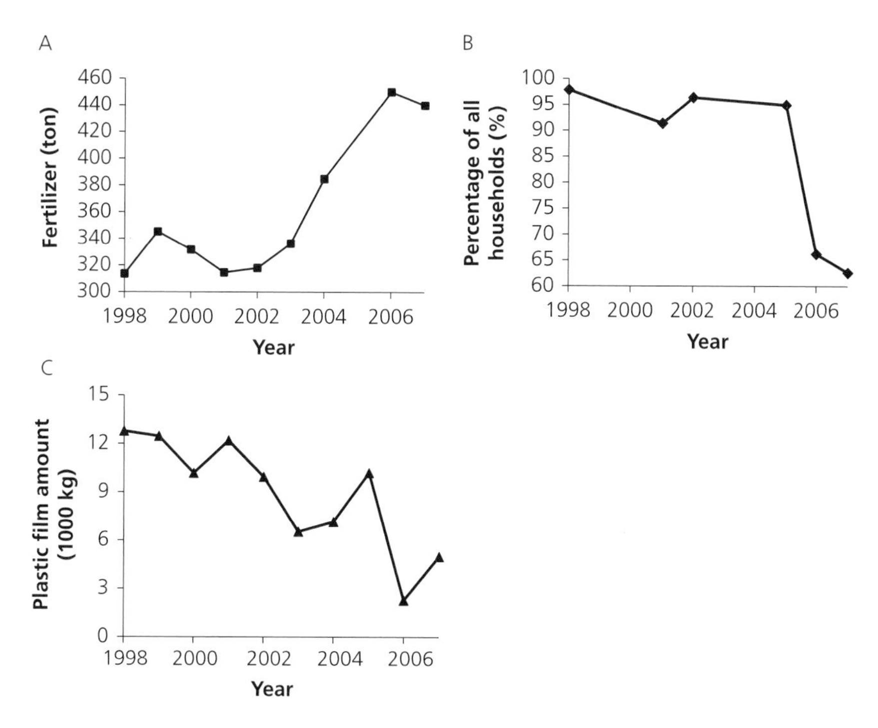

Figure 17.10 (A) Total amount of fertilizers purchased by Wolong farmers from outside markets over time; (B) percentage of sampled households purchasing fertilizers over time; (C) total amount of plastic film purchased by Wolong farmers over time. Adapted from Liu et al. (2015a).

mitigate the contraction of the area used as cropland due to reconversion to forest.

Effects

Using fertilizers made in cities has increased agricultural production but also polluted the local environment (e.g., soil and water) and affected organisms. In 2007, Wolong farmers in 177 surveyed households spent 114 000 yuan on fertilizers (Liu et al., 2013b). The effects of fertilizers extend to spillover systems. For instance, nitrogenous fertilizer production in China is known for having a large impact on GHGs. Chinese companies mainly use coal as an energy source to produce nitrogenous fertilizers, while the rest of the world mainly uses natural gas (Kahrl et al., 2010). GHGs emitted from fertilizer production impact distant systems all over the world via contributions to climate change. For example, the 440 tons of fertilizer applied in Wolong in 2007 (see Figure 17.10A) could emit as much as 4092 tons of CO_2 (9.3 tCO_2 tN^{-1}, including the fertilizer production process; Kahrl et al., 2010).

A positive feedback drives this telecoupling. The use of industrial products helps increase agricultural yield, thus generating more income that could be used to purchase more industrial products including agricultural technology.

17.3 Similarities, differences, and relationships among telecouplings

Among the telecoupling processes analyzed above, all but one increased in strength over time until 2008 when the Wenchuan earthquake occurred (Figure 17.11). The telecoupling process with the most pronounced change is the information spread about Wolong (occurrence of words "Wolong Nature Reserve" in published English books and articles). The only telecoupling that has decreased somewhat is the purchase of plastic film from outside. This change is most likely attributed to the decrease in cropland area due to the implementation of NFCP and GTGP. All telecouplings except NFCP and GTGP occurred in or before 1998. The

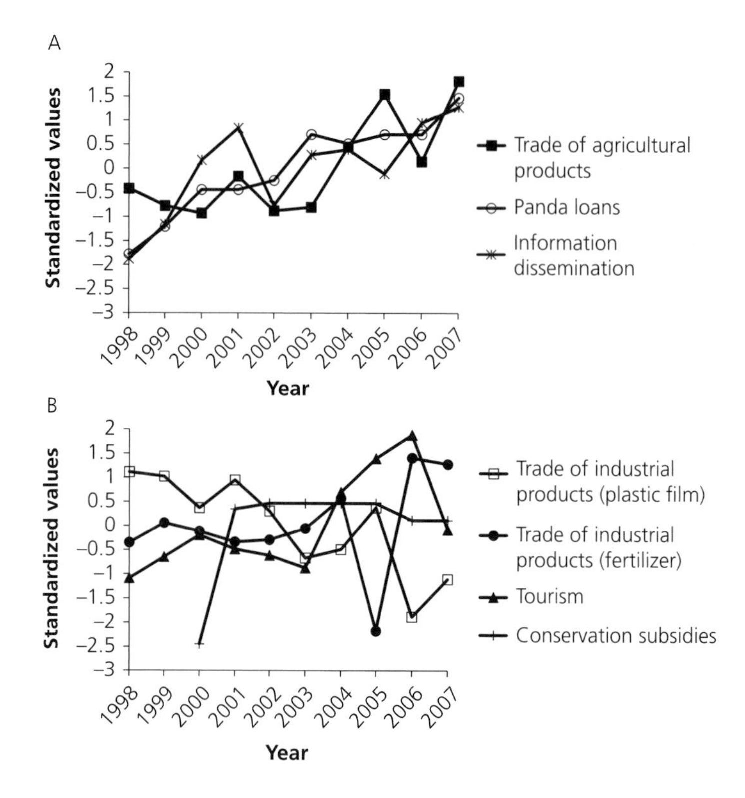

Figure 17.11 Changes in telecouplings for those with flows (A) leaving and (B) entering Wolong Nature Reserve from 1998 to 2007. Values shown are standardized across years. Trade in agricultural products refers to total income from cash crops and livestock in Wolong. Panda loans are the number of pandas sent from Wolong to outside zoos on loan. Information dissemination is the number of international news articles containing the words "Wolong Nature Reserve" published in the English language. Trade of industrial products is presented in two parts: total amount of fertilizers and total amount of plastic film purchased by Wolong residents. Tourism is represented by the total number of incoming tourists visiting Wolong. Conservation subsidies are total amounts of funding that Wolong received for the Natural Forest Conservation Program (NFCP) and the Grain to Green Program (GTGP). Adapted from Liu et al. (2015a).

implementation of GTGP and NFCP began in 2000 and 2001, respectively (Figure 17.11).

The relationships among the telecouplings are complex. They may enhance each other. Information spread may be a key driver for many other telecouplings, since it increased to a greater degree and occurred earlier than other telecouplings (Figure 17.11). For instance, information spread through media exposure has historically been a key driver of the increase in tourism to Wolong (Liu, 2012). During the early years of tourism development in Wolong, there was a spike in tourists' arrival in 1983 (from less than 10 000 to over 20 000 visitors). This sudden increase was triggered by one of the first international media reports on Wolong regarding a suspected panda starvation resulting from a mass bamboo flowering and die-off (Liu, 2012). Since then, the Chinese government has designated Wolong as a tourist attraction, invested in tourism facilities and promoted it through the global news media. Thus, the government has increased the exposure of Wolong to the rest of the world. For instance, 24% (n = 70) of the visitors to the Wolong breeding center whom we interviewed in 2005 said they had previously read print media reports on Wolong, and 29% had seen television programs (Liu, 2012). In addition, increased investment in tourism and increases in tourist volumes have in turn increased media exposure about Wolong. For instance, a Google search for "Wolong" and "tour" in English yielded over 80 000 hits as of September 2014, many of which were promotional tourist packages and tourist recommendations.

Panda loans have significantly affected the international information spread about Wolong. Around 20% of all media reports found in LexisNexis® about Wolong concern panda loans, which is more than any other topic. Such loans allow people from around the world who do not have the opportunity to visit Wolong to still learn about Wolong. Loaned pandas often require multimillion dollar investments in exhibits, many of which incorporate educational materials to share information about Wolong with visitors. Loaned pandas are also featured on webcams that are geared toward disseminating information to a wide audience. For instance, more than 237 000 hits were recorded on the panda cam website of the National Zoo in Washington DC within one day after the birth of a baby panda (*Associated Press*, 2012). This website thus serves as an information portal for Internet users worldwide. Information spread may also encourage donations to Wolong's breeding center. In recent years, such funds have helped pay for the operating costs of a giant panda reintroduction program geared toward releasing pandas to the wild (Pandas International, 2012).

Another example is conservation subsidies and tourism. When interviewed, local people in Wolong perceived a potential benefit from NFCP for tourism development, presumably due to their perception that protecting the forest would help attract nature-based tourists (Yang et al., 2013c). Another related conservation program called the Grain to Bamboo Program (GTBP) has an even greater link to tourism. The GTBP pays local people to plant bamboo mainly along the roadways, in part to improve the roadside aesthetic appeal of the reserve to visiting tourists (Yang et al., 2013c).

Telecouplings may also offset each other. The trade of agricultural and industrial products is at odds with off-farm income-generation activities. For example, the decrease in the amount of plastic film used may in part be due to an over 50% reduction in an average household's agricultural land as a result of GTGP. Tourism also provides a source of non-farm income that may discourage agricultural development, and in turn, agricultural trade. For example, households with tourist-related income had on average 50% less farmland compared to households not engaged in tourism in 2006 (average of 0.63 vs 1.25 mu, 1 mu = 0.0667 ha; Liu et al., 2012). Households with more initial cropland are also significantly less likely to take up tourism as an alternative income source (Yang, 2013). Off-farm telecouplings also affect each other. For instance, tourism had a significant positive effect on total household income, but the effect of tourism was also significantly dependent upon the household's participation in NFCP (Yang, 2013; see also Chapter 9). Tourism and NFCP participation were antagonistic in affecting total household income because those with lower tourism income only had higher total income if their income from NFCP was also high (Yang, 2013). Although tourism provides a potentially higher economic gain for locals who participate, it is less stable compared to agriculture, NFCP, and GTGP. The instability leaves many locals wary about investing in tourism activities (e.g., see fluctuations in Figure 17.6A; Yang, 2013).

There are spatial overlaps between some telecoupling processes. Perhaps this is best illustrated via the relationship between panda loans and tourists (e.g., destinations of panda loans and origins of tourists, Figures 17.3–17.5). After excluding the tourists coming from within Sichuan and the nearby Chongqing municipality, there is a significant positive correlation between panda loan destinations and origins of tourists across China ($R = 0.72$, $p < 0.05$). Three of the top four locations are shared between panda loans and tourists (Beijing, Guangzhou, and Shanghai). Internationally, the top two countries for both panda loans and tourists are also shared: Japan and the USA. These similarities reflect the locations with high human populations in developed areas that would be interested and have resources to undertake both international travel and hosting pandas in zoos. Loans began to increase several years prior to the boom in tourism (Figure 17.11), largely due to the rapid increase in breeding technologies that preceded tourism development. Loans may have encouraged subsequent tourism by raising the international status of Wolong.

Telecouplings may also lead to the formation of other telecouplings. The panda loans are a good example because they induce and are induced by other telecouplings such as trade agreements between China and other countries, including procurement of cars, renewable energy, and other resources by China (Buckingham et al., 2013). For instance, the loan of pandas to the Edinburgh Zoo in 2011 was part of a £2.6 bn (US$3.94 billion) collection of business deals, including China securing rights to oil from a Scottish oil refinery (*The Guardian*, 2011).

17.4 Discussion

This chapter represents the first effort to study multiple telecouplings across borders and across local to global scales. It uncovers many similar and different spatiotemporal patterns and relationships among multiple telecouplings. The distance-defying patterns illustrated by tourism and panda loans suggest that geographical proximity is not necessarily the only determinant of telecouplings. For example, tourist destination choices are mainly determined by distance and cost, but this process is also heavily shaped by tourist motivations such as discovering new places and experiencing other cultures (Fesenmaier et al., 2006, Nicolau and Mas, 2006).

Telecouplings illustrated in this study are quite common around the world. Some telecouplings may be independent of scale and context, behaving in a similar manner across systems at different scales and in different contexts. For example, almost all rural areas import industrial products such as fertilizers produced elsewhere. On the other hand, many farmers sell agricultural products to outside markets (Jacoby, 2000). Many rural areas are destinations for tourists who live in cities (Lane, 1994). Information about various places is disseminated worldwide through publications, mass media, the Internet, and other communication channels. Financial support from external sources for conservation (e.g., payments for ecosystem services) is increasingly common worldwide (Chen et al., 2009, Yang et al., 2013c). Pandas are endemic to China and panda loans are relatively limited at the global scale. But many countries or places offer other wildlife species such as tigers, zebras, alligators, lions, and wolves to numerous zoos (Braverman, 2010). In many ways, the presence of other wildlife species in zoos plays roles (e.g., education) similar to the role of loaned pandas.

It is much more challenging to study telecouplings than local couplings (human–nature interactions within a system) because telecouplings involve many components that go beyond a single location, across multiple scales, and across administrative boundaries. Naturally it is even more challenging to study and quantify multiple telecouplings simultaneously than one telecoupling at a time. As a result, many research gaps exist. The biggest unknown ones are the spillover systems. In some cases, it is even unclear where the spillover systems are. Furthermore, many other environmental and socioeconomic effects across the telecoupled systems are not measured quantitatively. Feedbacks and relationships among multiple telecouplings require further quantification. While much remains to be done, this study lays a good foundation for future research and management to enhance positive effects and reduce negative effects of telecouplings on environmental sustainability and human well-being.

17.5 Summary

Many studies have focused on human–nature interactions within a particular area. There is little research on multiple reciprocal interactions, simultaneous socioeconomic and environmental impacts, and relationships with other areas. This chapter addressed these important knowledge gaps by applying the new integrated framework of telecoupling (socioeconomic and environmental interactions among two or more areas over distances). Results show that even the small and remote model coupled system of Wolong Nature Reserve had multiple telecoupling processes with the rest of the world. These included panda loans, tourism, information dissemination, conservation subsidies, and trade of agricultural and industrial products. The telecoupling processes exhibited non-linear patterns and have varying socioeconomic and environmental effects in various areas across the world. For example, as of 2010, 85 pandas had been loaned from Wolong to zoos in several countries. These loans have diverse effects such as introducing considerable economic costs to the receiving zoos and creating diffuse CO_2 emissions via animal transport. The chapter explored the substantial similarities, differences, and relationships among different telecouplings, which cannot be detected by traditional separate studies. For instance, most of the telecouplings examined have been increasing over time. Telecouplings may also offset one another. Tourism was at odds with agricultural production and trade. Households in Wolong that participated in tourism had on average 50% less farmland compared to households not engaged in tourism in 2006. Such an integrated study leads to a more comprehensive understanding of distant human–nature interactions and has important implications for global sustainability and human well-being.

References

Associated Press (2012) After a Surprise Panda Birth in DC, Anxiety Awaits. http://www.wivb.com/dpp/news/nation/After-a-surprise-panda-birth-in-DC-anxiety-awaits_79060825.

Avissar, R. and Werth, D. (2005) Global hydroclimatological teleconnections resulting from tropical deforestation. *Journal of Hydrometeorology*, **6**, 134–45.

Balmford, A., Beresford, J., Green, J., et al. (2009) A global perspective on trends in nature-based tourism. *PLoS Biology*, 7, e1000144.

BBC (2011) Who, What, Why: how do you fly a panda 5,000 miles? *BBC News Magazine*. http://www.bbc.com/news/magazine-16000130.

Blueskymodel (2014) *1 Air Mile*. http://blueskymodel.org/air-mile#average-aircraft.

Brants, T. and Franz, A. (2006) *Web 1T 5-gram Version 1*. Linguistic Data Consortium, Philadelphia, PA.

Braverman, I. (2010) Zoo registrars: a bewildering bureaucracy. *Duke Environmental Law and Policy Forum*, 21, 165.

Brown, J. (2011) Zoo Orders Chinese Food Delivery from Holland: Edinburgh-based panda breeding plan described as "madness" after bamboo is sourced in Netherlands.*Independent*. http://www.independent.co.uk/environment/nature/zoo-orders-chinese-food-delivery-from-holland-6258681.html.

Buckingham, K.C., David, J.N.W., and Jepson, P. (2013) Environmental reviews and case studies: diplomats and refugees: panda diplomacy, soft "cuddly" power, and the new trajectory in panda conservation. *Environmental Practice*, **15**, 262–70.

Carter, N.H., Viña, A., Hull, V., et al. (2014) Coupled human and natural systems approach to wildlife research and conservation. *Ecology and Society*, 19, 43.

Chen, X., Lupi, F., He, G., and Liu, J. (2009) Linking social norms to efficient conservation investment in payments for ecosystem services. *Proceedings of the National Academy of Sciences of the United States of America*, 106, 11812–17.

Chen, X., Frank, K., Dietz, T., and Liu, J. (2012) Weak ties, labor migration, and environmental impacts: towards a sociology of sustainability. *Organization & Environment*, **25**, 3–24.

Chinese National Committee for Man and the Biosphere (1998) *Nature Reserves and Ecotourism*. Science and Technology Press of China, Beijing, China (in Chinese).

Conservation International (2011) *Critical Ecosystem Partnership Fund*. http://www.conservation.org/how/pages/hotspots.aspx.

Daily, G. (1997) *Nature's Services: Societal Dependence on Natural Ecosystems*. Island Press, Washington, DC.

DeFries, R.S., Rudel, T.K., Uriarte, M., and Hansen, M. (2010) Deforestation driven by urban population growth and agricultural trade in the twenty-first century. *Nature Geoscience*, **3**, 178–181.

Eakin, H., DeFries, R., Kerr, S., et al. (2014) Significance of telecoupling for exploration of land-use change. In K.C. Seto and A. Reenberg, eds, *Rethinking Global Land Use in an Urban Era*, pp. 141–62. MIT Press, Cambridge, MA.

Ellis, S., Pan, W., Xie, Z., and Wildt, D. (2006) The giant panda as a social, biological, and conservation phenomenon. In D.E. Wildt, A. Zhang, H. Zhang, D.L. Janssen, and S. Ellis, eds, *Giant Pandas: Biology, Veterinary Medicine, and Management*, pp. 1–16. Cambridge University Press, Cambridge, UK.

Feldon, A. (2008) *Pandamonium*. http://www.imdb.com/title/tt2842562/?ref_=nm_flmg_dr_3.

Fesenmaier, D.R., Wöber, K.W., and Werthner, H. (2006) *Destination Recommendation Systems: Behavioral Foundations and Applications*. CABI Pub., Cambridge, MA.

Folke, C., Jansson, Å., Rockström, J., et al. (2011). Reconnecting to the biosphere. *Ambio*, **40**, 719–38.

Garrett, R.D., Rueda, X., and Lambin, E.F. (2013) Globalization's unexpected impact on soybean production in South America: linkages between preferences for non-genetically modified crops, eco-certifications, and land use. *Environmental Research Letters*, 8, 044055.

Gasparri, N.I. and De Waroux, Y.L.P. (2014) The coupling of South American soybean and cattle production frontiers: new challenges for conservation policy and land change science. *Conservation Letters*. DOI: 10.1111/conl.12121.

Gasparri, N.I., Kuemmerle, T., Meyfroidt, P., et al. (2015) The emerging soybean production frontier in southern Africa: conservation challenges and the role of south-south telecouplings. *Conservation Letters*. DOI: 10.1111/conl.12173.

Ghimire, K.B. (1997) Conservation and social development: an assessment of Wolong and other panda reserves in China. In K.B. Ghimire and M.P. Pimbert, eds,

Environmental Politics and Impacts of National Parks and Protected Areas, pp. 187–213. Earthscan Publications, London, UK.

Gössling, S., Hansson, C.B., Hörstmeier, O., and Saggel, S. (2002) Ecological footprint analysis as a tool to assess tourism sustainability. *Ecological Economics*, **43**, 199–211.

The Guardian, (2011)Giant Pandas from China Help Seal New Business Deals with UK. http://www.theguardian.com/world/2011/jan/10/pandas-china-uk-oil-refinery-deal.

He, G., Chen, X., Liu, W., et al. (2008) Distribution of economic benefits from ecotourism: a case study of Wolong Nature Reserve for giant pandas in China. *Environmental Management*, **42**, 1017–25.

Hoekstra, A.Y. and Wiedmann, T.O. (2014) Humanity's unsustainable environmental footprint. *Science*, **344**, 1114–17.

Hong Kong SAR Government (2012) SDEV Inspects HK-funded Reconstruction Projects in Wolong. *Hong Kong Government News*. http://www.info.gov.hk/gia/general/201204/12/P201204120517.htm.

Hull, V., Xu, W., Liu, W., et al. (2011) Evaluating the efficacy of zoning designations for protected area management. *Biological Conservation*, **144**, 3028–37.

International Civil Aviation Organization (2014) *Carbon Emissions Calculator*. http://www.icao.int/environmental-protection/CarbonOffset/Pages/default.aspx.

Jacoby, H.G. (2000) Access to markets and the benefits of rural roads. *The Economic Journal*, **110**, 713–37.

Kahrl, F., Li, Y., Su, Y., Tennigkeit, T., et al. (2010) Greenhouse gas emissions from nitrogen fertilizer use in China. *Environmental Science & Policy*, **13**, 688–94.

Kastner, T., Erb, K.-H., and Nonhebel, S. (2011) International wood trade and forest change: a global analysis. *Global Environmental Change*, **21**, 947–56.

Konar, M., Dalin, C., Suweis, S., et al. (2011) Water for food: the global virtual water trade network. *Water Resources Research*, 47, W05520.

Lambin, E.F. and Meyfroidt, P. (2011) Global land use change, economic globalization, and the looming land scarcity. *Proceedings of the National Academy of Sciences of the United States of America*, **108**, 3465–72.

Lane, B. (1994) What is rural tourism? *Journal of Sustainable Tourism*, **2**, 7–21.

Levitt, T. (1982) *The Globalization of Markets*. Division of Research, Graduate School of Business Administration, Harvard University, Boston, MA.

Li, C., Zhou, S., Xiao, D., et al. (1992) A general description of Wolong Nature Reserve. In Wolong Nature Reserve, Sichuan Normal College, eds, *The Animal and Plant Resources and Protection of Wolong Nature Reserve*, pp. 313–25. Sichuan Publishing House of Science and Technology, Chengdu, China (in Chinese).

Li, W. and Han, N. (2001) Ecotourism management in China's nature reserves. *Ambio*, **30**, 62–63.

Liu, J. (2014) Forest sustainability in China and implications for a telecoupled world. *Asia & the Pacific Policy Studies*, **1**, 230–50.

Liu, J. (2015) Promises and perils for the panda. *Science*, **348**, 642.

Liu, J., Dietz, T., Carpenter, S.R., et al. (2007) Complexity of coupled human and natural systems. *Science*, **317**, 1513–16.

Liu, J., Hull, V., Batistella, M., et al. (2013a) Framing sustainability in a telecoupled world. *Ecology and Society*, **18**, 2.

Liu, J., Hull, V., Luo, J., et al. (2015a) Multiple telecouplings and their complex interrelationships. *Ecology and Society*. **20**, 44.

Liu, J., Hull, V., Moran, E., et al. (2014) Applications of the telecoupling framework to land-change science. In K.C. Seto and A. Reenberg, eds, *Rethinking Global Land Use in an Urban Era*, pp. 119–39. MIT Press, Cambridge, MA.

Liu, J., Li, S., Ouyang, Z., et al. (2008) Ecological and socioeconomic effects of China's policies for ecosystem services. *Proceedings of the National Academy of Sciences of the United States of America*, **105**, 9477–82.

Liu, J., Linderman, M., Ouyang, Z., et al. (2001) Ecological degradation in protected areas: the case of Wolong Nature Reserve for giant pandas. *Science*, **292**, 98–101.

Liu, J., McConnell, W., and Luo, J. (2013b) *Wolong Household Study [China]*. ICPSR34365-v1. Inter-university Consortium for Political and Social Research, Ann Arbor, MI. http://doi.org/10.3886/ICPSR34365.v1.

Liu, J., Mooney, H., Hull, V., et al. (2015b) Systems integration for global sustainability. *Science*, 347(6225), 1258832.

Liu, J., Ouyang, Z., Pimm, S.L., et al. (2003) Protecting China's biodiversity. *Science*, **300**, 1240–41.

Liu, J. and Yang, W. (2012) Water sustainability for China and beyond. *Science*, **337**, 649–50.

Liu, J. and Yang, W. (2013) Integrated assessments of payments for ecosystem services programs. *Proceedings of the National Academy of Sciences of the United States of America*, **110**, 16297–98.

Liu, W. (2012) *Patterns and Impacts of Tourism Development in a Coupled Human and Natural System*. Doctoral Dissertation, Michigan State University, East Lansing, MI.

Liu, W., Vogt, C.A., Luo, J., et al. (2012) Drivers and socioeconomic impacts of tourism participation in protected areas. *PLoS ONE*, 7, e35420.

Ministry of Forestry (1998) *Wolong Nature Reserve Master Plan*. Beijing, China(in Chinese).

Mouland, B. (2001) Hope at Last as the Giant Panda Enjoys a Baby Boom. *Daily Mail*. http://www.highbeam.com/doc/1G1-78409889.html.

Myers, N., Mittermeier, R.A., Mittermeier, C.G., et al. (2000) Biodiversity hotspots for conservation priorities. *Nature*, **403**, 853–58.

National Post (2010) Corporations Cuddling up to Pandas. http://www.canada.com/story.html?id=25701c3a-5e13–14bf8-b8d8–918b3b9204b6.

Neal, J.D., Uysal, M., and Sirgy, M.J. (2007) The effect of tourism services on travelers' quality of life. *Journal of Travel Research*, **46**, 154–63.

News.Gov.Hk (2014) Paul Chan Concludes Sichuan Visit. http://archive.news.gov.hk/en/categories/admin/html/2014/03/20140305_172605.shtml.

Newsome, D., Moore, S.A., and Dowling, R.K. (2002) *Natural Area Tourism: ecology, impacts, and management*. Channel View Publications, Clevedon, England.

Nicolau, J.L. and Mas, F.J. (2006) The influence of distance and prices on the choice of tourist destinations: the moderating role of motivations. *Tourism Management*, **27**, 982–96.

Pan, P.P. (2000) Meet the Pandas; Zoo Bringing Home Pick of China's Litter. *The Washington Post*. http://www.highbeam.com/doc/1P2–556874.html.

Pandas International (2012) *Reintroduction Program: 2012*. http://www.pandasinternational.org/wptemp/program-areas-2/reintroduction-program/reintroduction-program-2012/.

Peters, G.P., Minx, J.C., Weber, C.L., and Edenhofer, O. (2011) Growth in emission transfers via international trade from 1990 to 2008. *Proceedings of the National Academy of Sciences of the Untied States of America*, **108**, 8903–08.

Schaller, G.B. (1994) *The Last Panda*. University of Chicago Press, Chicago, IL.

Sichuan Academy of Forestry (2014) *The Overall Planning of Wolong National Nature Reserve, Sichuan, China*. Sichuan Department of Forestry, Chengdu, China (in Chinese).

Sichuan Provincial Bureau of Statistics (2007) *Sichuan Statistical Yearbook. 1998–2007*. http://www.sc.stats.gov.cn/tjcbw/tjnj/index.html (in Chinese).

Simons, C. (2003) Where Little Pandas Come From. *The New York Times*. http://www.nytimes.com/learning/students/pop/20030106snapmonday.html.

Smithsonian National Zoological Park (2012) *Smithsonian Institution National Zoo: 125 years*. http://www.americanbison.si.edu/the-national-zoo-at-125/.

State Council Information Office of China (2015) *Press Conference on the Fourth National Panda Survey Results*. http://www.scio.gov.cn/xwfbh/gbwxwfbh/fbh/Document/1395514/1395514.htm (in Chinese).

State Forestry Administration (2006) *The Third National Survey Report on the Giant Panda in China*. Science Publishing House, Beijing, China (in Chinese).

Stevens, K., Irwin, B., Kramer, D., and Urquhart, G. (2014) Impact of increasing market access on a tropical small-scale fishery. *Marine Policy*, **50**, 46–52.

Swaisgood, R.R., Wei, F., Wildt, D.E., et al. (2009) Giant panda conservation science: how far we have come. *Biology Letters*, **6**, 143–45.

United Nations Statistics Division (2012) *United Nations Commodity Trade Statistics Database (UNcomtrade)*. http://comtrade.un.org/db.

Viña, A., Chen, X.D., McConnell, W.J., et al. (2011) Effects of natural disasters on conservation policies: the case of the 2008 Wenchuan Earthquake, China. *Ambio*, **40**, 274–84.

Wildt, D., Lu, X., Lam, M., et al. (2006) Partnerships and capacity building for securing giant pandas ex situ and in situ: how zoos are contributing to conservation. In D.E. Wildt, A. Zhang, H. Zhang, D.L. Janssen, and S. Ellis, eds, *Giant Pandas: Biology, Veterinary Medicine, and Management*, pp. 520–40. Cambridge University Press, Cambridge, UK.

Wolong Administration Bureau Department of Social and Economic Development (2008) *Annual Report on the Rural Economy of Wolong Nature Reserve*. Wolong Administration Bureau, Wolong, China (in Chinese).

Wolong Nature Reserve (1998–2010) *Wolong Socio-economic Report*. Wolong, China (in Chinese).

Wolong Nature Reserve (2005) *Chronicles of Wolong Nature Reserve* (in Chinese).

World Travel and Tourism Council (2014) *WTTC World Economic Impact Research Report 2014*. http://www.wttc.org/focus/research-for-action/economic-impact-analysis/.

Yang, W. (2013) *Ecosystem Services, Human Well-being, and Policies in Coupled Human and Natural Systems*. Doctoral Dissertation, Michigan State University, East Lansing, MI.

Yang, W., Dietz, T., Kramer, D.B., et al. (2015) An integrated approach to understanding the linkages between ecosystem services and human well-being. *Ecosystem Health and Sustainability*, **1**, 19.

Yang, W., Dietz, T., Kramer, D.B., et al. (2013a) Going beyond the Millennium Ecosystem Assessment: an index system of human well-being. *PLoS ONE*, **8**, e64582.

Yang, W., Dietz, T., Liu, W., et al. (2013b) Going beyond the Millennium Ecosystem Assessment: an index system of human dependence on ecosystem services. *PLoS ONE*, **8**, e64581.

Yang, W., Liu, W., Viña, A., et al. (2013c) Performance and prospects of payments for ecosystem services programs: evidence from China. *Journal of Environmental Management*, **127**, 86–95.

Zhang, Z., Zhang, A., Hou, R., et al. (2006) Historical perspective of breeding giant pandas ex situ in China and high priorities for the future. In D.E. Wildt, A. Zhang, H. Zhang, D.L. Janssen, and S. Ellis, eds, *Giant Pandas: Biology, Veterinary Medicine, and Management*, pp. 455–68. Cambridge University Press, Cambridge, UK.

CHAPTER 18

Lessons from Local Studies for Global Sustainability

Jianguo Liu, Vanessa Hull, Wu Yang, Andrés Viña, Li An, Neil Carter, Xiaodong Chen, Wei Liu, Zhiyun Ouyang, and Hemin Zhang

18.1 Introduction

An important feature of a model system is that the results from the system can be applicable to many other systems (Chapter 2). However, unlike model organisms such as the fruit fly (*Drosophila melanogaster*) and model ecosystems (Chapter 2), model coupled human and natural systems are more complex because they consist of both natural and human components as well as their interactions. Thus, generalizable findings from model coupled systems are often more difficult to obtain.

Fortunately, our team's long-term research in the model coupled system of Wolong Nature Reserve has been fruitful. Our work in this real-world laboratory has produced ideas and methods for understanding how coupled systems work and what this might mean for sustainability in Wolong and beyond. Besides providing useful information for conservation and sustainability policy making, these ideas and methods have also contributed to existing theories. Examples include theories on complexity (An et al., 2005), social norms (Chen et al., 2009), collective action (Yang et al., 2013a), and forest transition (Viña et al., 2011). Our work in this model coupled system has also contributed to the development of new theories such as those about telecoupling (Liu, 2014, Liu et al., 2013a, 2014, 2015, Liu and Yang, 2013; Chapter 17).

In the past two decades, our team has also learned many lessons through research in Wolong and through applying some of the results from Wolong to other coupled systems worldwide, as illustrated in previous chapters. Here we highlight ten of them. It is our hope that these lessons may be useful for studying other coupled systems and for helping achieve global sustainability.

18.2 Insights from model coupled systems are applicable to other systems

It has been documented that some aspects of coupled systems are different at different scales (scale dependent; Cumming et al., 2006, Zurlini et al., 2006) and in different contexts (context dependent; Abel et al., 2006, McDaniels et al., 2008). However, our team's work indicates that there are also some features of coupled systems that are the same or similar at different scales (scale independent) and in different contexts (context independent), in addition to scale-dependent and context-dependent features. This demonstrates the utility of model coupled systems like Wolong. In other words, methods and findings from a model coupled system can be applicable to other systems, including those in which the focal system is embedded and those that are far away from the focal system.

For instance, our finding that the number of households increased faster than population size in Wolong (Figure 18.1A) led to the discovery of a similar pattern in 141 countries (Figure 18.1B). This

Pandas and People. Edited by Jianguo Liu, Vanessa Hull, Wu Yang, Andrés Viña, Xiaodong Chen, Zhiyun Ouyang, and Hemin Zhang. © Oxford University Press 2016. Published 2016 by Oxford University Press.

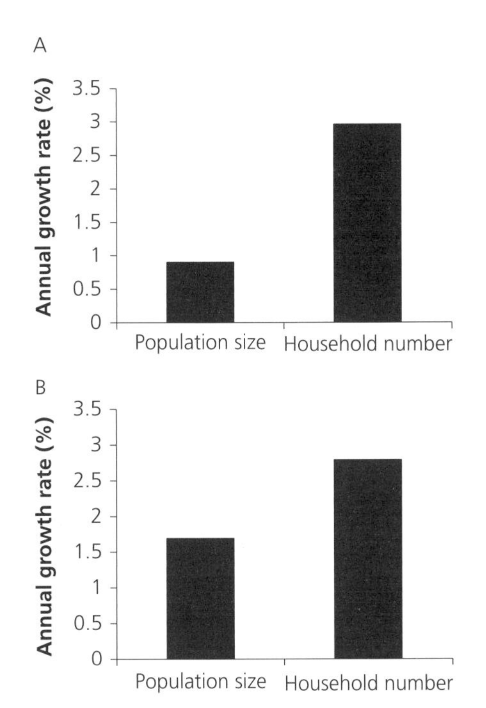

Figure 18.1 Increases in household numbers vs population sizes from 1985 to 2000 in (A) Wolong and in (B) 141 countries around the world.

pattern occurred not only at the reserve scale, but also at national and global scales. It occurred not only in the context of an agropastoral community in a protected area, but also in other contexts such as urban and suburban communities, in Western and non-Western cultures, and in forest, wetland, and grassland ecosystems (Liu et al., 2003a).

Because of these and other similarities across scales and contexts, some of the results and methods from work in Wolong have been scaled up to the entire giant panda range, the country of China, and the entire planet Earth (Figure 18.2). For example, they have contributed to broader studies on giant panda habitat dynamics across the species' range (Chapter 15; Li et al., 2013, Viña et al., 2010). They have also helped to understand China's environmental challenges (Liu, 2014, Liu et al., 2003b, Liu and Raven, 2010), ecosystem services programs (Liu et al., 2008, 2013b), and water sustainability issues (Liu and Yang, 2012). Insights from Wolong have broad implications for managing China's nature reserve system (Liu, 2013, Liu et al., 2003b) made up of approximately 2600 nature reserves (Liu et al., 2013b). For example, the decline of forest in Wolong Nature Reserve after its establishment (Liu et al., 2001) is similar to what happened in many other nature reserves (Curran et al., 2004, Maiorano et al., 2008), although their contexts differ.

Some methods that our team developed in Wolong have been applied to other coupled systems. For example, the agent-based model (ABM) developed for Wolong (An et al., 2005) has been adapted to Chitwan National Park in Nepal (An et al., 2014). Although Wolong and Chitwan are in different contexts (e.g., different countries), both have a

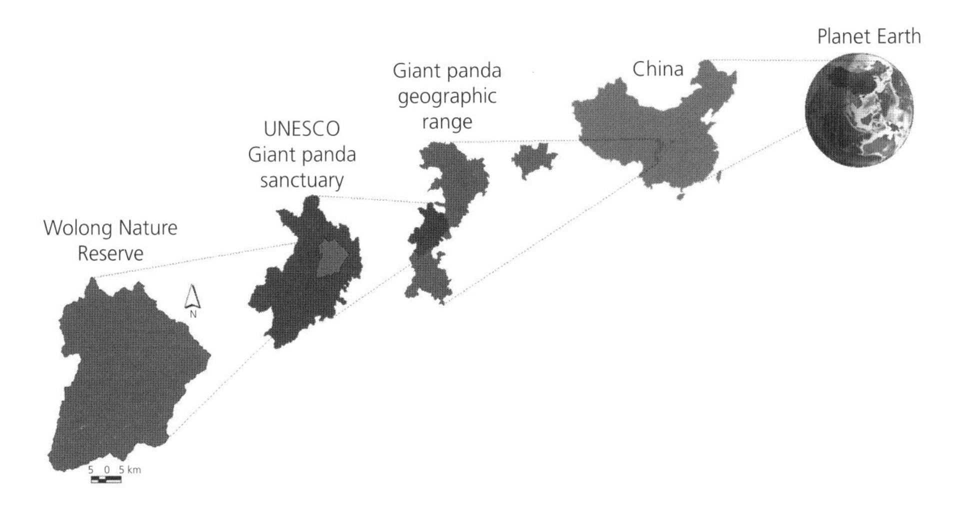

Figure 18.2 Scaling up from our model system in Wolong Nature Reserve to the entire planet.

similar hierarchical structure of agents (e.g., individuals, households, and communities) and similar types of human–nature interactions (e.g., marriage impacts household formation and, in turn, land cover; An et al., 2014). Insights into the complexity of Wolong have been used for cross-site syntheses with Chitwan (Carter et al., 2014; Chapter 16). Wolong and Chitwan have similarities in coupled system properties and dynamics. These similarities exist even though the two systems have different geographic settings (mountainous area vs plain), biological composition (e.g., pandas vs tigers), and demographic distribution (residents inside vs outside the protected area).

Wolong also shares many complex attributes (e.g., time lags, non-linearity, legacy effects, and heterogeneity) with other coupled systems around the world (Liu et al., 2007a). Such similarities exist even though the systems are located on different continents (Africa, Europe, North America, and South America). Those systems also differ in ecosystem types (tropical forests, agricultural highlands, lakes, urban, and wetlands), political structure (e.g., socialist or capitalist), human population density (from very sparse to dense), and economic conditions (developing vs developed).

Research in Wolong was the catalyst for the formation of the International Network of Research on Coupled Human and Natural Systems (CHANS-Net.org). This network serves as a platform for more than 1500 scholars from all over the globe interested in coupled systems research and applications to "see, be seen, find, and connect." Users share their project methods and ideas and alert the community to milestones and special events. The website received 6200 page views in its first year, 2009, and increased in the years following, reaching 17 900 page views in 2014. The network also has organized, convened, and/or sponsored 27 workshops and symposia on coupled systems research at major national and international scientific meetings (Table 18.1). In addition, a major component of the mission of CHANS-Net is to foster the careers of students and early-career scientists (CHANS Fellows). CHANS-Net supports them to attend, present, network, and learn from senior scholars at the various events. This program has sponsored 74 young scholars from all over the globe. The events have led to many fruitful interdisciplinary collaborations (e.g., Feola et al., 2015, Roy et al., 2013).

18.3 Humans and nature go hand in hand

The second lesson is fundamental to the core of coupled systems. It is that humans and nature go hand in hand. Our team's work in Wolong indicates that it is not possible for a given event in a system to affect only nature or only people, as there are cascading effects and everything is connected. Often, the interactions between humans and nature seem like a seesaw with the two subsystems on opposing sides but tightly coupled. Humans might degrade nature for their own gain in the short term, but might suffer in the long term due to feedbacks from changes in nature. On the other hand, nature might flourish under conditions when humans have moved away or their population has declined in a given area. Achieving sustainability of the entire coupled system requires a good balance of both sides. While a perfectly balanced "happy medium" between the two may appear to be a utopian fantasy, it is possible to achieve some degree of balance when both sides of the seesaw are adequately considered.

Human–nature conflicts and coexistence are common in Wolong (Chapter 4), as in many other parts of the world (e.g., Chapter 16). For example, collection of fuelwood in the reserve met an important need for people. But this practice was in clear conflict with giant panda conservation (Chapters 4, 7, and 10; An et al., 2001, Bearer et al., 2008). Emerging livestock-raising practices in the reserve also represent points of conflict for pandas and people. Livestock are a primary income source for people but can also directly compete with pandas for food and space (Chapter 4; Hull et al., 2014). On the other hand, there is also human–nature coexistence. Improvements in electricity from hydropower together with forest monitoring and payments from the Natural Forest Conservation Program (NFCP) helped change energy sources for people and minimize effects of fuelwood collection on panda habitat (Chapters 5, 7, and 13; An et al., 2002, Yang et al., 2013b). Participation in the Grain to Green Program and the Grain to Bamboo Program (GTGB) showed

Table 18.1 List of symposia and workshops organized, convened, and/or sponsored by CHANS-Net.

Name of symposium or workshop	Affiliate institution	Year and location	URL
Symposium on "Complexity in Human–Nature Interactions across Landscapes"	US Regional Association (International Association for Landscape Ecology)	2009, Snowbird, UT	http://www.usiale.org/snowbird2009/session_presentation.php?id=A1
Workshop on "Challenges and Opportunities in Research on Complexity of Coupled Human and Natural Systems"	US Regional Association (International Association for Landscape Ecology)	2009, Snowbird, UT	http://www.usiale.org/snowbird2009/index.php?id=halfWorkshops#complexity
Symposium on "Frontiers in Research on Coupled Human and Natural Systems (CHANS): Current Progress and Future Opportunities"	National Science Foundation headquarters	2010, Arlington, VA	http://chans-net.org/events/frontiers-research-coupled-human-and-natural-systems-chans-current-progress-and-future-opport
Symposium on "Coupled Human and Natural Systems in China and Nepal"	Global Land Project—Open Science Meeting	2010, Tempe, AZ	http://www.glp2010.org/
Seven different symposia on Coupled Human and Natural Systems Research	Annual Meeting of the Association of American Geographers	2010, Washington DC	http://meridian.aag.org/callforpapers/program/calendar.cfm?dn=1&mtgID=55
Six different symposia on Coupled Human and Natural Systems Research	American Association for the Advancement of Science	2011, Washington DC	http://chans-net.org/events/chans_aaas_2011
Workshop on "Land Change Meta-analysis"	Global Land Project	2012, Amsterdam, The Netherlands	http://chans-net.org/news/mapping-landscape-land-change-synthesis
Symposium on "Disentangling Diverse Drivers and Complex Dynamics of Coupled Human and Natural Systems (CHANS)"	US Regional Association (International Association for Landscape Ecology)	2012, Newport, RI	http://www.usiale.org/newport2012/sessions/symposium-nasa-msu-symposium-disentangling-diverse-drivers-and-complex-dynamics-coupled
CHANS-Net Workshop	American Geophysical Union	2012, San Francisco, CA	http://chans-net.org/news/chans-agu-2012
Four different Symposia on "Coupled Human and Natural Systems and Global Change"	American Geophysical Union	2012, San Francisco, CA	http://chans-net.org/news/chans-agu-2012
Two different Symposia on Coupled Human and Natural Systems	American Geophysical Union	2013, San Francisco, CA	http://chans-net.org/news/chans-net-members-gear-agu-again
Symposium on "Ecological Sustainability in a Telecoupled World"	Ecological Society of America	2013, Minneapolis, MN	https://eco.confex.com/eco/2013/webprogram/Session8773.html

promise for fostering gains for both people and pandas (Chapters 5 and 13; Liu et al., 2008, Yang et al., 2013b). These programs converted cropland to forest plantations and bamboo plantations, respectively. Ultimately, the seesaw is constantly shifting as new effects emerge and the two sides provide feedback to each other. For example, there is a potential increase in crop raiding by wildlife as a result of forest recovery (Chapter 13), which may lead to an increase in human–wildlife conflict. As human–nature interactions are dynamic, it is important to continue monitoring them and managing them adaptively.

18.4 Good intentions sometimes lead to bad outcomes

The third lesson is that good intentions can sometimes lead to bad outcomes. Managers working in difficult political climates try to put their best foot forward and make the best decision possible about how to manage a complex coupled system based on the information available. But sometimes not all of the information needed is available, especially in systems where not enough coupled system research has been conducted to inform such decisions. The dearth of information is particularly a challenge for human attitudes and mechanisms behind resource-use behaviors. These human dimensions are often overlooked in the management of natural resources (Leys and Vanclay, 2011, Liu et al., 2010).

Perhaps this issue can be illustrated with one of the first key findings of our research. We found that a faster decline in forest cover and panda habitat occurred after Wolong was established as a nature reserve (Liu et al., 2001). This finding had implications for protected areas worldwide by demonstrating that despite the best of intentions, a protected area designation does not necessarily protect biodiversity. The intention itself was not enough to produce a good outcome in the context of an ever-growing human population inside the reserve (not subject to China's national one-child policy). Other challenges that were difficult to overcome included a rapidly developing local economy, increasing tourism, and a lack of available energy source aside from fuelwood extracted from the forests (Liu et al., 2001). What the government overlooked in this protected area designation was the local context and the livelihoods of people living inside the reserve, which required other policies to manage them effectively.

Another example is a 1980s project of the Chinese government and the World Food Programme. This project sought to relocate farmers from the core panda habitat area inside the reserve to the main road by providing subsidized housing (Ghimire, 1997, Liu et al., 1999). The intention behind the project to relieve pressure on core panda habitat was good, but the outcome again was bad. The farmers did not move into the housing. The project sponsors did not recognize that if the farmers had relocated to the housing complex, they would be too far from their cropland (their primary lifeline). At that time, the farmers also would have had no other income-earning opportunities. These examples illustrate the importance of going beyond intentions. It is crucial to strive for sound conservation planning that considers all major aspects of the coupled system and anticipates all important potential outcomes in advance.

18.5 Put researchers in the local residents' shoes

The fourth lesson is to put ourselves in the local residents' shoes. Over the years, we took the time to get to know local community members by spending time in their homes and participating in their everyday lives and family events. By doing so, we began to see Wolong residents through their eyes and soak in the culture from their perspective. Our most meaningful insights came as a result of conducting thought exercises in which we put ourselves in the shoes of local farmers. We imagined ourselves navigating the complex labyrinth of changing policies, sudden disturbances, and shifting community and ecological dynamics in Wolong. We asked ourselves questions such as: Would we cut trees for fuelwood if we had no other way to cook food for our families and heat our homes? Would we raise horses as a new livelihood strategy if we were struggling to make ends meet? Asking ourselves questions like these and honestly commiserating over our answers helped us to see the coupled system more clearly. Ultimately, our answers showed that we would do things very similarly to what the locals have actually done. As a result, we stopped seeing local residents as objects to blame and began to understand that meeting their livelihood needs is at the very core of the sustainability of the coupled system as a whole.

This realization is what spurred us to focus on conservation incentives as ways of achieving conservation while also helping local livelihoods. It is not a fluke that incentive-based conservation programs have surged in popularity in recent years around the globe (Bulte et al., 2008). With a progressively developed global cash economy, humans residing in and around vulnerable coupled systems worldwide have substantial and often urgent economic needs (McShane et al., 2011, Turner et al., 2012). Putting oneself in the shoes of a local resident makes one realize that people need cash to feed their families, send children to school, pay for health care, and save for an uncertain future. Programs aimed at conserving nature by limiting the behaviors of people and their access to natural resources are not likely to succeed if these basic needs are not met (Swallow et al., 2009).

In Wolong, we saw the success of incentive-based conservation practices first hand. Where establishing the protected area alone failed to curb forest degradation, the incentive-based methods that prioritized local livelihoods appeared to succeed

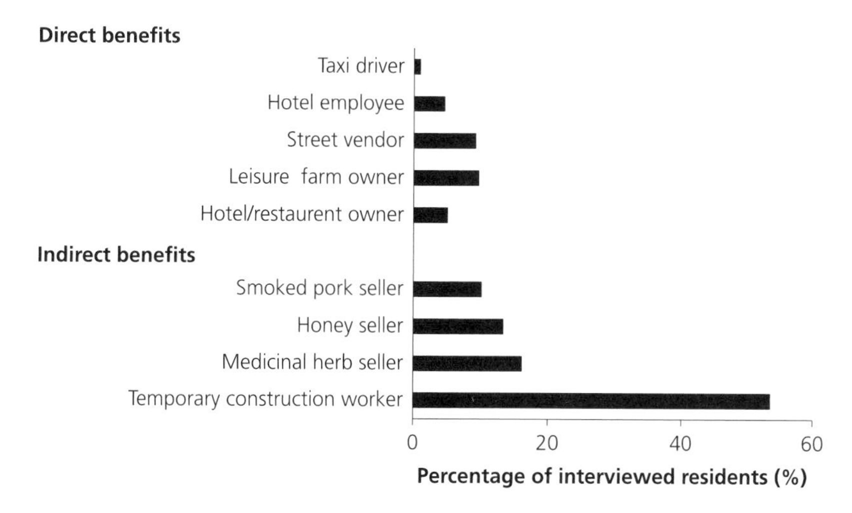

Figure 18.3 Percentage of interviewed local residents (n = 217) receiving direct and indirect benefits from tourism in Wolong in 2005–2007. Figure drawn from data in Liu et al. (2012).

(Chapter 6; Liu et al., 2008, Yang et al., 2013b). In addition to forest-cover increases, local people perceived and reported personal benefits from NFCP and GTGP/GTBP and expressed interest in re-enrolling if the programs were to be renewed (Chapters 11 and 13). One of the reasons that the incentive-based approach has worked well in Wolong is that the programs appeared to be well managed, and funds were fairly distributed with transparency. This sound management was not the case in many other areas where breakdowns in equity have occurred (Daw et al., 2011, Ferraro and Simpson, 2002). On the other hand, when we put ourselves in the shoes of the local people, we also saw that tourism in Wolong was suffering from an unequal distribution of benefits. Much less income was distributed among community members relative to outside companies (He et al., 2008). Few tourism jobs were available for local residents (Figure 18.3). Tourism was less successful than other strategies at helping both people and the environment (He et al., 2008, Liu et al., 2012).

18.6 Money isn't everything

The fifth lesson is that money is not the only factor that matters. Although money is important in shaping human behaviors, people are multidimensional. They have different cultural and family histories, religious and spiritual values, and ethical standards. People also have diverse sources of identity that relate to natural resources in different ways (Barrera-Bassols and Toledo, 2005, Colding and Folke, 2001). If these factors are not also considered, the coupled system cannot be fully understood or managed well.

The local residents that our team interviewed expressed strong cultural values and traditions, many of which were tied to behaviors we were studying. For instance, smoking pork using fuelwood was a strong cultural tradition of local residents that intersected with many important holidays (Yang et al., 2013c, Chapter 10). Many people with whom our team talked (especially older people) also expressed an inability to adapt to life outside the reserve in cities (even if given the opportunity). They preferred a more relaxed rural life against a fast-paced city atmosphere and preferred the higher environmental quality in the reserve to polluted cities. Our work on social networks also showed the strong family and community bonds among local residents, which often affected decision-making (Chapter 11; Chen et al., 2009, Yang et al., 2013b). Overall, local residents' attitudes and behaviors could not be easily predicted using a simple mathematical formula because they weighed multiple, sometimes competing interests at any given time.

18.7 Expect the unexpected

The sixth lesson we learned is to expect the unexpected. Surprises are a key component of coupled systems (Liu et al., 2007b). They happen because the systems are complex. Often a surprise may occur as a result of a past event in the system that was not previously considered, manifesting later as a legacy effect or a lagged effect. Other times a surprise may occur because the system has passed a threshold that was not anticipated, a sudden tipping point that throws the system into a new state. The challenge of anticipating or predicting the surprises may not always be possible, but can be helped by amassing long-term data and using them to develop systems models with scenario analyses (see Chapter 14 for examples).

In Wolong, one of the biggest surprises was how local residents responded to conservation policies in ways that were not anticipated and in conflict with policy goals. For example, NFCP involved a ban on timber harvesting. Many residents initially responded to this ban by cutting off branches of trees instead of the entire trunk for fuelwood. This behavior was not expected and was not addressed in the policy guidelines, but could compromise forest structure and suitability for wildlife (e.g., birds). Another related example is that some residents split up their households into smaller ones to receive more NFCP subsidies, which were handed out at the household level (see Figure 18.4 for the sudden jump in number of households in 2001 when NFCP started). Such behavior could work against the policy because more households and reduction in household size also increase the per-capita energy consumption, thus requiring more fuelwood (Liu et al., 2003a).

The emergence of raising horses was also a surprise because that had not been a part of local agricultural livelihoods in previous decades (Hull et al., 2014). The horses ate large amounts of bamboo in panda habitat, but the reserve managers were unaware of their presence for over a year. After our team informed the reserve managers about our research results, raising horses was banned. However, there was a surprising increase in raising other livestock species (e.g., sheep and yaks; Chapter 4). Our experience with studying the livestock suggests that a chain of surprises can occur. They can often emerge faster than the managers can come up with effective policies to deal with them. This phenomenon highlights the need for long-term research, monitoring, and adaptive management.

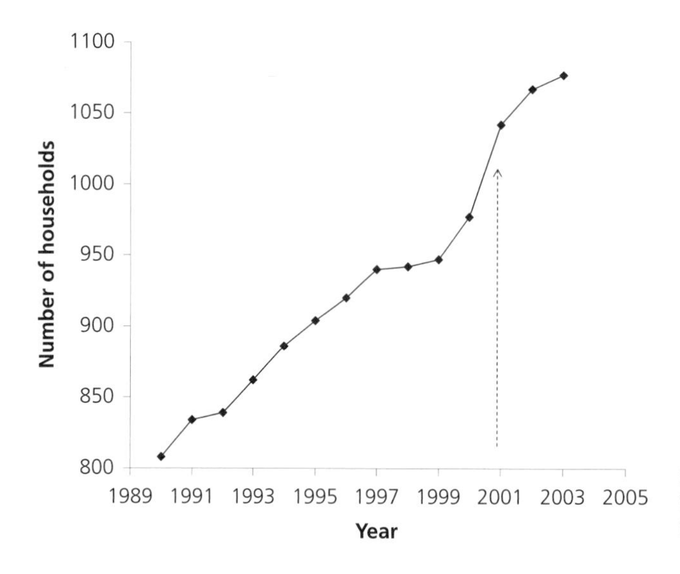

Figure 18.4 Household numbers in Wolong Nature Reserve over time, including a surge after the implementation of new conservation programs in 2001.

18.8 Diversify toolbox to meet complex challenges

The seventh lesson we learned is that researchers and managers should bring a diverse toolbox of ideas and approaches to the table when tackling sustainability in coupled systems. In our many years of work, we never found a "quick fix" or a simple solution to the challenges we encountered. From perusal of the literature, the same appears to be true in many other systems (e.g., Doremus, 2003, Underwood, 2011). This observation speaks to the complexity of coupled systems. We found instead that a mix of different factors helped explain a given phenomenon and a combination of different policies or management measures were needed to solve conservation challenges. This approach requires an ability to think creatively and integrate diverse perspectives.

In the early years of research on Wolong, it would have been easy to assume that the main problem was the number of people in the reserve. One might think that if the population could be controlled or limited, then perhaps the system could be sustained. However, our work suggests that it is not so simple. Given the failure of past household relocation attempts (Chapter 3), the best way to shrink population numbers in the reserve is to improve the education of young people so they can find jobs outside the reserve after graduation (Liu et al., 2001). That said, a single population regulation policy is unlikely to be a "quick fix" because our work has shown that other factors aside from population growth drive system dynamics. For instance, household proliferation (due to factors such as cultural shifts away from multigenerational households) is an even more important driver of environmental degradation than population sizes alone (Liu et al., 2003a). Education level, demographic makeup of the household (An et al., 2001), and social norms (Chen et al., 2009) are other factors that may also affect how people behave.

Ultimately, multiple different policies need to come together to achieve sustainability in Wolong. In addition to the education incentive, other approaches include incentive-based PES programs. We discussed many of the benefits from these programs in some previous chapters (e.g., Chapters 5, 7, and 13), and we believe they should be included as integral parts of the toolbox. However, the money earned from these policies cannot support a household by itself. It is simply not enough. Money earned from PES programs are supplements to household income (averaging US$143 per household/year from NFCP and US$571 per ha/year from GTGP; see Chapter 5, Yang et al., 2013b). These funds cannot serve as a household's entire income. Tourism is another item that should be in the toolbox of approaches to achieve sustainability, because tourism helps to diversify income generation for local people. However, tourism alone is not a solution because, as discussed in Chapter 12, the tourism industry is volatile and most local households benefit little from tourism.

Zoning is another policy that should be a part of the toolbox. Zoning allows protected areas with people living inside them to spatially segregate human development from core conservation areas (Hull et al., 2011). Without zoning, it is possible that human impacts could spread throughout the entire area and fragment key habitats. However, as discussed in Chapter 13, zoning is not a "quick fix." Zoning has only protected against infrastructural development in core conservation areas, but not against on-the-ground human activities such as livestock penetration into forests in Wolong that occur without an awareness or appreciation of zoning boundaries. The take-home message here is that each policy has strengths and weaknesses. Thus, a policy portfolio with multiple complementary policies is needed to tackle sustainability from different angles.

18.9 It's a marathon, not a sprint

The eighth lesson learned is that coupled system research is a marathon and not a sprint. The complexity involved in this type of research necessitates a long-term commitment. Scientific discovery is a non-linear process in that the most meaningful insights may not come to light until many years of research have been conducted. For instance, we learned a great deal about our model coupled system during its recovery after the unexpected earthquake hit the study area. This catastrophic event occurred after we had already been studying Wolong for 13 years. If we had stopped our study prior to this event, we would have missed the opportunity to learn many issues about the coupled system. Example research topics include vulnerability, resilience, and adaptability (Chapter 12), as well as the effectiveness of policies (Viña et al., 2011). In addition, the 13 years of data prior to the earthquake helped to strengthen our understanding of the profound shifts in the system after the earthquake (Yang et al., 2013c, 2015, Zhang et al., 2011, 2014). Taking a long-term approach also allowed us to examine the trajectories of various human and natural components. Furthermore, it created causal chains explaining how and why different phenomena occurred, such as how conservation policies shifted labor allocation over time or why people decided to out-migrate or engage in new income-generation activities.

There are considerable barriers to obtaining funding to support such long-term research via traditional funding channels. It is ironic that sustainability is a long-term issue and coupled systems are very dynamic, but funding for research on sustainability is often short lived. The trend in the research community with increasingly limited funding appears to reward short-term, low-risk projects aimed at producing rapid results. However, this trend is at odds with another major funding goal to support transformative research. Transformative research is "research creating radically innovative insights and effectively contributing to solutions for the severe sustainability problems facing our society" (Wiek, 2010). In our experience, most transformative research on coupled systems requires a long-term effort.

One solution to this disconnection between long-term transformative research needs and short-term funding limitations is to conduct coordinated relay research under a comprehensive long-term framework. Like a relay race in sports, relay research is a central part of our philosophy of long-term work on the model coupled system. It involves splitting up a large body of planned work into multiple, smaller projects assigned to different team members (e.g., students or postdoctoral researchers). The projects should all fall under the same general framework, but can be distinct from one another and fit together like unique but integrated puzzle pieces. For instance, early in our research, we had one student working on socioeconomics and another student working on remote-sensing approaches to understand forest-cover change. Both students collaborated to produce integrated results (An et al., 2001, Linderman et al., 2006). Housing the data in a common platform (e.g., the Inter-university Consortium for Political and Social Research, ICPSR; ICPSR, 2014) can assist in integrating findings. Such an arrangement can in turn help foster collaborative work among different generations of team members over long time periods.

18.10 Think "outside the box"

The ninth lesson we learned is that it is important to consider any given coupled system in a broader context that extends outside the perceived study area boundaries. In other words, think "outside the box." Any given coupled system is actually a part of a network of multiple coupled systems that are connected to one another by flows (e.g., movement of information, energy, people, goods, or organisms). No system can be fully understood in isolation (Liu et al., 2015). Considering outside influences as drivers of change or impacts on other systems as externalities only does not do the system justice because it ignores the feedbacks between the focal system and other systems. The telecoupling framework allows for these interactions among coupled systems to be explicitly integrated (Chapter 17; Liu et al., 2013a).

As documented in Chapter 17 and several other chapters, Wolong may be small in area, but its impact is large and stretches across the entire globe. Besides internal factors, dynamics in Wolong are driven by agents, flows, and causes originating in other systems via telecouplings. Examples include information flow, national conservation policies, tourism, agricultural markets, and panda loans (Chapter 17). Wolong is a receiving system for tourists and money from agricultural trade, tourism, and conservation policies. In turn, Wolong is a sending system for pandas, agricultural products, and information. The importance of telecoupling as a means of thinking "outside the box" cannot be overstated. In fact, our work in Wolong demonstrates that telecouplings sometimes can be even more important than local couplings in driving system dynamics.

This observation is perhaps best appreciated by our findings documenting the impact of the earthquake, which cut off many telecouplings between Wolong and the outside world and in turn threatened human livelihoods inside Wolong (Chapter 12). Researchers need to explicitly account for and quantify telecoupling agents, causes, flows, and effects. Such quantification would help to better anticipate responses to such perturbations and provide guidance for how to overcome sustainability challenges (Liu et al., 2015).

18.11 Engaging stakeholders can foster positive change

The tenth lesson we learned through our research is that reaching out to stakeholders to share findings and make recommendations for the future can foster positive change. Our team has worked closely alongside the managers of Wolong over the years and has had meaningful reciprocal interactions with them that have benefited both sides. This relationship was worthwhile even when research findings did not reflect positively on the management. For instance, the government agencies at first did not believe our finding that there was faster habitat degradation in Wolong after the reserve was established (Liu et al., 2001). Through sharing objective results and expressing genuine respect for the challenges they faced, we are delighted that several management changes were made and panda habitat has been recovering (Viña et al., 2007).

We have also worked together with local residents, researchers, and government officials in Wolong over the years in other ways. On one hand, we have received logistical support and knowledge from the local residents, researchers, and administration. On the other hand, we have shared our techniques, approaches, and data through daily interactions, guest lectures, and workshops with the local researchers and managers. We have also hosted several local researchers for training and collaboration at our research center at Michigan State University in the USA. Overall, we have found that the management community especially appreciates the trust that has been fostered due to our long-term commitment year after year to send students and faculty to work in Wolong. This trust has been valuable in a time when many organizations appear for short-term periods only to disappear soon after—a pattern that prevents long-term trust and consistency from being developed.

We feel that it is also rewarding to reach out to the public to share our experiences and research findings. Over the years, our team has given countless guest lectures at university campuses, local K-12 classrooms, and other public forums. We have also engaged in interactions with the news media (e.g., *The New York Times*, BBC, *Time* magazine, and China Central Television—CCTV) and social media. We have found that members of the public are eager for information and also ask pointed questions that challenge us to think more broadly and deeply about our work. Because coupled systems are complex, it is all the more fruitful for us to reach out and share our findings.

The story of Wolong is an engaging one that captures people's attention. We try to harness people's fascination for Wolong to convey broader ideas about sustainability. We want the public to understand that humans are not isolated but are a part of a larger system that includes nature. It is our hope that the story of Wolong helps people comprehend that everything humans do affects nature and affects the sustainability of the planet Earth, which in turn affects humans. It is important to create win–win solutions to sustain the environment and improve human well-being. Ultimately, we want to convey a hopeful message emphasizing that when there are deleterious patterns, it is essential to change human behaviors to ensure that resources are sustainably managed for future generations.

18.12 Summary

Our team's two decades of interdisciplinary research in Wolong Nature Reserve and beyond has revealed many complexities of human–nature interactions. Here we have summarized ten overarching lessons from our research in Wolong and its applications to other coupled human and natural systems. Our results indicate that Wolong is a valuable model coupled system as methods and findings from Wolong can be applicable to larger systems in which it is embedded and to other systems worldwide. Seeing the system from the multifaceted perspective of local residents in Wolong allowed us to better appreciate the difficulties that residents face and the diverse and dynamic factors (e.g., not only economic, but also cultural, social, and political) that they consider in making decisions. We also found that good intentions can sometimes lead to bad socioeconomic and environmental outcomes. It is essential to take a long-term approach to reveal complex attributes such as time lags, legacy effects, and feedbacks. Working with managers has given us an appreciation for the management problems they face and a platform from which to use our research to foster positive change. For example, we found that multiple interacting policies were needed to effectively manage human–nature interactions for sustainability. Other lessons include thinking "outside the box," diversifying the toolbox of ideas and approaches to meet complex challenges, and expecting the unexpected. We hope these lessons are useful for others undertaking the challenging, yet rewarding, task of conducting long-term, interdisciplinary research on coupled systems around the globe.

References

Abel, N., Cumming, D.H., and Anderies, J.M. (2006) Collapse and reorganization in social-ecological systems: questions, some ideas, and policy implications. *Ecology and Society*, **11**, 17.

An, L., Linderman, M., Qi, J., et al. (2005) Exploring complexity in a human-environment system: an agent-based spatial model for multidisciplinary and multiscale integration. *Annals of the Association of American Geographers*, **95**, 54–79.

An, L., Liu, J., Ouyang, Z., et al. (2001) Simulating demographic and socioeconomic processes on household level and implications for giant panda habitats. *Ecological Modelling*, **140**, 31–49.

An, L., Lupi, F., Liu, J., et al. (2002) Modeling the choice to switch from fuelwood to electricity: implications for giant panda habitat conservation. *Ecological Economics*, **42**, 445–57.

An, L., Zvoleff, A., Liu, J., and Axinn, W. (2014) Agent-based modeling in coupled human and natural systems (CHANS): lessons from a comparative analysis. *Annals of the Association of American Geographers*, **104**, 723–45.

Barrera-Bassols, N. and Toledo, V.M. (2005) Ethnoecology of the Yucatec Maya: symbolism, knowledge, and management of natural resources. *Journal of Latin American Geography*, **4**, 9–41.

Bearer, S., Linderman, M., Huang, J., et al. (2008) Effects of fuelwood collection and timber harvesting on giant panda habitat use. *Biological Conservation*, **141**, 385–93.

Bulte, E.H., Lipper, L., Stringer, R., and Zilberman, D. (2008) Payments for ecosystem services and poverty reduction: concepts, issues, and empirical perspectives. *Environment and Development Economics*, **13**, 245–54.

Carter, N.H., Viña, A., Hull, V., et al. (2014) Coupled human and natural systems approach to wildlife research and conservation. *Ecology and Society*, **19**, 43.

Chen, X., Lupi, F., He, G., and Liu, J. (2009) Linking social norms to efficient conservation investment in payments for ecosystem services. *Proceedings of the National Academy of Sciences of the United States of America*, **106**, 11812–17.

Colding, J. and Folke, C. (2001) Social taboos: "invisible" systems of local resource management and biological conservation. *Ecological Applications*, **11**, 584–600.

Cumming, G.S., Cumming, D.H., and Redman, C.L. (2006) Scale mismatches in social-ecological systems: causes, consequences, and solutions. *Ecology and Society*, **11**, 14.

Curran, L.M., Trigg, S.N., McDonald, A.K., et al. (2004) Lowland forest loss in protected areas of Indonesian Borneo. *Science*, **303**, 1000–03.

Daw, T., Brown, K., Rosendo, S., and Pomeroy, R. (2011) Applying the ecosystem services concept to poverty alleviation: the need to disaggregate human well-being. *Environmental Conservation*, **38**, 370–79.

Doremus, H. (2003) A policy portfolio approach to biodiversity protection on private lands. *Environmental Science & Policy*, **6**, 217–32.

Feola, G., Lerner, A.M., Jain, M., et al. (2015) Researching farmer behaviour in climate change adaptation and sustainable agriculture: lessons learned from five case studies. *Journal of Rural Studies*, **39**, 74–84.

Ferraro, P.J. and Simpson, R.D. (2002) The cost-effectiveness of conservation payments. *Land Economics*, **78**, 339–53.

Ghimire, K.B. (1997) Conservation and social development: an assessment of Wolong and other panda reserves in China. In K.B. Ghimire and M.P. Pimbert, eds, *Environmental Politics and Impacts of National Parks and Protected Areas*, pp. 187–213. Earthscan Publications, London, UK.

He, G., Chen, X., Liu, W., et al. (2008) Distribution of economic benefits from ecotourism: a case study of Wolong Nature Reserve for giant pandas in China. *Environmental Management*, **42**, 1017–25.

Hull, V., Xu, W., Liu, W., et al. (2011) Evaluating the efficacy of zoning designations for protected area management. *Biological Conservation*, **144**, 3028–37.

Hull, V., Zhang, J., Zhou, S., et al. (2014) Impact of livestock on giant pandas and their habitat. *Journal for Nature Conservation*, **22**, 256–64.

ICPSR (2014) Inter-University Consortium for Political and Social Research. http://www.icpsr.umich.edu/icpsrweb/ICPSR/.

Leys, A.J. and Vanclay, J.K. (2011) Social learning: a knowledge and capacity building approach for adaptive co-management of contested landscapes. *Land Use Policy*, **28**, 574–84.

Li, Y., Viña, A., Yang, W., et al. (2013) Effects of conservation policies on forest cover change in giant panda habitat regions, China. *Land Use Policy*, **33**, 42–53.

Linderman, M.A., An, L., Bearer, S., et al. (2006) Interactive effects of natural and human disturbances on vegetation dynamics across landscapes. *Ecological Applications*, **16**, 452–63.

Liu, J. (2013) Complex forces affect China's biodiversity. In N.S. Sodhi, L. Gibson, and P. Raven, eds, *Conservation Biology: Voices from the Tropics*, pp. 207–15. Wiley-Blackwell, Oxford, UK.

Liu, J. (2014) Forest sustainability in China and implications for a telecoupled world. *Asia & the Pacific Policy Studies*, **1**, 230–50.

Liu, J., Daily, G.C., Ehrlich, P.R., and Luck, G.W. (2003a) Effects of household dynamics on resource consumption and biodiversity. *Nature*, **421**, 530–33.

Liu, J., Dietz, T., Carpenter, S.R., et al. (2007a) Complexity of coupled human and natural systems. *Science*, **317**, 1513–16.

Liu, J., Dietz, T., Carpenter, S.R., et al. (2007b) Coupled human and natural systems. *Ambio*, **36**, 639–49.

Liu, J., Hull, V., Batistella, M., et al. (2013a) Framing sustainability in a telecoupled world. *Ecology and Society*, **18**, 26.

Liu, J., Hull, V., Moran, E., et al. (2014) Applications of the telecoupling framework to land-change science. In K.C. Seto and A. Reenberg, eds, *Rethinking Global Land Use in an Urban Era*, pp. 119–39. MIT Press, Cambridge, MA.

Liu, J., Li, S., Ouyang, Z., et al. (2008) Ecological and socioeconomic effects of China's policies for ecosystem services. *Proceedings of the National Academy of Sciences of the United States of America*, **105**, 9477–82.

Liu, J., Linderman, M., Ouyang, Z., et al. (2001) Ecological degradation in protected areas: the case of Wolong Nature Reserve for giant pandas. *Science*, **292**, 98–101.

Liu, J., Mooney, H., Hull, V., et al. (2015) Systems integration for global sustainability. *Science*, **347**, 1258832.

Liu, J., Ouyang, Z., and Miao, H. (2010) Environmental attitudes of stakeholders and their perceptions regarding protected area-community conflicts: a case study in China. *Journal of Environmental Management*, **91**, 2254–62.

Liu, J., Ouyang, Z., Pimm, S.L., et al. (2003b) Protecting China's biodiversity. *Science*, **300**, 1240–41.

Liu, J., Ouyang, Z., Tan, Y., et al. (1999) Changes in human population structure: implications for biodiversity conservation. *Population and Environment*, **21**, 45–58.

Liu, J., Ouyang, Z., Yang, W., et al. (2013b) Evaluation of ecosystem service policies from biophysical and social perspectives: the case of China. In S.A. Levin, ed., *Encyclopedia of Biodiversity*(second edition), **vol. 3**, pp. 372–84. Academic Press, Waltham, MA.

Liu, J. and Raven, P. (2010) China's environmental challenges and implications for the world. *Critical Reviews in Environmental Science and Technology*, **40**, 823–51.

Liu, J. and Yang, W. (2012) Water sustainability for China and beyond. *Science*, **337**, 649–50.

Liu, J. and Yang, W. (2013) Integrated assessments of payments for ecosystem services programs. *Proceedings of the National Academy of Sciences of the United States of America*, **110**, 16297–98.

Liu, W., Vogt, C.A., Luo, J., et al. (2012) Drivers and socioeconomic impacts of tourism participation in protected areas. *PLoS ONE*, **7**, e35420.

Maiorano, L., Falcucci, A., and Boitani, L. (2008) Size-dependent resistance of protected areas to land-use change. *Proceedings of the Royal Society B: Biological Sciences*, **275**, 1297–304.

McDaniels, T., Chang, S., Cole, D., et al. (2008) Fostering resilience to extreme events within infrastructure systems: characterizing decision contexts for mitigation and adaptation. *Global Environmental Change*, **18**, 310–18.

McShane, T.O., Hirsch, P.D., Trung, T.C., et al. (2011) Hard choices: Making trade-offs between biodiversity conservation and human well-being. *Biological Conservation*, **144**, 966–72.

Roy, E.D., Morzillo, A.T., Seijo, F., et al. (2013) The elusive pursuit of interdisciplinarity at the human-environment interface. *Bioscience*, **63**, 745–53.

Swallow, B.M., Kallesoe, M.F., Iftikhar, U.A., et al. (2009) Compensation and rewards for environmental services in the developing world: framing pan-tropical analysis and comparison. *Ecology and Society*, **14**, 26.

Turner, W.R., Brandon, K., Brooks, T.M., et al. (2012) Global biodiversity conservation and the alleviation of poverty. *Bioscience*, **62**, 85–92.

Underwood, J.G. (2011) Combining landscape-level conservation planning and biodiversity offset programs: a case study. *Environmental Management*, **47**, 121–29.

Viña, A., Bearer, S., Chen, X., et al. (2007) Temporal changes in giant panda habitat connectivity across boundaries of Wolong Nature Reserve, China. *Ecological Applications*, **17**, 1019–30.

Viña, A., Chen, X., McConnell, W.J., et al. (2011) Effects of natural disasters on conservation policies: the case of the 2008 Wenchuan Earthquake, China. *Ambio*, **40**, 274–84.

Viña, A., Tuanmu, M.-N., Xu, W., et al. (2010) Range-wide analysis of wildlife habitat: implications for conservation. *Biological Conservation*, **143**, 1960–69.

Wiek, A. (2010) *Sustainability Science: transformative research beyond scenario studies*. Symposium at the annual meeting of the American Association for the Advancement of Science. https://aaas.confex.com/aaas/2010/webprogram/Session1822.html.

Yang, W., Dietz, T., Kramer, D.B., et al. (2013c) Going beyond the Millennium Ecosystem Assessment: an index system of human well-being. *PLoS ONE*, **8**, e64582.

Yang, W., Dietz, T., Kramer, D.B., et al. (2015) An integrated approach to understanding the linkages between ecosystem services and human well-being. *Ecosystem Health and Sustainability*, **1**, 19.

Yang, W., Liu, W., Viña, A., et al. (2013a) Nonlinear effects of group size on collective action and resource outcomes. *Proceedings of the National Academy of Sciences of the United States of America*, **110**, 10916–21.

Yang, W., Liu, W., Viña, A., et al. (2013b) Performance and prospects of payments for ecosystem services programs: evidence from China. *Journal of Environmental Management*, **127**, 86–95.

Zhang, J., Hull, V., Huang, J., et al. (2014) Natural recovery and restoration in giant panda habitat after the Wenchuan earthquake. *Forest Ecology and Management*, **319**, 1–9.

Zhang, J., Hull, V., Xu, W., et al. (2011) Impact of the 2008 Wenchuan earthquake on biodiversity and giant panda habitat in Wolong Nature Reserve, China. *Ecological Research*, **26**, 523–31.

Zurlini, G., Riitters, K., Zaccarelli, N., et al. (2006) Disturbance patterns in a socio-ecological system at multiple scales. *Ecological Complexity*, **3**, 119–28.

PART IV

Perspectives

In the final part of the book (Chapter 19), Liu et al. offer a vision for future directions for this exciting field of coupled human and natural systems (CHANS) research. They outline several issues of future research that would take the field to new heights. These include: (1) conducting longer-term research; (2) bringing more components together; (3) integrating across multiple spatial scales; (4) exploring spillover systems; (5) developing better tools; and (6) translating research into policy and practice. Throughout, Liu et al. provide example ideas for future research on each of these issues in Wolong. Examples include: (1) investigating long-term recovery patterns from the earthquake; (2) linking humans to other components of biodiversity in the reserve; (3) exploring implications of climate change impacts in Wolong on broader range-wide patterns; (4) analyzing relationships between Wolong and other areas engaged in agricultural trading; (5) developing a "Digital CHANS" that would reproduce and test complex dynamics occurring in Wolong in an artificial environment; and (6) participating in the codesign, coproduction, and coimplementation of policies alongside diverse administrative units in the reserve. They also reflect more broadly on how these same topics of future research can be pursued in other coupled systems to transform coupled-systems research for global sustainability.

CHAPTER 19

Future Directions for Coupled Human and Natural Systems Research

Jianguo Liu, Vanessa Hull, Zhiyun Ouyang, and Hemin Zhang

19.1 Introduction

The world is facing unprecedented challenges, such as climate change, land-use change, human population increase, disease spread, food shortage, and social unrest (Future Earth, 2014, Liu et al., 2015, Reid et al., 2010, United Nations, 2014). With new challenges emerging every day, coupled human and natural systems (CHANS) are in constant flux. These challenges affect both humans and nature in complex ways, often destabilizing the balance between the two. In order to understand and manage such dynamic systems for global sustainability, CHANS research needs to be constantly improving and building on past findings over long-term time frames.

Now is an exciting time for coupled systems research, with many new advances on the horizon for the future development. But several improvements are needed to be more integrative (Liu et al., 2015). For example, research to date has tended to be conducted over short-term time frames, over limited spatial extents, and without a strong connection to policy and practice. Recent advances in individual fields such as computer science (e.g., "big data" approaches) have yet to be fully embraced in coupled systems research. Research on coupled systems needs to grow and evolve to be more informative to address pressing sustainability issues.

After conducting 20 years of coupled systems research in Wolong Nature Reserve, we have learned a great deal about our model system and coupled systems in general. The model coupled systems approach that we outlined in Chapter 2 proved worthwhile because it gave us the opportunity to delve deeply into the economic, environmental, political, and cultural context of a specific case in detail over a long-term time frame. As illustrated in Chapter 18, this approach led to a number of broadly applicable lessons. These include lessons on CHANS theories, research methods, drivers of change and system vulnerability, and application of research to management and outreach. Nonetheless, in many ways we feel that we are just getting started and may have uncovered only the tip of a large iceberg. Our hope is to continue to use Wolong as a laboratory to inform and elevate global coupled systems research to a new level.

In this chapter, we outline several areas of future research for more systems integration. Here we define systems integration as holistic approaches to integrating various components of coupled systems at organizational, spatial, and temporal scales (Liu et al., 2015). Areas of improvement include (1) longer-term research, (2) more cross-component integration, (3) tighter integration across multiple spatial scales, (4) exploration of spillover systems, (5) development of better tools, and (6) translation of research results into policy and practice (Figure 19.1; Liu et al., 2015). Throughout this chapter, we use ideas and plans for future research in our model system as an example to illustrate how to pursue research in these areas. Advances in these areas would allow coupled systems research to make significant contributions to addressing global sustainability challenges.

Pandas and People. Edited by Jianguo Liu, Vanessa Hull, Wu Yang, Andrés Viña, Xiaodong Chen, Zhiyun Ouyang, and Hemin Zhang. Published 2016 by Oxford University Press.

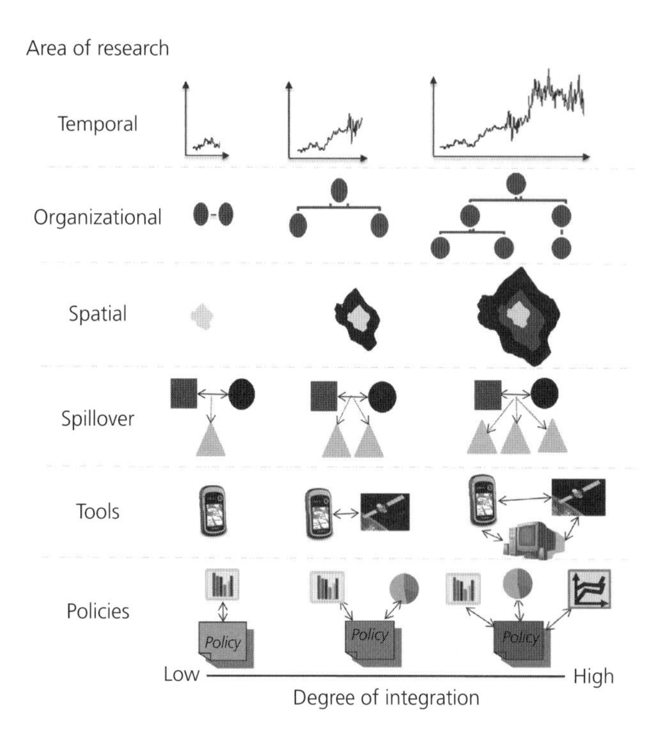

Figure 19.1 Degrees of integration across the six main areas in coupled systems research related to time, components, space, spillover systems, tools, and policies. Increased integration involves longer time frames, more components across organizational levels, and analysis across multiple spatial scales. In addition, it embraces complex effects occurring in spillover systems. Spillover systems (triangles) are those affected by interactions occurring between sending (square) and receiving (circle) systems. Integration also uses more advanced tools (e.g., digital software or cloud computing) and greater efforts to apply research findings to policies.

19.2 Conducting longer-term research

The first area in which coupled systems research has much room to grow is in embracing a longer-term way of approaching research. Some of the greatest scientific achievements occur as a result of a commitment to continuous study in the same system for decades. One example is the European Organization for Nuclear Research (CERN), founded in 1954 in France as a premier research center for fundamental physics research (European Organization for Nuclear Research, 2014). CERN has been operating world-class instruments to collect original data for 60 years (and counting). The long-term commitment has allowed the center to build upon past discoveries and achieve groundbreaking findings related to nuclear physics and astrophysics that would not have been revealed over a short time span; for instance, the discovery of new field particles W and Z, which won the Nobel Prize in Physics in 1984 (European Organization for Nuclear Research, 2014). Another example is the Long Term Ecological Research Network (LTER). This network consists of 26 long-term field research stations supported by the National Science Foundation (NSF), the first six of which were established in 1980 (The Long Term Ecological Research Network, 2014). The sites are distributed throughout the United States, Antarctica, and islands in the Caribbean and Pacific. Sites encompass diverse ecosystems. They support a wide variety of research topics including air quality, land-use change, soil nutrient cycling, forest succession, and stream ecology. Since researchers have studied the same sites year after year, they have been able to learn about long-term processes such as climate change, disturbances, evolution, and land-cover transitions. This emphasis has led the way for novel findings such as identification of key determinants dictating variability in above-ground net primary production across biomes represented in 11 LTER sites (Knapp and Smith, 2001). In 1993, LTER expanded to ILTER (International LTER), a global network of 40 national LTER networks (http://www.ilternet.edu). There are a few examples in coupled systems research, and the number is growing. Examples include some LTER sites (e.g., two urban LTER sites in the USA) that have incorporated social sciences (Redman et al., 2004). In 2003, European LTER launched platforms on Long-Term Socio-Ecological Research (LTSER) to conduct coupled systems research. The current 31 LTSER platforms are distributed across 21 countries in the European Union. LTSER sites are larger than LTER sites (in terms of physical space). They also encompass socioeconomic units with adequate information on socioeconomic and demographic features that allow coupled systems research (rather than just ecological research; Haberl et al., 2006, Singh et al., 2010).

Longer-term studies are especially needed for sustainability, because sustainability is a long-term issue itself and because human–nature interactions are becoming increasingly complex and dynamic in the globalized world. For example, there could be lag effects that do not emerge until after a delay or cumulative effects that build over time. Latent effects that are dormant for a long period and triggered unexpectedly by a second event due to synergistic effects also may occur. So may legacy effects that create path dependencies whereby the future is constrained by the past (Figure 19.2). Not accounting for these effects in management could threaten the sustainability of the system. For instance, the historical patterns of mosquito ditch construction in Massachusetts (USA) had little impact on fisheries for decades. Emerging impacts in the form of residential development and recreational fishing later interacted with this historical phenomenon. Ditches then unexpectedly became a key factor in the decline of marsh habitat for fisheries decades later (Coverdale et al., 2013).

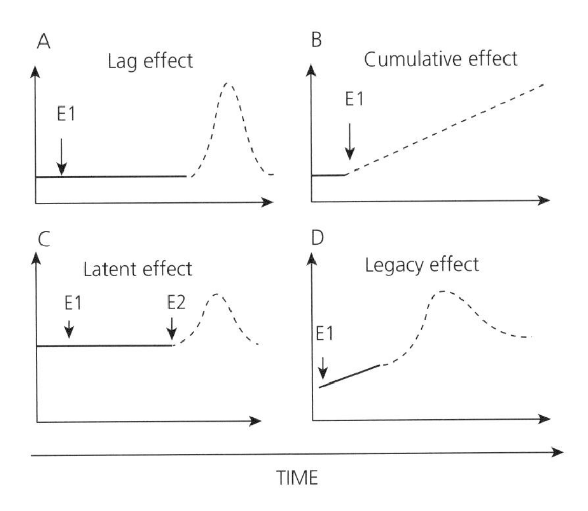

Figure 19.2 Pictorial representation of examples of types of long-term effects in coupled systems involving one or more events (E1, E2). Examples include (A) lag effect in which a punctuated event occurs and its effect is not observed for some time later; (B) cumulative effect that builds up over time; (C) latent effect that is minimal or dormant for a long period of time until a new event creates synergistic effects; (D) legacy effect in which a past event impacts the future trajectory. The vertical arrow points to the time when the event takes place.

Coupled systems researchers could partner with existing longitudinal studies such as LTER (Grimm et al., 2013, Mirtl et al., 2013). Integrated teams could introduce new research questions, data, and approaches that address human–nature interactions into the existing frameworks (Collins et al., 2006, Redman et al., 2004). On the other hand, existing longitudinal studies on the social side (e.g., the Panel Study of Income Dynamics; Institute for Social Research, 2014) could collaborate with ecologists to integrate ecological data into their existing frameworks. Coupled systems researchers could also collaborate with each other to conduct cross-site syntheses and telecoupling research. One example is the international community of coupled systems researchers established via the International Network of Research on Coupled Human and Natural Systems (CHANS-Net.org; Center for Systems Integration and Sustainability, 2014; see Chapters 16 and 18). Answering fundamental questions (e.g., the examples in Table 19.1) is essential to developing generalizable theories about coupled systems, identifying sources of variability, and understanding telecouplings in diverse coupled systems across the globe.

One topic of study that needs to be further developed over longer time frames is resilience, or the ability of systems to recover from disturbance. An extensive body of literature has been developed on human–nature interactions and resilience (Folke et al., 2002, Levin and Lubchenco, 2008, Walker et al., 2009). However, much of the research to date has been conducted over short-term time frames. In Wolong, one topic that we would like to further develop concerns the long-term effects of the earthquake on human–nature interactions. We have conducted research detailing the short-term response of the reserve to this major disaster (Viña et al., 2011, Zhang et al., 2011, 2014), and our data suggest

Table 19.1 Example questions for future long-term coupled human and natural systems research.

Topic	Example questions
Feedbacks	• How do feedbacks emerge, evolve, and dissolve? • Which ecological and socioeconomic factors contribute to making a feedback "tighter" or "looser"? • How do feedbacks affect the resilience of a system? • Are coupled human–nature feedbacks different in magnitude and duration from feedbacks occurring only within human or within natural systems?
Thresholds	• How do thresholds vary over time and space? • What types of models are best able to predict coupled human–nature thresholds? • How do thresholds impact humans and nature differently? • What factors are more easily manipulated to prevent a negative threshold from being reached?
Vulnerability	• How does vulnerability vary over time and space? • How is the vulnerability of a system affected by different durations and frequencies of disturbances? • Is a system more vulnerable when it has fewer linkages between humans and nature? • How does the vulnerability of individual actors manifest itself at higher hierarchical levels of the system?
Time lags	• How does the duration of time lags vary over time and space? • What types of human–nature interactions have short vs long time lags? • How do time lags affect resilience? • Are time lags between human and natural subsystems different from those within human or natural subsystems?
Telecoupling	• How do telecouplings interact with local couplings? • How are spillover systems affected by telecouplings in ways that have not been detected in place-based research? • How do multiple, different telecouplings interact with one another? • Under which conditions is a system with stronger telecouplings than local couplings more resilient?
Management implications	• What factors make institutions more resilient in the face of unexpected human–nature interactions? • What types of frameworks are needed to detect and account for tradeoffs in coupled systems? • How can institutions better anticipate time lags and implement management measures before negative effects unfold? • How can institutions better collaborate across telecoupled systems? • How do we govern telecoupled systems through flow-based management?

that the response so far may be just part of a larger ripple effect that may not be seen until after a time lag (Yang et al., 2015). There are many uncertainties in this present period of reconstruction, such as whether and how the main road will be permanently repaired and to what extent the new infrastructure will be functional for local people. These uncertainties may cause the system to shift into a new, altered state with negative impacts on the natural system over the long term if unregulated (e.g., with increased livestock grazing, see Chapter 4). The movement of people to government housing structures along the main road after the earthquake may also significantly alter human–nature interactions in ways that are not yet apparent (Chapter 14). Future research may involve tracking the pattern of vegetation regrowth in earthquake-affected areas to ascertain long-term trends in biodiversity. In addition, it is worthwhile to consider how such long-term recovery may interact with conservation policies. Short-term studies demonstrated the importance of conservation programs for offsetting negative earthquake impacts on forests (Viña et al., 2011). Nonetheless, there was only a marginal positive gain from artificial tree and bamboo planting restoration programs in earthquake-affected sites (Zhang et al., 2014). The long-term picture may differ greatly because of a lag effect.

Another related topic that requires a long-term approach is human–nature feedbacks. Such feedbacks occur when effects of one of the two subsystems (humans or nature) affect the other, which in turn feed back to exert effects on the first subsystem. Feedbacks are a central component of coupled systems but are understudied (Hull et al., 2015). The dearth of research could be because they may take a long time to emerge (Liu et al., 2015). We have identified several examples of feedbacks

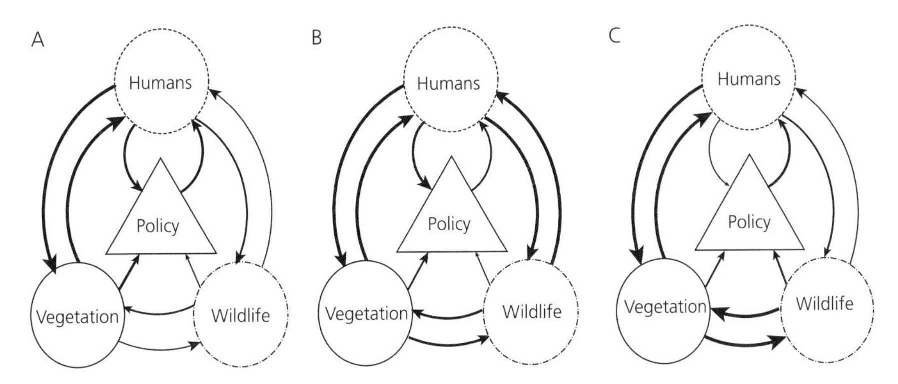

Figure 19.3 Hypothetical dynamics of feedback loops over three time periods. Arrows indicate directions of feedback, with line thickness representing relative strength. (A) Policies emerge to discourage vegetation degradation. (B) Cropland damage by wildlife increases as a function of vegetation recovery, inducing changes in socioeconomic activities. (C) New policies emerge (e.g., compensation for wildlife-derived crop losses), reducing the strength of the feedback depicted in (B).

occurring in Wolong (Figure 19.3). They include policies implemented to protect forests (Chapters 5 and 13), increased crop raiding by wildlife in response to improved forest, and policies enacted to control wildlife damage (Chapters 13 and 16). There is a need to delve deeper into understanding the mechanisms behind these feedbacks and the long-term consequences, in addition to exploring other feedbacks that we have yet to uncover or quantify in Wolong. One example could involve potential increased poaching due to the negative effects of landslides on access to other forms of income (e.g., agricultural markets). Another example is the evolution of new livestock policies (Hull et al., 2011a) in response to environmental impacts caused by people's innovation in the livestock sector.

19.3 Bringing more components together

Most coupled systems research has involved a few key variables or sectors of interest. Common variables of interest include population size, income, and land use on the human side, and animal population size, plant abundance, and land cover on the natural side. To be more integrative and obtain a more holistic picture of coupled systems, it is important to bring more components together. This integration would illustrate how many seemingly disconnected components are linked with each other. This task can be accomplished by integrating across different organizational levels or units (e.g., trophic levels or institutions). Examples are featured in emerging research on human–nature nexuses (e.g., energy–water nexus and food–energy–water nexus). In this approach, researchers integrate across diverse issues (or nodes), including water, food security, national defense, air quality, energy, and climate (Liu et al., 2015; Figure 19.4). Interest in the nexus approach has increased rapidly since 2010. Nonetheless, there was a long period of little or no activity after the first paper on the nexus approach (a food–energy nexus) was published in 1982 (Figure 19.4A). The vast majority of studies so far have analyzed two-node nexuses (80%), with only 16% and 4% of the studies on nexuses with three and four nodes, respectively (Figure 19.4B). Several types of two-node nexuses have received substantial attention, including water–energy nexus, water–food nexus, energy–food nexus, and air–climate nexus (Figure 19.4C). Among the three- and four-node nexuses are energy–environment–land nexus and climate–energy–food–water nexus (Figure 19.4D). These studies not only shed light on new patterns and processes in coupled systems, but also inform sustainability efforts. For example, one study coupled global energy security policy with policies related to other sectors such as climate change and air pollution (via an air–climate–energy

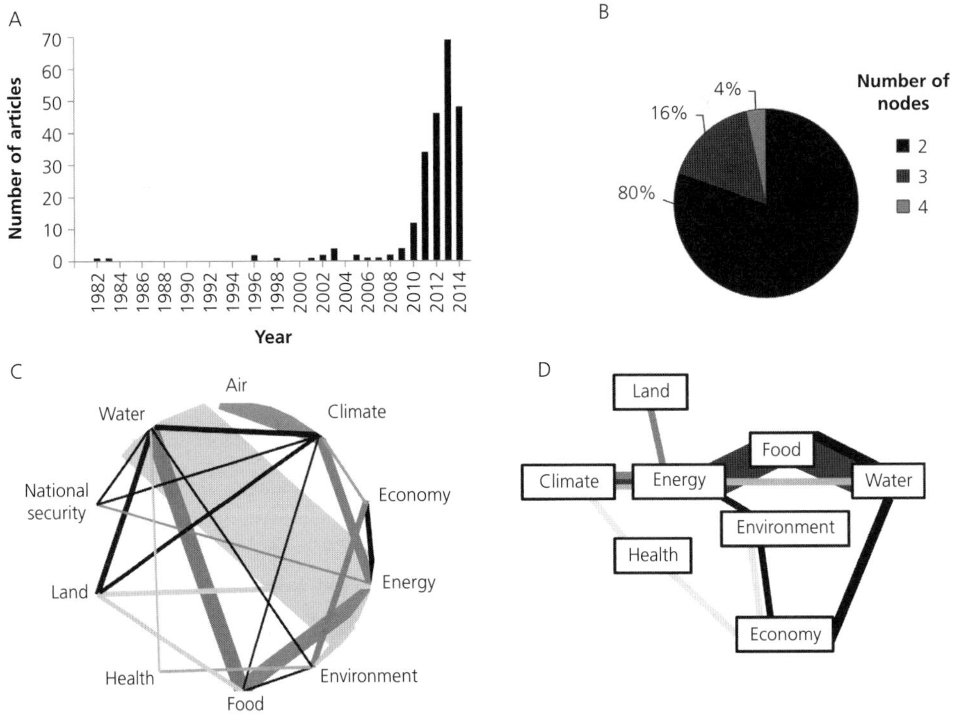

Figure 19.4 (A) The rising annual number of papers using the nexus approach (e.g., water–food nexus, land–economy–environment nexus, food–energy–water nexus, or food–energy–water–climate nexus) recorded in the Web of Sciences (as of August 14, 2014). (B) Percentages of two-, three- and four-node nexuses in all the papers in (A). (C) Frequency of two-node nexuses (e.g., the energy–water nexus) investigated in two or more papers in (A). (D) Nexuses with three or more nodes in two or more papers in (A). In (C) and (D), each nexus is represented by a unique shade (e.g., in (D) the lightest gray refers to a climate–health–economy–environment nexus), thickness of line corresponds to frequency of the types of nexuses studied, and nexuses in only one study are omitted for simplicity.

nexus; Bollen et al., 2010). Such an approach improved global oil sustainability compared to implementing energy policy alone (Bollen et al., 2010).

Umbrella concepts such as ecosystem services (Daily, 1997, Mace et al., 2012), environmental footprints (Hoekstra and Wiedmann, 2014), and planetary boundaries (Rockstrom et al., 2009, Steffen et al., 2015) also bring human–nature interactions together. The study of ecosystem services (discussed in Chapter 5) involves quantifying the benefits of nature to humans, e.g., clean water, nutrient cycling, and recreation (Millennium Ecosystem Assessment, 2005). The environmental footprints are measures of human resource use and waste generation (Hoekstra and Wiedmann, 2014). Planetary boundaries are thresholds for necessary Earth system functions (e.g., stratospheric ozone, global freshwater, and nitrogen cycling) beyond which humanity cannot sustain itself (Rockstrom et al., 2009). Among nine boundaries evaluated, five have already crossed the thresholds (Steffen et al., 2015). So far, these three umbrella concepts have been studied separately despite the fact that they are interconnected (Figure 19.5). Humans have increased consumption of various ecosystem services (and other natural capital), and most non-market ecosystem services have decreased (Figures 19.5A and 19.5B). In contrast, environmental footprints have increased, in some instances, past sustainable levels (Figure 19.5C). These impacts also have either approached or gone beyond the planetary boundaries (Figure 19.5D). Negative feedbacks are needed to more effectively manage these threats to sustainability.

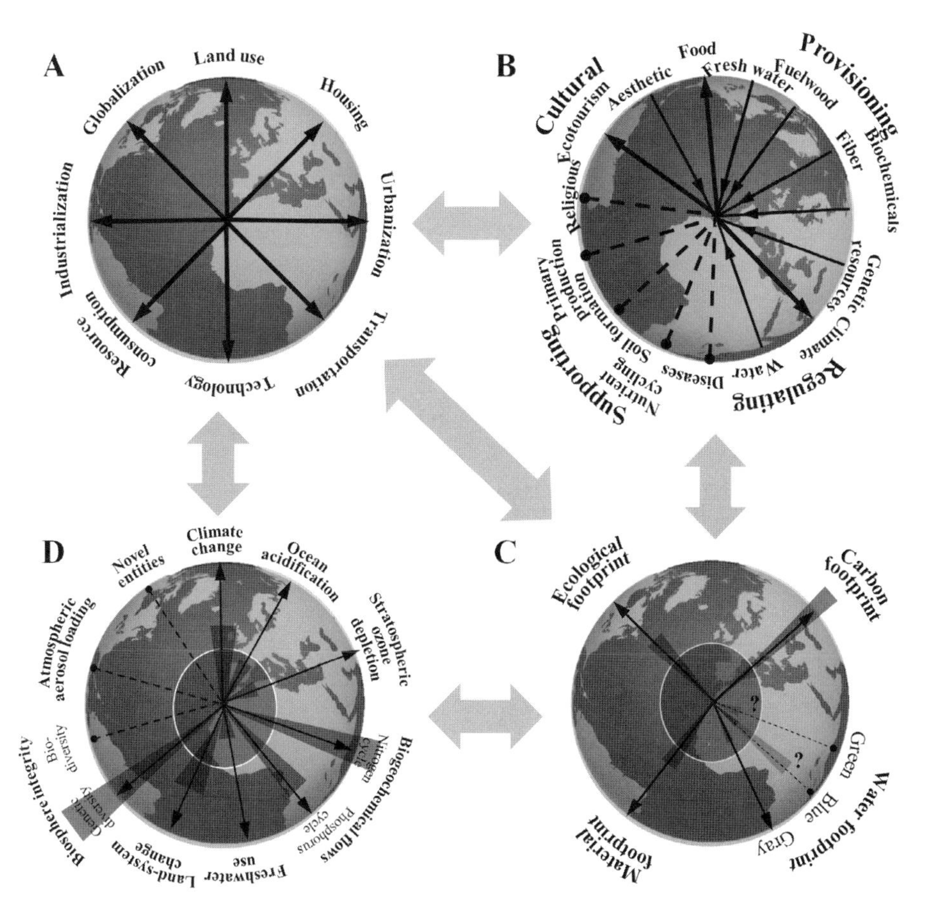

Figure 19.5 Examples of (A) human activities, (B) ecosystem services, (C) environmental footprints, and (D) planetary boundaries. Outward arrows correspond to increases in values, inward arrows decreases, and dashed lines no data. In (C) and (D), the inner circle represents maximum sustainable footprints (Hoekstra and Wiedmann, 2014) and safe operating space for nine planetary boundaries (Rockström et al., 2009), respectively. Wedges indicate estimated current values for the variables. In (C), at least three types of footprints (ecological, carbon, and material) have exceeded maximum sustainable footprints (Hoekstra and Wiedmann, 2014). Question marks (?) mean the information is uncertain. For (D), all planetary system variables have increased and five boundaries have been crossed (Steffen et al., 2015). Figures are modified and reproduced from Liu et al. (2015).

There are several major understudied components of our model coupled system (Wolong) that we hope to delve into in the future. One is the interactions between individual pandas and people. This topic has never been quantitatively studied to date. Quantification is strongly needed, considering the complexities of human impacts on panda habitat (Chapter 4) and differences in individual pandas (e.g., age, sex, weight, and behavior). Current studies involve complex agent-based modeling of human behaviors at the individual level (e.g., An and Liu, 2010, An et al., 2014). But on the panda side, research has been largely limited to population-level studies. Because pandas avoid humans and live in dense habitats with low visibility, it is difficult to directly observe pandas in the wild (Hull et al., 2014). One way to overcome such a difficulty is to use advanced sensors to track individual pandas as they move through the complex landscape. We have started to track a small number of giant pandas using global positioning system (GPS) collars (Hull et al., 2014). But it is necessary to track more pandas to draw more definite conclusions. Promising leads include three-axis accelerometers

(Sikka et al., 2004) and infrared camera traps (Carter et al., 2012), both of which provide detailed information while having a low impact on the animal. Linking individual panda behaviors to frequently studied human impacts such as timber harvesting is needed. But it is also likely that pandas are sensitive to other impacts such as sounds that are difficult for researchers to "see" on the ground. One novel avenue to explore might be soundscape research, which involves cutting-edge approaches to quantifying sounds across landscapes (Pijanowski et al., 2011).

Another area of inquiry is the investigation of interactions between humans and other aspects of biodiversity in the reserve. Our prior work has focused mainly on giant pandas because they are a charismatic flagship species that drives conservation and management, but Wolong also supports thousands of other animal and plant species (Chapter 3). In addition to serving important roles in the natural subsystem, these species also offer humans with a wide range of ecosystem services (Yang et al., 2013a, b, 2015). For example, medicinal herbs provide relief from health-related ailments, trees prevent soil erosion and landslides, and insects serve as pollinators. Recent research has documented the overlap in the niche of pandas with many other endemic species of concern (Xu et al., 2014). This finding validates the focus on pandas as being meaningful for the ecosystem as a whole. Nonetheless, pandas cannot serve as a surrogate for all species because some species may be more or less sensitive to human impacts than pandas and may interact with humans in different ways. For instance, some species may be the targets of human hunters (e.g., sambar *Rusa unicolor*) and others may be more likely to engage in crop raiding (e.g., wild boar *Sus scrofa*). Further research is needed to ascertain how changes in their population sizes and spatial distribution may create cascading effects on multiple other aspects of the system (e.g., other animal species, plants, and people).

We also need to understand aquatic systems and their linkages with terrestrial systems in the reserve. We have considered water as a resource for pandas (Hull et al., 2011b) and as a source of alternative energy for people (via hydropower; An et al., 2001, 2002), but we have not yet delved into the dynamics of these interactions and relationships between them. For instance, it is necessary to understand the impact of hydropower plants on the aquatic system and tie this into our existing analysis of energy transition in the reserve. In addition, it is important to further investigate dynamics and influencing factors (e.g., climate change) of perennial water sources. These resources are important for the survival of pandas throughout their mountainous habitat and may also act as a reservoir for disease transfer with livestock (Hull et al., 2011a).

19.4 Integrating across multiple spatial scales

Most coupled systems research to date has been conducted at one or two spatial scales. More efforts to link coupled human and natural patterns and processes across multiple spatial scales are needed because causes and effects may be different and may interact in unpredictable ways across scales (Zurlini et al., 2006). For example, the Natural Forest Conservation Program in China is implemented in different ways across China as a result of differences in the structure of local institutions. Some local governments reward local people for participation and others do not (Chapter 11). Furthermore, it is important to avoid scale mismatches (Cumming et al., 2006). A classic example is the decline of marine great whale populations. These animals have a large home range but have been monitored at small spatial scales. Scales of study have been inadequate for understanding and predicting population crashes influenced by diverse harvesting policies across multiple institutions and countries (Cumming et al., 2006). In Chapter 17, we discussed the telecoupling approach as one way to conduct coupled systems research that goes beyond the focal study site and takes into account processes occurring at broader scales. Much work is needed in telecoupling research and other related issues to quantify complex relationships between different system components across different scales.

One future area of research that we hope to undertake in Wolong across multiple spatial scales deals with climate change. Climate change is a hot topic of study in many disciplines, but few studies have integrated climate change analyses within coupled systems across multiple spatial scales (Liu et al., 2015). To fill this gap, we have recently received a four-year grant from NSF, with plans to study complex effects of climate change on nature reserve networks across the entire panda range. We will build on our previous studies in another area of panda habitat (the Qinling Mountains) that simulated climate impacts on bamboo using phenological signatures extracted from remote sensing data (Tuanmu et al., 2013). We hope to explicate the complex interactions and feedbacks between humans and nature under climate change, such as the potential shifts in land use and government policies in response to panda habitat loss across scales.

Wolong will continue to serve as an important testing ground for this range-wide analysis and allow us to integrate data at local scales with processes occurring at broader scales (Figure 19.6). We plan to integrate our existing data with state-of-the-art remote sensing technologies to project the future state of Wolong under different climate change scenarios (e.g., extremes). We intend to map the potential future distribution of major bamboo species (and, by proxy, giant pandas) using multispectral approaches and state-of-the-art algorithms (e.g., Hilker et al., 2009a, b). These projections could also be integrated with the human community via agent-based models that capture changes in the behavior of people as they experience such shifts over time. Such shifts may result in complex human–nature feedbacks. For example, people could take up more agricultural land if panda habitat shifts to higher elevations. Another example is that people could collect more fuelwood if the hydrological flows and hydropower plants are negatively affected by warming trends. Results could then be scaled up to account for the relationships between Wolong and other coupled systems throughout the panda range. Of particular interest is how Wolong contributes to networks and corridors of remaining panda habitat and how policies in response to climate change influence policy decisions elsewhere. Processes related to climate, wildlife habitat, and land use in Wolong will be integrated with other processes across a reserve network and up to reserve meta-networks (Figure 19.6). Cross-scale interactions will also be quantified. Such an approach will allow us to better understand the sensitivity of coupled systems to climate change and to inform climate change policies in coupled systems across the globe.

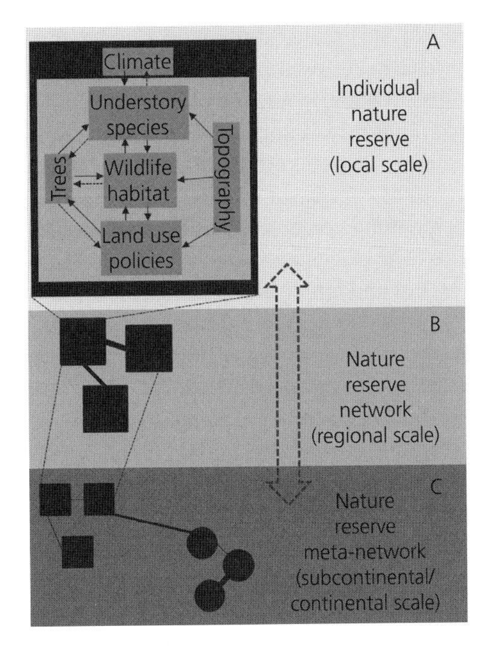

Figure 19.6 Conceptual framework of a forested macrosystem depicting interrelationships among (A) individual nature reserves at the local scale, (B) reserve networks at the regional scale, and (C) reserve meta-networks at subcontinental and continental scales. Rectangles and circles represent nature reserves in two different reserve networks, and solid lines refer to corridors between nature reserves and reserve networks. In (A) understory species and wildlife habitat are affected by climate change and other factors. Cross-scale interactions are illustrated by a vertical block arrow (from Liu et al., 2013b).

19.5 Exploring spillover systems

Spillover systems are one of the main types of systems involved in telecoupling (socioeconomic and environmental interactions over distances) but are the least studied (Chapter 17; Liu et al., 2015). These systems are affected by flows between the sending and receiving systems in diverse ways. For example, spillover systems are affected by CO_2 emissions from producing and transporting products between sending and receiving systems. Spillover systems also serve as stopovers for migratory animals (Liu et al., 2013a; Chapter 17). As spillover systems have been ignored in most research, key system interactions are missed, and effects may go unnoticed, leading to surprising consequences that threaten sustainability. An example is dead zones (low oxygen areas in water bodies) that form in the Gulf of Mexico. These dead zones are caused by fertilizer application for agricultural production and trade in the Mississippi basin in the USA (Rabalais et al., 2002). Identification and quantification of spillover systems (e.g., causes and effects) is needed to better understand and anticipate such complex processes.

In our research in Wolong, there are many opportunities to explore spillover systems. Wolong is a spillover system for some telecouplings and creates spillover effects in other systems by serving as either a sending or receiving system for other telecouplings (Chapter 17). In a broader sense, Wolong is a spillover system influenced by climate change due to emissions of greenhouse gases throughout China and the world. The climate change impact on bamboo in Wolong involves agents and flows far beyond its borders (Tuanmu et al., 2013). A spillover effect that we have not yet investigated is the impact on Wolong's local agricultural market of changing market prices due to trade among buyers and sellers in different regions or countries. An important set of questions we have not answered include: what are the ecosystem services generated from the protection of Wolong (or the entire panda range) for places outside? How does Wolong's (or the panda range's) provision of ecosystem services affect outsiders' well-being? Another area for future study of spillovers is the impact of tourism advertising and infrastructure for other nearby tourist sites on future tourist volume in Wolong. The interactions between Wolong and other areas also create spillover systems. For example, tourists from foreign countries often stop in Beijing or Chengdu before arriving in Wolong. In these cases, Beijing and Chengdu are spillover systems. Understanding and quantifying these spillover systems will help place Wolong in clearer focus with respect to how it affects and is affected by other coupled systems.

19.6 Developing better tools—tackling big data

Addressing issues such as those discussed in the previous sections of this chapter will require new tools to manage the large volumes of diverse coupled human and natural data. The emerging realm of "big data" provides some clues about where to start. The term "big data" refers to "the increasing volume, variety, and velocity of data streams across sectors" of scientific research (Hampton et al., 2013: 156). Major discoveries are expected across scientific disciplines in the coming decades due to such data, which come in large volumes that cannot be handled by traditional data tools, thus requiring new databases and software. Central to the premise of "big science" is a culture of data sharing and adequate data curation to foster long-term collaboration across institutes (Hampton et al., 2013). The goal is for the data to have a life that lasts beyond a single manuscript, extends outside of a single study site, and is used by multiple research organizations. This objective can be achieved using data-sharing and analyzing software, such as cloud computing and massively parallel processing, which could allow for online databases to be implemented and shared across networks of servers globally.

This technology is useful for creating digital organisms—computer software programs that can self-replicate and evolve over time. The digital realm has come to the forefront of many scientific fields including physics, management, medicine, sociology, and ecology. It is popular because it allows for flexibility in designing diverse types of systems in a controlled environment to test hypotheses and examine scenarios. An example is the software program Avida (Ofria and Bryson, 2015) maintained under the NSF Center for the Study of Evolution in Action (BEACON) at Michigan State University. Organisms in this novel program can reproduce at rates thousands of times faster than bacteria, thus allowing fundamental theories about evolution to be tested in a way never possible before (Zimmer, 2005). Digital organisms have been useful for understanding evolution of host–parasite interactions, mutualisms, and predator–prey interactions in ecology (Fortuna et al., 2013). On the sociology side, digital organisms have informed work on knowledge transfer, social intelligence, and social interactions (Erickson, 2009).

At an even larger scale, a related concept of "Digital Earth" was put forth by Al Gore in 1998 in his visionary speech outlining ideas on the future of technology (Gore, 1998). "Digital Earth" was envisioned as a computerized, multidimensional representation of the entire globe, with human and natural data on each location stored as layers and accessible to all. The vision has not yet been realized, although parts of it have come to fruition, such as the detailed, high-resolution maps of the Earth's surface now available on Google Earth. The potential for these maps to be integrated with diverse layers of data such as socioeconomic data and field sampling data has not yet been tapped but should be kept on the horizon as a future goal. At a smaller scale, the Wolong administration is currently developing a "Digital Wolong." This program is a shared online database where all administrative and conservation work is to be recorded and shared to facilitate cross-department collaboration. One particular component would involve posting monitoring data from infrared cameras distributed throughout the reserve.

In order to better understand and predict an entire coupled system using the increasing volumes of data that are available about the model system of Wolong, we are planning to create a "Digital CHANS." It will begin with a digital replica or avatar of Wolong—an artificial, virtual, self-replicating, and spatially explicit system that represents the real Wolong. Some of its components may include something like the aforementioned self-replicating and evolving digital organisms with networking databases such as "Digital Earth" or "Digital Wolong." Besides representing the abiotic factors (e.g., topography, weather, and rivers), Digital CHANS will be made up of interacting biotic components. Examples include animals, plants, people, and institutions such as households and communities. Human activities such as farming, tourism, livestock grazing, transportation, and conservation will also be included. It will be a three-dimensional visualization of Wolong similar to those shown on Google Earth. More importantly, it will enable users to understand and model interactions within and across the system boundaries, and run a variety of scenarios. The Digital CHANS will be linked to distant systems via telecoupling processes. For example, it will allow local residents to interact with tourists from around the world, and to respond to changes such as job opportunities in cities and conservation payments from the central government. It will give users a virtual laboratory to understand the workings of a coupled system in a way that has not been done before and to allow us to systematically and comprehensively address pressing questions related to sustainability (Table 19.1) that we have not been able to address before using only smaller pieces of the puzzle. While we have already built various models that focus on a number of specific issues (e.g., Chapter 14), this laboratory could be used to holistically experiment with an entire CHANS, further help develop sustainability theories, and guide future research and practice. Such simulation experiments will allow us to simulate possible policy scenarios for the future and present projections to stakeholders on sustainability outcomes for various management measures. Building on the Digital CHANS of Wolong, a Digital CHANS of the entire world would ultimately be possible.

19.7 Translating research into policy and practice

To make research more productive and more meaningful to society (e.g., to help achieve the United Nations Sustainable Development Goals; United Nations, 2014), it is important that the coupled systems community collaborate better with stakeholders (e.g., conservation organizations, resource managers, and development agencies). The conventional approach of "translating" science into policy and practice tends to be fraught with lack of interactions between researchers and stakeholders until the research is done (Liu et al., 2015). Coupled systems researchers can be transdisciplinary leaders in adopting a new paradigm in which scientists and stakeholders codesign research, coproduce results, and coimplement new measures (e.g., strategies and decision-making; Future Earth, 2014). Coupled systems research is uniquely positioned to help weigh tradeoffs and competing objectives that inevitably accompany the management of coupled systems involving multiple stakeholders. Agencies often lack the infrastructure to anticipate and effectively respond to complexities in human–nature interactions such as feedbacks, time lags, and surprises. Coupled systems researchers can contribute by sharing knowledge and data about unexpected human–nature tradeoffs, such as invasive species and disease increasing with global trade (Hulme, 2009,

Karesh et al., 2005). The coupled systems framework also can help to evaluate policies that are not working as they were intended by illuminating complex human–nature interactions that produce unintended consequences. For example, as a result of land-use change, policies promoting biofuels have increased CO_2 emissions in some places despite their intended goal to have the inverse effect (Searchinger et al., 2008). This surprising effect occurred due to unanticipated land-use changes (e.g., conversion of grassland to crops) to make way for biofuels.

The coupled systems research community can also initiate dialogue about ways to restructure institutions so they are better equipped to address increasing threats to global sustainability (Liu et al., 2013a). This restructuring may include ways to better account for interactions and feedbacks between people and nature, such as by creating new interdisciplinary departments or interagency working groups. In addition, coupled systems researchers can help strategize ways to design institutions that are better integrated across space and time. Ostrom (2012) first championed the idea of "polycentric" institutions to alleviate threats that are nested within different levels of a system. For instance, connecting the hierarchical structure of organizations (e.g., at national, regional, and local levels) could help account for telecouplings among distant coupled systems. Such systems may be managed by different agencies that may not otherwise communicate with one another effectively (Chapter 17). Furthermore, institutions can be made more flexible so that they can adapt to new human–nature interactions and feedbacks that emerge over time, such as climate change and emerging infectious diseases.

In Wolong, we hope to further strengthen our existing relationships with stakeholders in the future so we may continue to generate research results that are not only theoretically interesting but also practical for planning and policy making. It would be worthwhile for us to initiate roundtable discussions with stakeholders to identify their concerns and information needs for future planning so we can design studies and scenario analyses that directly address these needs. In turn, we could share our findings, and nuanced uncertainties and surprises in our data that may be important for decision-making. Examples might include a sudden threshold of fuelwood collection impacts or a surprising twist in practices of raising livestock. Codesign, coproduction, and coimplementation could also help assess functionality and efficiency of various administrative departments (e.g., tourism, agricultural, and natural resource management departments) to improve communication and collaboration. They would be useful in finding ways in which different departments could build off one another and avoid conflicts regarding sustainability goals.

19.8 Summary

Future development in several areas is needed for the science of coupled human and natural systems (CHANS) to grow to new heights. Longer-term research is crucial to help explicate complexities such as feedbacks and time lags in coupled systems. More cross-component integration can help draw linkages between more diverse sectors of coupled systems through human–nature nexus approaches such as the food–energy–water nexus. Another area of future research is the need for tighter integration across multiple spatial scales. Spillover systems are particularly understudied and deserve special attention. Better tools are key to tackling the challenges in coupled systems research. For example, cloud computing and massively parallel processing can accommodate "big data" needs and the development of Digital CHANS for whole system experiments. Furthermore, future research on coupled systems should place greater emphasis on new ways of translating research results into policy and practice through working even closer with stakeholders to codesign, cocreate, and coimplement research and management measures. This transdisciplinary approach will allow opportunities to develop new theories, prepare for uncertainties, and anticipate unintended effects of policies. Research on these diverse themes will enrich our understanding of our model coupled system (Wolong Nature Reserve) and other coupled systems, and take us in new directions as we continue to unravel additional layers of complexity in human–nature interactions. By doing so, we hope to help lead the field of coupled systems research into the next era of transdisciplinary research for sustainability across local to global levels.

References

An, L. and Liu, J. (2010) Long-term effects of family planning and other determinants of fertility on population and environment: agent-based modeling evidence from Wolong Nature Reserve, China. *Population and Environment*, **31**, 427–59.

An, L., Liu, J., Ouyang, Z., et al. (2001) Simulating demographic and socioeconomic processes on household level and implications for giant panda habitats. *Ecological Modelling*, **140**, 31–49.

An, L., Lupi, F., Liu, J., et al. (2002) Modeling the choice to switch from fuelwood to electricity: implications for giant panda habitat conservation. *Ecological Economics*, **42**, 445–57.

An, L., Zvoleff, A., Liu, J., and Axinn, W. (2014) Agent-based modeling in coupled human and natural systems (CHANS): lessons from a comparative analysis. *Annals of the Association of American Geographers*, **104**, 723–45.

Bollen, J.C., Hers, S., and van der Zwaan, B. (2010) An integrated assessment of climate change, air pollution, and energy security policy. *Energy Policy*, **38**, 4021–30.

Carter, N.H., Shrestha, B.K., Karki, J.B., et al. (2012) Coexistence between wildlife and humans at fine spatial scales. *Proceedings of the National Academy of Sciences of the United States of America*, **109**, 15360–65.

Center for Systems Integration and Sustainability (2014) *CHANS-Net.org*.

Collins, S.L., Bettencourt, L.M.A., Hagberg, A., et al. (2006) New opportunities in ecological sensing using wireless sensor networks. *Frontiers in Ecology and the Environment*, **4**, 402–07.

Coverdale, T.C., Herrmann, N.C., Altieri, A.H., and Bertness, M.D. (2013) Latent impacts: the role of historical human activity in coastal habitat loss. *Frontiers in Ecology and the Environment*, **11**, 69–74.

Cumming, G.S., Cumming, D.H. and Redman, C.L. (2006) Scale mismatches in social–ecological systems: causes, consequences, and solutions. *Ecology and Society*, **11**, 14.

Daily, G. (1997) *Nature's Services: Societal Dependence on Natural Ecosystems*. Island Press, Washington, DC.

Erickson, T. (2009) "Social" systems: designing digital systems that support social intelligence. *AI & Society*, **23**, 147–66.

European Organization for Nuclear Research (2014) *CERN*. http://home.web.cern.ch/.

Folke, C., Carpenter, S., Elmqvist, T., et al. (2002) Resilience and sustainable development: building adaptive capacity in a world of transformations. *Ambio*, **31**, 437–40.

Fortuna, M.A., Zaman, L., Wagner, A.P., and Ofria, C. (2013) Evolving digital ecological networks. *PLoS Computational Biology*, **9**, e1002928.

Future Earth (2014) http://www.futureearth.info/.

Gore, A. (1998) The digital earth: understanding our planet in the 21st century. *Australian Surveyor*, **43**, 89–91.

Grimm, N.B., Redman, C.L., Boone, C.G., et al. (2013) Viewing the urban socio-ecological system through a sustainability lens: lessons and prospects from the Central Arizona-Phoenix LTER Programme. In S.J. Singh, H. Haberl, M. Chertow, M. Mirtl, and M. Schmid, eds, *Long Term Socio-Ecological Research*, pp. 217–46. Springer, Netherlands.

Haberl, H., Winiwarter, V., Andersson, K., et al. (2006) From LTER to LTSER: conceptualizing the socioeconomic dimension of long-term socioecological research. *Ecology and Society*, **11**, 13.

Hampton, S.E., Strasser, C.A., Tewksbury, J.J., et al. (2013) Big data and the future of ecology. *Frontiers in Ecology and the Environment*, **11**, 156–62.

Hilker, T., Wulder, M.A., Coops, N.C., et al. (2009a) A new data fusion model for high spatial-and temporal-resolution mapping of forest disturbance based on Landsat and MODIS. *Remote Sensing of Environment*, **113**, 1613–27.

Hilker, T., Wulder, M.A., Coops, N.C., et al. (2009b) Generation of dense time series synthetic Landsat data through data blending with MODIS using a spatial and temporal adaptive reflectance fusion model. *Remote Sensing of Environment*, **113**, 1988–99.

Hoekstra, A.Y. and Wiedmann, T.O. (2014) Humanity's unsustainable environmental footprint. *Science*, **344**, 1114–17.

Hull, V., Shortridge, A., Liu, B., et al. (2011b) The impact of giant panda foraging on bamboo dynamics in an isolated environment. *Plant Ecology*, **212**, 43–54.

Hull, V., Tuanmu, M.-N., and Liu, J. (2015) Synthesis of human-nature feedbacks. *Ecology and Society*, **20**, 17.

Hull, V., Xu, W., Liu, W., et al. (2011a) Evaluating the efficacy of zoning designations for protected area management. *Biological Conservation*, **144**, 3028–37.

Hull, V., Zhang, J., Zhou, S., et al. (2014) Impact of livestock on giant pandas and their habitat. *Journal for Nature Conservation*, **22**, 256–64.

Hulme, P.E. (2009) Trade, transport, and trouble: managing invasive species pathways in an era of globalization. *Journal of Applied Ecology*, **46**, 10–18.

Institute for Social Research (2014) *Panel Study of Income Dynamics*. http://psidonline.isr.umich.edu.

Karesh, W.B., Cook, R.A., Bennett, E.L., and Newcomb, J. (2005) Wildlife trade and global disease emergence. *Emerging Infectious Diseases*, **11**, 1000–02.

Knapp, A.K. and Smith, M.D. (2001) Variation among biomes in temporal dynamics of aboveground primary production. *Science*, **291**, 481–84.

Levin, S.A. and Lubchenco, J. (2008) Resilience, robustness, and marine ecosystem-based management. *BioScience*, **58**, 27–32.

Liu, J., Hull, V., Batistella, M., et al. (2013a) Framing sustainability in a telecoupled world. *Ecology and Society*, **18**, 26.

Liu, J., Mooney, H., Hull, V., et al. (2015) Systems integration for global sustainability. *Science*, **347**(6225), 1258832.

Liu, J., Viña, A., and Winkler, J. (2013b) *Complex Effects of Climate Change on Nature Reserve Networks at Macroscales*. Proposal to the National Science Foundation.

Mace, G.M., Norris, K., and Fitter, A.H. (2012) Biodiversity and ecosystem services: a multilayered relationship. *Trends in Ecology & Evolution*, **27**, 19–26.

Millennium Ecosystem Assessment (2005) *Ecosystems & Human Well-being: Synthesis*. Island Press, Washington, DC.

Mirtl, M., Orenstein, D.E., Wildenberg, M., et al. (2013) Development of LTSER platforms in LTER-Europe: challenges and experiences in implementing place-based long-term socio-ecological research in selected regions. In S.J. Singh, H. Haberl, M. Chertow, M. Mirtl, and M. Schmid, eds, *Long Term Socio-Ecological Research*, pp. 409–42. Springer, Netherlands.

Ofria, C. and Bryson, D.M. (2015) *Avida Digital Evolution Platform*. http://avida.devosoft.org/.

Ostrom, E. (2012) Nested externalities and polycentric institutions: must we wait for global solutions to climate change before taking actions at other scales? *Economic Theory*, **49**, 353–69.

Pijanowski, B.C., Villanueva-Rivera, L.J., Dumyahn, S.L., et al. (2011) Soundscape ecology: the science of sound in the landscape. *Bioscience*, **61**, 203–16.

Rabalais, N.N., Turner, R.E., and Wiseman, W.J.,Jr. (2002) Gulf of Mexico hypoxia, a.k.a. "the dead zone." *Annual Review of Ecology and Systematics*, **33**, 235–63.

Redman, C.L., Grove, J.M., and Kuby, L.H. (2004) Integrating social science into the long-term ecological research (LTER) network: social dimensions of ecological change and ecological dimensions of social change. *Ecosystems*, **7**, 161–71.

Reid, W., Chen, D., Goldfarb, L., et al. (2010) Earth system science for global sustainability: grand challenges. *Science*, **330**, 916–17.

Rockström, J., Steffen, W., Noone, K., et al. (2009) A safe operating space for humanity. *Nature*, **461**, 472–75.

Searchinger, T., Heimlich, R., Houghton, R.A., et al. (2008) Use of U.S. croplands for biofuels increases greenhouse gases through emissions from land-use change. *Science*, **319**, 1238–40.

Sikka, P., Corke, P., and Overs, L. (2004) Wireless sensor devices for animal tracking and control. *Local Computer Networks, 2004*. 29th Annual International Conference on IEEE.

Singh, S.J., Haberl, H., Gaube, V., et al. (2010) Conceptualising long-term socio-ecological research (LTSER): integrating the social dimension. In F. Müller, C. Baessler, H. Schubert, and S. Klotz, eds, *Long-term Ecological Research: Between Theory and Application*, pp. 377–98. Springer, Netherlands.

Steffen, W., Richardson, K., Rockström, J., et al. (2015) Planetary boundaries: guiding human development on a changing planet. *Science*, **347**(6223), 1259855.

The Long Term Ecological Research Network (2014) *LTER*, http://www.lternet.edu/.

Tuanmu, M.-N., Viña, A., Winkler, J.A., et al. (2013) Climate-change impacts on understorey bamboo species and giant pandas in China's Qinling Mountains. *Nature Climate Change*, **3**, 249–53.

United Nations (2014) *Sustainable Development Goals*. https://sustainabledevelopment.un.org/topics/sustainabledevelopmentgoals.

Viña, A., Chen, X.D., McConnell, W.J., et al. (2011) Effects of natural disasters on conservation policies: the case of the 2008 Wenchuan Earthquake, China. *Ambio*, **40**, 274–84.

Walker, B.H., Abel, N., Anderies, J.M., and Ryan, P. (2009) Resilience, adaptability, and transformability in the Goulburn-Broken Catchment, Australia. *Ecology and Society*, **14**, 12.

Xu, W., Viña, A., Qi, Z., et al. (2014) Evaluating conservation effectiveness of nature reserves established for surrogate species: case of a giant panda nature reserve in Qinling Mountains, China. *Chinese Geographical Science*, **24**, 60–70.

Yang, W., Dietz, T., Kramer, D.B., et al. (2015) An integrated approach to understanding the linkages between ecosystem services and human well-being. *Ecosystem Health and Sustainability*, **1**, 19.

Yang, W., Dietz, T., Liu, W., et al. (2013a) Going beyond the Millennium Ecosystem Assessment: an index system of human dependence on ecosystem services. *PLoS ONE*, **8**, e64581.

Yang, W., Liu, W., Viña, A., et al. (2013b) Performance and prospects of payments for ecosystem services programs: evidence from China. *Journal of Environmental Management*, **127**, 86–95.

Zhang, J., Hull, V., Huang, J., et al. (2014) Natural recovery and restoration in giant panda habitat after the Wenchuan earthquake. *Forest Ecology and Management*, **319**, 1–9.

Zhang, J., Hull, V., Xu, W., et al. (2011) Impact of the 2008 Wenchuan earthquake on biodiversity and giant panda habitat in Wolong Nature Reserve, China. *Ecological Research*, **26**, 523–31.

Zimmer, C. (2005) Testing Darwin. *Discover Magazine*. http://discovermagazine.com/2005/feb/cover.

Zurlini, G., Riitters, K., Zaccarelli, N., et al. (2006) Disturbance patterns in a socio-ecological system at multiple scales. *Ecological Complexity*, **3**, 119–28.

Index

D

T

U